PRECALCULUS MATHEMATICS

PRECALCULUS MATHEMATICS
Algebra, Trigonometry, and Analytic Geometry

Margaret L. Lial
Charles D. Miller

American River College
Sacramento, California

Scott, Foresman and Company
Glenview, Illinois   Brighton, England

Library of Congress Catalog Card Number: 72-97195.
ISBN: 0-673-07863-9.
AMS 1970 Subject Classification 98A10.

Regional offices of Scott, Foresman and Company are located in Dallas, Texas;
Glenview, Illinois; Oakland, New Jersey; Palo Alto, California; Tucker, Georgia;
and Brighton, England.

Preface

This book is designed for students who need a good, solid foundation in algebra, trigonometry, and analytic geometry before going on to calculus or other work in mathematics. We assume that students come to this text with at least an introductory course in algebra. The book progresses in a gradually increasing level of sophistication so that a student taking a course from this text is thoroughly and completely prepared to enter a more advanced course. Since there are currently several different approaches to the teaching of calculus, we have been careful to insure that the text provides good preparation for all these different calculus courses.

Too often precalculus texts maintain a high level of rigor throughout the entire text. We, however, prefer to begin at a reasonable level, and increase in rigor so that at the end of the course the student has reached the level of rigor expected in calculus.

There is now a trend away from the excesses of rigor, sophistication, and notational complexities that were creeping into mathematics texts a few years ago. The trend now is to intuition and understanding—appeal to the intuition of the student and give careful guidance so that learning and understanding follow. Mathematics is a lively, exciting and fascinating subject, and it has always seemed a shame to hide this behind a cloud of logical connectives, set symbols, and complicated axiom structures.

There is ample analytic geometry in this text—all topics necessary for success in calculus are included. Functions are emphasized throughout. The $a + bi$ definition of complex number is used, rather than the (a, b) definition so confusing to many students. Much material is included on inequalities, which are so important in the definitions of limits and continuity found in calculus. There are many problem sets and exercises, enough for almost any kind of course.

The student-oriented style of writing that we use has now been completely tested in many, many classes at a very large number of schools of all sizes and types. We are highly gratified at the student reaction to our efforts. We have tried to write for the students, in a way they can understand. Topics have been explained as simply and clearly as possible and many examples of each new topic are included. Judging by the reactions to our earlier books, we feel that we have succeeded in writing some of the most readable texts on the market.

Several people have helped to produce this text. Vern Heeren of American River College provided a thorough, careful reading of the manuscript. Pamela Conaghan of Scott, Foresman was very helpful. Zallia Todd and Sharlene Wills helped in the technical preparation of the manuscript.

Margaret L. Lial

Sacramento, California   Charles D. Miller

Table of Contents

Chapter One

The Real Number System

Algebra is primarily the study of numbers and the general properties of numbers. In this chapter, we introduce the idea of set and then concentrate on one of the most important sets of numbers, the set of real numbers. In Section 1.1 we discuss the necessary definitions needed to begin the study of the set of real numbers.

1.1 Sets of Numbers

A **set** is a collection of **elements** or **members.** We say a set is **well-defined** if its elements are clearly determined, that is, if we can decide whether or not a given element belongs to the set. We may describe a set by listing the elements of the set between braces (called **set braces**). For example, we can write the set containing the numbers 1, 2, 3, and 4 as

$$\{1, 2, 3, 4\}.$$

Since the order in which the elements are listed is unimportant, we could write instead $\{1, 2, 4, 3\}$ or any other arrangement of the four numbers. It is customary to use capital letters to refer to sets. Thus, we might refer to the set described above as S, where

$$S = \{1, 2, 3, 4\}.$$

To indicate that the number 1 is an element of the set S, we write $1 \in S$ (read "1 is an element of the set S" or "1 belongs to the set S"), while we write $5 \notin S$ to indicate that 5 is not an element of S.

It is sometimes more convenient to describe a set with a verbal phrase. Thus, we might describe the set S above as the set consisting of the first four counting numbers. In this example the notation $\{1, 2, 3, 4\}$, where we listed the elements between set braces, is briefer and clearer than the verbal description. However, the set F consisting of all the numbers between 0 and 1 could not be described by a listing of the elements. (Try it.) Set F is an example of an *infinite set*, one which has an unending list of distinct elements. On the other hand, a *finite set* is one which has a limited number of elements. Some infinite sets, unlike F, can be described by a listing process. For example, the set of numbers used for counting, called the **natural numbers,** can be written as

$$N = \{1, 2, 3, 4, \cdots\}.$$

A set whose elements are all elements of a second set is said to be a **subset** of that second set. Each element of the set $S = \{1, 2, 3, 4\}$ described above is an element of N, and so S is a subset of N, which we write as $S \subset N$. In this case, N is not a subset of S, written $N \not\subset S$, since there are elements in N which are not in S (the elements 5 and 6 for example).

Two sets, A and B, are said to be **equal** whenever $A \subset B$ and $B \subset A$. Thus, by $A = B$ we mean that the two sets contain exactly the same elements. If two sets contain the same number of elements (but not necessarily the same elements), they are called **equivalent** sets. From this we see that the two sets

$$S = \{1, 2, 3, 4\} \qquad \text{and} \qquad W = \{a, b, c, d\}$$

are equivalent sets, but not equal sets.

Two sets can be combined to form a third set. The set of all elements which belong to *either* of the sets A *or* B is called the **union** of A and B, written $A \cup B$. The set of all elements which belong to *both* A *and* B is called the **intersection** of A and B and is written $A \cap B$. For example, given

$$A = \{0, 1, 2, 3\} \qquad \text{and} \qquad B = \{0, 2, 4, 6\},$$

we have $A \cup B = \{0, 1, 2, 3, 4, 6\}$ while $A \cap B = \{0, 2\}$.

Example 1 Find $M \cup N$ and $M \cap N$ if $M = \{2, 4, 6\}$ and $N = \{1, 3\}$.

By the definitions, we have $M \cup N = \{1, 2, 3, 4, 6\}$. The set $M \cap N$ consists of all elements common to both M and N, but there are no such elements. To express this fact, we introduce a set called the **empty set** (or **null set**), written $\varnothing$. Thus, $M \cap N = \varnothing$.

Two sets, such as M and N in Example 1, whose intersection is the empty set, are called **disjoint sets.** Hence, the set of all even numbers and the set of all odd numbers are disjoint sets.

The set U of all elements to be used in a particular discussion is called the **universe of discourse** or **universal set.** For a given universe U and set A, the set of all elements in U which are not in set A is called the **complement** of A, written A'. For example, if U is the set of natural numbers N, and E is the set of even natural numbers, then E' is the set of odd natural numbers. Note that

$$E \cup E' = U \qquad \text{and} \qquad E \cap E' = \varnothing,$$

by the definition of complementary sets.

A convenient way to represent sets requires the use of a **variable,** a letter used to represent one unspecified element of a set of numbers. The set of numbers is called the **domain** or **replacement set** of the variable. For example, we can write $\{x \mid x \in N\}$ (read "set of all x such that x is an element of N") to represent the set of natural numbers. In this case N is the domain of the variable x.

At this point, let us list some of the sets of numbers which we shall be using throughout this book.

▶ *natural numbers* $\qquad \{1, 2, 3, 4, \cdots\}$

▶ *whole numbers* $\qquad \{0, 1, 2, 3, 4, \cdots\}$

▶ *integers* $\qquad \{\cdots, -2, -1, 0, 1, 2, \cdots\}$

▶ *rational numbers* $\qquad \left\{\dfrac{p}{q} \,\middle|\, p \text{ and } q \text{ are integers, } q \neq 0\right\}$ or $\{x \mid x \text{ is a repeating or terminating decimal}\}$

▶ *irrational numbers* $\quad \{x \mid x \text{ is a nonrepeating, nonterminating decimal}\}$

▶ *real numbers* $\qquad \{x \mid x \text{ is a rational or irrational number}\}$

▶ *imaginary numbers* $\quad \{x \mid x = a + bi \text{ where } a \text{ and } b \text{ are real numbers, } b \neq 0, \, i^2 = -1\}$

▶ *complex numbers* $\qquad \{x \mid x = a + bi, \, a \text{ and } b \text{ are real numbers}\}$

In this chapter we shall make a thorough study of the set of real numbers. Then we shall study the imaginary and complex numbers in a later chapter.

1.1 Exercises

List the elements of the following sets between set braces.

1. $\{x \mid x \text{ is less than 5 and } x \text{ is a natural number}\}$
2. $\{x \mid x \text{ is a whole number between } -2\frac{1}{2} \text{ and } 2\frac{1}{2}\}$
3. $\{x \mid x \text{ is an integer between } -2\frac{1}{2} \text{ and } 2\frac{1}{2}\}$
4. $\{x \mid x \text{ is a natural number between } -2\frac{1}{2} \text{ and } 2\frac{1}{2}\}$
5. $\{x \mid x \text{ is a positive odd integer greater than 5}\}$

Given $U = \{x \mid x$ is an integer$\}$, $M = \{0, 2, 4, 6, 8\}$, $N = \{x \mid x$ is an even integer$\}$, $P = \{x \mid x$ is an odd integer$\}$, $Q = \{-1, 0, 1, 2, 3, 4\}$, and $R = \{-3, -2, -1\}$, find each of the following sets.

6. $M \cap Q$
7. $M \cup Q$
8. $M \cap P$
9. $N \cup P$
10. $N \cap P$

11. $Q \cap R$
12. N'
13. $M' \cap P$
14. $Q \cap R'$
15. $U' \cup R$

16. $(M \cap Q) \cap N$
17. $(Q \cap R) \cup M$
18. $(R \cap N) \cup Q$
19. $R \cap (P \cap Q)$
20. $(M' \cap Q) \cup R$

Let N represent the set of natural numbers, W the set of whole numbers, I the integers, F the rational numbers, H the irrational numbers, and R the real numbers. Complete each of the following statements with the appropriate symbol: $\subset$, $=$, $\cap$, or $\cup$.

21. $N \underline{\hspace{1cm}} I$
22. $N \underline{\hspace{1cm}} F$
23. $N \underline{\hspace{1cm}} W = W$
24. $N \underline{\hspace{1cm}} R = R$
25. $I \underline{\hspace{1cm}} R = I$

26. $F \underline{\hspace{1cm}} H = \emptyset$
27. $F \underline{\hspace{1cm}} H = R$
28. $\emptyset \underline{\hspace{1cm}} N$
29. $F' \underline{\hspace{1cm}} H$
30. $I \underline{\hspace{1cm}} (N \cup I)$

State the conditions on sets X, Y, and Z for which the following are true.

31. $X \cap Y = \emptyset$
32. $X \cup Y = \emptyset$
33. $X \cap Y = X$
34. $X \cup Y = X$
35. $X \cap \emptyset = X$

36. $X \cup \emptyset = X$
37. $X \cap Y' = \emptyset$
38. $X \cap Y = Y \cup X$
39. $X \cap (Y \cup Z) = (X \cap Y) \cup (X \cap Z)$
40. $X \cap (Y \cap Z) = (X \cap Y) \cap Z$

Use the definitions given in this section to show that the following are true for all sets X, Y, and Z.

41. $X \cap X' = \emptyset$
42. $X \cup X' = U$
43. $X \cap Y = Y \cap X$

44. $X \cup (Y \cup Z) = (X \cup Y) \cup Z$
45. If $X \subset Y$ and $Y \subset Z$, then $X \subset Z$.

1.2 Axioms of the Real Numbers

A **mathematical system** requires (1) a set of elements, (2) relations between the elements, (3) operations on the elements, and (4) properties, the rules which govern the manipulation of the elements. In the remainder of this chapter, we shall investigate the mathematical system which has the real numbers for elements.

The most important relation between two real numbers is that of equality. To indicate the relation "**a is equal to b**" we write $a = b$. The symbol $\neq$

represents the relation "is not equal to." (In a later section, we shall discuss another relation, "is less than.") Two expressions which represent the same real number are said to be **equal.** The axioms of equality which are stated below describe the assumed properties of the relation "equals."

Axioms of Equality

For all real numbers a, b, and c:

$a = a$.	*reflexive axiom*
If $a = b$, then $b = a$.	*symmetric axiom*
If $a = b$ and $b = c$, then $a = c$.	*transitive axiom*
If $a = b$, then a may replace b in any expression.	*substitution axiom*

A **binary** operation on a set is an operation on two elements of the set which produces a third element. (The three elements need not be distinct, nor need the third element be within the set.) There are two fundamental binary operations on real numbers, addition, indicated by $+$, and multiplication, indicated by $\cdot$, as in $2 \cdot 3$. When indicating a product with variables, the elements are written with no symbol between them, as ab or $2x$. Multiplication is also indicated by parentheses, as in the expression $2(x - y)$ which indicates that the quantity in the parentheses is to be multiplied by 2. Thus, grouping by parentheses determines the order in which operations must be performed.

The set of real numbers, together with the relation of equality and the operations of addition and multiplication form the **real number system.** This system is an example of a **field.** The rules which govern the operations of addition and multiplication of a field are called the **field axioms.**

Field Axioms for Addition and Multiplication

For all real numbers a, b, and c:

$a + b = b + a$.	
$ab = ba$.	*commutative axioms*
$(a + b) + c = a + (b + c)$.	
$(ab)c = a(bc)$.	*associative axioms*

There exists a unique real number 0, the **additive identity,** such that
$$a + 0 = 0 + a = a.$$

There exists a unique real number 1, the **multiplicative identity,** such that
$$a \cdot 1 = 1 \cdot a = a.$$

identity axioms

There exists a unique real number $-a$, the
additive inverse of a, such that

$$a + (-a) = (-a) + a = 0.$$

If $a \neq 0$, there exists a unique real number $1/a$,

the **multiplicative inverse** of a, such that

$$a \cdot \frac{1}{a} = \frac{1}{a} \cdot a = 1.$$

inverse axioms

$a + b$ is a real number.
ab is a real number.

closure axioms

$a(b + c) = ab + ac.$

distributive axiom

The closure axioms guarantee that the addition or multiplication of two real numbers will be a real number. The commutative axioms assure us that we can add two numbers or multiply two numbers without regard to order, while the associative axioms show that the way in which we group, when using only one operation with more than two numbers, has no effect on the final result.

The identity axioms establish the existence of the identity elements, 0 for addition and 1 for multiplication, which preserve the identity of any element with which they are combined. The additive inverse axiom notes the existence of the *additive inverse* element $-a$ for each element a, which when added to a, yields the *additive identity*, 0. And the multiplicative inverse axiom notes the existence of the *multiplicative inverse* element $1/a$ for each nonzero element a, which when multiplied by a, yields the *multiplicative identity*, 1. The additive inverse element $-a$ is also called the **negative** of a, while the multiplicative inverse element $1/a$ is also called the **reciprocal** of a.

The distributive axiom describes a relationship between multiplication and addition, which, in effect, changes an indicated product to an indicated sum, or an indicated sum to an indicated product. By the commutative axiom, the distributive axiom can also be written as

$$(a + b)c = ac + bc.$$

Repeated application of the distributive axiom gives the **extended distributive axiom,**

$$a(b + c + d + \ldots + n) = ab + ac + ad + \ldots + an,$$

which can also be rewritten using the commutative axiom.

We shall define **subtraction** for any real numbers a and b as

$$\blacktriangleright \qquad a - b = a + (-b).$$

Thus, the subtraction $a - b$ is defined as the addition of a and the additive inverse of b. **Division** of any real number a by any nonzero real number b

(denoted $a \div b$, $\dfrac{a}{b}$, or a/b) is defined analogously as follows.

▶
$$a \div b = \frac{a}{b} = a \cdot \frac{1}{b}.$$

That is, to find the quotient a/b, we find the product of a and the multiplicative inverse of b.

1.2 Exercises

Answer true or false. If false, tell why.

1. The set of integers is closed with respect to addition.
2. The set of irrational numbers is closed with respect to multiplication.
3. The set $\{0, 1\}$ is closed with respect to subtraction.
4. The set $\{0, 1\}$ is closed with respect to multiplication.
5. The set $\{1, -1\}$ is closed with respect to division.
6. The set of irrational numbers contains an identity element for multiplication.
7. The set of natural numbers contains an identity element for addition.
8. The set of even numbers satisfies the inverse axiom for addition.
9. The set of rational numbers between 0 and 1 inclusive satisfies the inverse axiom for multiplication.
10. Any subset of the real numbers satisfies the commutative axioms.

Identify the axiom(s) which is illustrated in each of the following.

11. $3\left(\dfrac{1}{3}\right) = \left(\dfrac{1}{3}\right)3$

12. $\dfrac{3}{4} + \left(-\dfrac{3}{4}\right) = 0$

13. $2\left(\dfrac{1}{4}\right) + 2\left(\dfrac{3}{4}\right) = 2(1)$

14. $x + y = y + x$
15. $(y + x) + 2x = y + (x + 2x)$
16. If $(x + y) + 2x = y + (x + 2x)$, and $y + (x + 2x) = y + 3x$, then $(x + y) + 2x = y + 3x$.

17. $(x + 6)\left(\dfrac{1}{x + 6}\right) = 1, (x \neq -6)$

18. If x is a real number, $x + 2$ is a real number.
19. If $x + 2 = 3$ and $(x + 2) - y = z$, then $3 - y = z$.
20. $(7 - y) + 0 = 7 - y$

In this section we stated that the field of real numbers forms a mathematical system. Another example of a mathematical system is one in which the elements are sets. The following exercises formulate this system.

21. State two relations between sets.
22. State two binary operations on sets.
23. State five axioms for sets. (Hint: refer to Exercise Set 1.1, Exercises 41–45.)

Use the axioms and definitions of real numbers to complete the following proofs.

24. Prove: $a[b + (c + d)] = ab + ac + ad$.

$$a[b + (c + d)] = ab + a(c + d) \qquad \rule{3cm}{0.4pt}$$
$$ab + a(c + d) = ab + ac + ad \qquad \rule{3cm}{0.4pt}$$
$$a[b + (c + d)] = ab + ac + ad \qquad \rule{3cm}{0.4pt}$$

25. Prove: $0 - a = -a$.

$$0 - a = 0 + (-a) \qquad \rule{3cm}{0.4pt}$$
$$0 + (-a) = -a \qquad \rule{3cm}{0.4pt}$$
$$0 - a = -a \qquad \rule{3cm}{0.4pt}$$

26. Prove: $a - a = 0$.

$$a - a = a + (-a) \qquad \rule{3cm}{0.4pt}$$
$$a + (-a) = 0 \qquad \rule{3cm}{0.4pt}$$
$$a - a = 0 \qquad \rule{3cm}{0.4pt}$$

27. Prove: $\dfrac{1}{a}(ba) = b$.

$$\frac{1}{a}(ba) = \frac{1}{a}(ab) \qquad \rule{3cm}{0.4pt}$$
$$\frac{1}{a}(ab) = \left(\frac{1}{a} \cdot a\right)b \qquad \rule{3cm}{0.4pt}$$
$$\left(\frac{1}{a} \cdot a\right)b = 1 \cdot b \qquad \rule{3cm}{0.4pt}$$
$$1 \cdot b = b \qquad \rule{3cm}{0.4pt}$$
$$\frac{1}{a}(ba) = b \qquad \rule{3cm}{0.4pt}$$

28. Prove: $(a + b)c = ac + bc$.

$$(a + b)c = c(a + b) \qquad \rule{3cm}{0.4pt}$$
$$c(a + b) = ca + cb \qquad \rule{3cm}{0.4pt}$$
$$(a + b)c = ca + cb \qquad \rule{3cm}{0.4pt}$$
$$ca + cb = ac + bc \qquad \rule{3cm}{0.4pt}$$
$$(a + b)c = ac + bc \qquad \rule{3cm}{0.4pt}$$

29. Prove: $(m + n) + (p + q) = (m + p) + (n + q)$.

$(m + n) + (p + q) = [(m + n) + p] + q$ _____

$[(m + n) + p] + q = [m + (n + p)] + q$ _____

$[m + (n + p)] + q = [m + (p + n)] + q$ _____

$[m + (p + n)] + q = [(m + p) + n] + q$ _____

$[(m + p) + n] + q = (m + p) + (n + q)$ _____

$(m + n) + (p + q) = (m + p) + (n + q)$ _____

30. Prove: $(a + b) + (-a) = b$.

$(a + b) + (-a) = a + [b + (-a)]$ _____

$a + [b + (-a)] = a + [-a + b]$ _____

$a + [-a + b] = [a + (-a)] + b$ _____

$[a + (-a)] + b = 0 + b$ _____

$0 + b = b$ _____

$(a + b) + (-a) = b$ _____

1.3 Consequences of the Field Axioms

From the basic definitions and axioms stated in Section 1.2, we can demonstrate many further properties of the real numbers. These consequences of the axioms and definitions are called **theorems** and are distinguished by name from the axioms, since they follow deductively from the basic properties which we accept without proof. In this section, we shall demonstrate the proofs of several important and useful properties which are consequences of the field axioms and the definitions.

The first theorem we shall prove is used so much that we give it a name.

THEOREM 1.1   ADDITION PROPERTY OF EQUALITY
For all real numbers a, b, and c,
if $a = b$, then $a + c = b + c$.

Proof

$a + c$ is a real number	closure axiom for addition
$a + c = a + c$	reflexive axiom
$a = b$	given
$a + c = b + c$	substitution axiom

The conclusion of this theorem can also be stated as $c + a = c + b$, by the commutative axiom. There is a comparable theorem for multiplication.

THEOREM 1.2   MULTIPLICATION PROPERTY OF EQUALITY
For all real numbers, a, b, and c,
if $a = b$, then $ac = bc$.

The conclusion here could also be stated as $ca = cb$. The proof of the multiplication property of equality is similar to that for Theorem 1.1 and is left for the exercises.

THEOREM 1.3   CANCELLATION PROPERTY OF ADDITION
For all real numbers a, b, and c,
if $a + c = b + c$, then $a = b$.

Proof

$a + c = b + c$	given
$(a + c) + (-c) = (b + c) + (-c)$	addition property of equality
$a + [c + (-c)] = b + [c + (-c)]$	associative axiom
$c + (-c) = 0$	additive inverse axiom
$a + 0 = b + 0$	substitution axiom
$a = b$	additive identity axiom
	and substitution axiom

Note that Theorem 1.3 is the converse of Theorem 1.1—that is, the hypothesis and conclusion of Theorem 1.1 are respectively the conclusion and hypothesis of Theorem 1.3. It is convenient to combine these two theorems into one statement as follows.

THEOREM 1.3 (A)   For all real numbers a, b, and c,
$a = b$ if and only if $a + c = b + c$.

The words "if and only if" here tell us that the theorem is true with either part as hypothesis and the other as conclusion. Thus, in general, the phrase "*p* if and only if *q*" means "if *p*, then *q* *and* if *q*, then *p*." Proof of such a theorem requires proof of both "if *p*, then *q*" and "if *q*, then *p*."

THEOREM 1.4   CANCELLATION PROPERTY OF
MULTIPLICATION
For all real numbers a, b, and $c \neq 0$,
$a = b$ if and only if $ac = bc$.

To prove this theorem, we must prove two statements: (1) if $a = b$, then $ac = bc$, and (2) if $ac = bc$, then $a = b$. The proof is left for the exercises.

Another important and useful theorem is the multiplication property of zero.

THEOREM 1.5   MULTIPLICATION PROPERTY OF ZERO
For all real numbers a,
$a \cdot 0 = 0 \cdot a = 0$.

Proof

$a = a \cdot 1$	identity axiom of multiplication
$a = a(1 + 0)$	identity axiom of addition and substitution axiom
$a = a \cdot 1 + a \cdot 0$	distributive axiom
$a = a + a \cdot 0$	identity axiom of multiplication and substitution axiom
$-a + a = -a + (a + a \cdot 0)$	addition property of equality
$-a + a = (-a + a) + a \cdot 0$	associative axiom of addition
$0 = 0 + a \cdot 0$	additive inverse axiom and substitution
$0 = a \cdot 0$	additive identity axiom and substitution
$a \cdot 0 = 0$	symmetric axiom
$a \cdot 0 = 0 \cdot a$	commutative axiom of multiplication
$0 = 0 \cdot a$	transitive axiom of equality
$0 \cdot a = 0$	symmetric axiom

The rules for operating with quotients are summarized in Theorem 1.6.

THEOREM 1.6   For all real numbers a, $b \neq 0$, c, and $d \neq 0$:

(a) $\dfrac{a}{b} = \dfrac{c}{d}$ if and only if $ad = bc$

(b) $\dfrac{ad}{bd} = \dfrac{a}{b}$ *

(c) $\dfrac{1}{b} \cdot \dfrac{1}{d} = \dfrac{1}{bd}$

(d) $\dfrac{a}{b} \cdot \dfrac{c}{d} = \dfrac{ac}{bd}$

(e) $\dfrac{a}{b} + \dfrac{c}{b} = \dfrac{a + c}{b}$.

Theorem 1.6(a) describes the useful **equality test for rational numbers,** while (b) states the **fundamental theorem of rational numbers** which is the basis for reducing fractions to lowest terms or building fractions to higher terms. Theorem 1.6(c) states that the product of the multiplicative inverses of b and d equals the multiplicative inverse of the product bd. Parts (d) and (e) show how to perform the operations of multiplication and addition for rational numbers. The proofs of these statements are included in the exercises.

Since rational numbers are real numbers, the definition of subtraction for rational numbers follows from the definition of subtraction of real numbers.

$$\frac{a}{b} - \frac{c}{d} = \frac{a}{b} + \left(-\frac{c}{d}\right)$$

* Here $bd \neq 0$ follows from the two restrictions $b \neq 0$ and $d \neq 0$.

Similarly, the definition of division for rational numbers follows from the definition of division of real numbers and the definition of multiplication of rational numbers.

$$\frac{\frac{a}{b}}{\frac{c}{d}} = \frac{a}{b} \cdot \frac{1}{\frac{c}{d}} = \frac{a}{b} \cdot \frac{d}{c} = \frac{ad}{bc} \qquad \left(\frac{c}{d} \neq 0\right)$$

The next theorem includes several facts about operations with real numbers.

THEOREM 1.7   For all real numbers a and b:

(a) $-(-a) = a$

(b) $(-a) + (-b) = -(a + b)$

(c) $(-a)b = -(ab)$

(d) $(-a)(-b) = ab$

(e) $\dfrac{-a}{b} = \dfrac{a}{-b} = -\dfrac{a}{b},\ (b \neq 0)$

(f) $\dfrac{-a}{-b} = \dfrac{a}{b},\ (b \neq 0)$.

Theorem 1.7(a) states that the additive inverse of the additive inverse of a is a, while (b) states that the sum of the additive inverses of a and b is the additive inverse of the sum of a and b. Theorem 1.7(c) and (d) deal with products of additive inverses. Theorem 1.7(e) and (f) are consequences of the definition of division and Theorem 1.7(c) and (d).

The proof of Theorem 1.7(a) follows from the fact that both a and $-(-a)$ are additive inverses of $-a$ by definition. That is, $a + (-a) = 0$ and $-a + [-(-a)] = 0$. However, by the additive inverse axiom, for each real number, there is a *unique* additive inverse, so that we must have $-(-a) = a$. The proofs for the remainder of Theorem 1.7 are left as exercises.

1.3 Exercises

Complete the following proofs by giving reasons for each step, using axioms, definitions, or previously proved theorems. Some steps require more than one reason. Assume a, b, c, and d represent real numbers.

1. Prove: $(-a) + (-b) = -(a + b)$.

$[(-a) + (-b)] + (a + b) = (-a + a) + (-b + b)$ _____

$= 0 + 0$ _____

$= 0$ _____

$(-a) + (-b)$ is the additive inverse of $(a + b)$ _____

$-(a + b)$ is the additive inverse of $(a + b)$ _____

$(-a) + (-b) = -(a + b)$ _____

2. Prove: $-(a - b) = b - a$.

$$(b - a) + (a - b) = [b + (-a)] + [a + (-b)]$$
$$= [a + (-a)] + [b + (-b)]$$
$$= 0 + 0$$
$$= 0$$

$b - a$ is the additive inverse of $a - b$

$-(a - b)$ is the additive inverse of $a - b$

$-(a - b) = b - a$

3. Prove: If $x + a = b$, then $x = b - a$.

$$x + a = b \qquad\qquad\qquad\qquad\qquad \text{given}$$
$$(x + a) + (-a) = b + (-a)$$
$$x + [a + (-a)] = b + (-a)$$
$$x + 0 = b + (-a)$$
$$x = b + (-a)$$
$$x = b - a$$

4. Prove: $(-a)b = -(ab)$.

$$[(-a)b] + ab = [-a + a]b$$
$$= 0 \cdot b$$
$$= 0$$

$(-a)b$ is the additive inverse of ab

$-(ab)$ is the additive inverse of ab

$(-a)b = -(ab)$

5. Prove: If $b \neq 0$, $\dfrac{-a}{b} = -\dfrac{a}{b}$.

$$\frac{-a}{b} = -a\left(\frac{1}{b}\right)$$
$$= -\left(a \cdot \frac{1}{b}\right)$$
$$= -\frac{a}{b}$$

6. Prove: For $b \neq 0$ and $d \neq 0$, $\dfrac{a}{b} = \dfrac{c}{d}$ if and only if $ad = bc$.

PART I

$$\frac{a}{b} = \frac{c}{d} \qquad\qquad\qquad\qquad\qquad \text{given}$$

$$\frac{a}{b}(bd) = \frac{c}{d}(bd)$$

$$\left(a \cdot \frac{1}{b}\right)(bd) = \left(c \cdot \frac{1}{d}\right)(bd)$$

$$ad\left(\frac{1}{b} \cdot b\right) = cb\left(\frac{1}{d} \cdot d\right)$$

$$ad(1) = cb(1)$$

$$ad = bc$$

$ad = cb$ .... given

$$ad\left(\frac{1}{b} \cdot \frac{1}{d}\right) = cb\left(\frac{1}{b} \cdot \frac{1}{d}\right)$$ _____

$$\left(a \cdot \frac{1}{b}\right)\left(d \cdot \frac{1}{d}\right) = \left(c \cdot \frac{1}{d}\right)\left(b \cdot \frac{1}{b}\right)$$ _____

$$\left(a \cdot \frac{1}{b}\right)1 = \left(c \cdot \frac{1}{d}\right)1$$ _____

$$a \cdot \frac{1}{b} = c \cdot \frac{1}{d}$$ _____

$$\frac{a}{b} = \frac{c}{d}$$ _____

From parts I and II, we have $\frac{a}{b} = \frac{c}{d}$ if and only if $ad = bc$.

7. Prove: If $b \neq 0$ and $d \neq 0$, $\frac{ad}{bd} = \frac{a}{b}$.

$(ad)b = (ad)b$ _____

$(ad)b = (bd)a$ _____

$\frac{ad}{bd} = \frac{a}{b}$ _____

8. Prove: If $a \neq 0$ and $b \neq 0$, then $\frac{1}{a} \cdot \frac{1}{b} = \frac{1}{ab}$.

$$\left(\frac{1}{a} \cdot \frac{1}{b}\right)(ab) = \left(\frac{1}{a} \cdot a\right)\left(\frac{1}{b} \cdot b\right)$$ _____

$$= 1 \cdot 1$$ _____

$$= 1$$ _____

$\frac{1}{a} \cdot \frac{1}{b}$ is the multiplicative inverse of ab _____

$\frac{1}{ab}$ is the multiplicative inverse of ab _____

$\frac{1}{a} \cdot \frac{1}{b} = \frac{1}{ab}$ _____

Prove the following theorems. Assume a, b, and c are real numbers.

9. $(-1)a = -a$

10. If $x = \frac{a}{b}$, $b \neq 0$, then $a = bx$.

11. $a = b$ if and only if $ac = bc$, $(c \neq 0)$

12. $(-a)(-b) = ab$

13. $\frac{-a}{-b} = \frac{a}{b}$, $(b \neq 0)$

14. $\frac{a}{b} \cdot \frac{c}{d} = \frac{ac}{bd}$, $(b \neq 0, d \neq 0)$

15. $\frac{a}{b} + \frac{c}{b} = \frac{a + c}{b}$, $(b \neq 0)$

16. $a(b - c) = ab - ac$

1.4 Axioms of Order and Completeness

We saw in Section 1.2 that "is equal to" and "is not equal to" are relations between real numbers. Another important relation between real numbers is the order relation "is less than." To indicate the relation "a is less than b" we write $a < b$.

To discuss the ordering of the real numbers it is convenient to consider the correspondence which can be established between the set of real numbers and the set of points on a line. By choosing some point on a line as the point which we shall identify as 0, and then choosing a unit of measure, we can identify the number 1 with the point one unit to the right of 0, -1 with the point one unit to the left, and so on. We shall call this line the **number line,** and abbreviate the phrase "the point corresponding to the number x" as "the point x."

On the number line, points to the right of 0 represent positive numbers, and points to the left of 0 represent negative numbers. Note that the positive numbers increase toward the right, so that for positive numbers a and b, $a < b$ means a is to the left of b on the number line. To maintain this order, we shall define the relation "is less than" as follows.

▶   For all real numbers a and b,
     $a < b$ if and only if, for some positive real number c, $a + c = b$.

This relationship is illustrated in Figure 1.1.

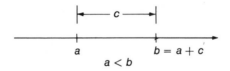

$$a < b$$

Figure 1.1

Several other order relations can be derived from the definitions of $<$ and $=$. We can restate the relation "a is less than b" as "b is greater than a," written $b > a$. The following variations of these relations are frequently used.

$\leq$	is less than or equal to
$\geq$	is greater than or equal to
$\not<$	is not less than
$\not>$	is not greater than

Sentences using these order relations, along with $<$, $>$, and $\neq$, are sometimes called inequalities.

From the definition of $a < b$, and by referring to the number line, we can see that the following order axioms hold.

Axioms of Order

For all real numbers a, b, and c:

Exactly one of these three conditions holds: *trichotomy axiom*
$a = b$, $a < b$, or $b < a$.
If $a > 0$ and $b > 0$, then $a + b > 0$ and $ab > 0$.   *closure axiom*

We can now restate the definition of positive and negative numbers in terms of inequalities.

▶   The number a is **positive** if and only if $a > 0$,
and a is **negative** if and only if $a < 0$.

Several further properties of the relation $<$ are given in Theorem 1.8.

THEOREM 1.8   For all real numbers a, b, and c:
(a)   if $a < b$, then $a + c < b + c$
(b)   if $a < b$ and $c > 0$, then $ac < bc$
(c)   if $a < b$ and $c < 0$, then $ac > bc$
(d)   if $a < b$ and $b < c$, then $a < c$
(e)   if $a > 0$, then $-a < 0$.

Proof of (a)

$a < b$	given
For some positive number m, $a + m = b$.	definition of $<$
$(a + m) + c = b + c$	addition property of equality
$(a + c) + m = b + c$	commutative and associative axioms
$a + c < b + c$	definition of $<$

The proofs of parts (b), (c), (d), and (e) are similar.

A nonempty set S of real numbers is said to have an **upper bound** b if b is a real number and for every element x in S, $x \le b$. If b is less than every other upper bound of S, it is called the **least upper bound.** Similarly, S has a lower bound b', if for every element x of S, $x \ge b'$. Also, b' is the **greatest lower bound** of S if it is larger than every other lower bound of S. For example, the set $M = \{0, 1, 2\}$ has as upper bounds the numbers 2, 3.5, $\sqrt{15}$, and so on, with 0 and all negative numbers as lower bounds. The number 2 is the least upper bound while 0 is the greatest lower bound. The set $\{x \,|\, x \le 2\}$ also has 2 as a least upper bound, but has no lower bound. The least upper bound and greatest lower bound need not belong to the set. For example, the set $\{x \,|\, x > 5\}$ has no upper bound, but has a greatest lower bound of 5, which is not an element of the set.

The last property needed to describe the set of real numbers is called the axiom of completeness. This axiom guarantees that the number line representing the set of real numbers will have no "gaps."

Axiom of Completeness

If a nonempty set S of real numbers has an upper bound, then S has a least upper bound. Also, if S has a lower bound, then S has a greatest lower bound.

With this axiom, we have completed the description of the real numbers as a mathematical system. The set of real numbers is a **complete ordered field,** since it satisfies the completeness axiom, the order axioms, and the field axioms. In fact, the set of real numbers is the only set which satisfies all of these axioms and thus is the only complete ordered field. The set of rational numbers is an ordered field but is not complete. To see why, consider the set $\{x \,|\, x \text{ is a rational number}, x^2 < 2\}$. This set has an upper bound in the set of rational numbers, for example, 1.5, but it has no least upper bound in the set of rational numbers. The field of complex numbers is not ordered, as is shown in a later chapter.

1.4 Exercises

Mark each of the following true or false.

1. $4 < -2$
2. $-7 < -5$
3. $-\dfrac{1}{2} > -\dfrac{1}{3}$

4. $\dfrac{1}{4} + \dfrac{1}{4} \le \dfrac{1}{2}$
5. $\dfrac{2}{3} \ge -\dfrac{2}{3}$

6. $\dfrac{23}{4} \not< \dfrac{15}{4}$

Identify the axiom or theorem of this section which is illustrated by each of the following statements. Assume x and y represent real numbers.

7. If $x < 2$ and $2 < y$, then $x < y$.
8. If $x < 4$ then $x + 7 < 11$.
9. If $x < -2$ then $4x < -8$.
10. If $x < -2$ then $-4x > 8$.
11. If $x + 3 < y$, then $x < y - 3$.
12. Either $x < 0$, $x = 0$, or $x > 0$.

For each of the following sets of real numbers give both the greatest lower bound and the least upper bound.

13. $\{x \,|\, x \le 6\}$
14. $\{x \,|\, -2 \le x < 5\}$
15. $\{x \,|\, x < 5 \text{ and } x \text{ is a natural number}\}$
16. $\{x \,|\, x < 5 \text{ and } x \text{ is a rational number}\}$

17. $\{\cdots, 10, 1, .1, .01, .001, \cdots\}$

18. $\left\{1, \dfrac{1}{2}, \dfrac{1}{3}, \dfrac{1}{4}, \cdots\right\}$

Complete the following proofs for all real numbers a, b, c, and d.

19. Prove: If $a < b$ and $c > 0$, then $ac < bc$.

$a < b$	given
For some $m > 0$, $a + m = b$.	_____
$(a + m)c = bc$	_____
$ac + mc = bc$	_____
$c > 0$	_____
$mc > 0$	_____
$ac < bc$	_____

20. Prove: If $a < b$ and $c < 0$, then $ac > bc$.

$a < b$	given
For some $m > 0$, $a + m = b$.	_____
$(a + m)c = bc$	_____
$ac + mc = bc$	_____
$c < 0$	_____
$mc < 0$	_____
$-(mc) > 0$	_____
$(ac + mc) + (-mc) = bc + (-mc)$	_____
$ac = bc + (-mc)$	_____
$bc < ac$	_____

21. Prove: $a > b$ if and only if $-a < -b$.

PART I

$a > b$	_____
$a(-1) < b(-1)$	_____
$a(-1) = -a$ and $b(-1) = -b$	_____
$-a < -b$	_____

PART II

$-a < -b$	given
_____	_____
_____	_____
_____	_____

22. Prove: If $a > 1$, then $a^2 > 1$.

$a > 1$	given
$a(a) > 1(a)$	_____
$1(a) = a$	_____
$1(a) > 1$	_____
$a^2 = a(a) > 1$	_____

23. Prove: If $a < b$ and $c < d$, then $a + c < b + d$.

$a < b$ and $c < d$	given
For some $m > 0$, $a + m = b$.	_____
For some $n > 0$, $c + n = d$.	_____
$(a + m) + (c + n) = b + d$	_____
$(a + c) + (m + n) = b + d$	_____
$m + n > 0$	_____
$a + c < b + d$	_____

24. Prove: For positive numbers a, b, c, and d, if $a < b$ and $c < d$, then $ac < bd$.

For some positive number m, $a + m = b$.	_____
For some positive number n, $c + n = d$.	_____
$(a + m)(c + n) = bd$	_____
$a(c + n) + m(c + n) = bd$	_____
$ac + an + mc + mn = bd$	_____
$an + mc + mn > 0$	_____
$ac < bd$	_____

25. Prove: If $a > 1$, then $a^3 > 1$.

1.5 Absolute Value

Associated with each real number there is a nonnegative quantity which represents the number's distance from 0 on the number line. This is called the **absolute value** of the number, written $|x|$, and is defined as follows.

▶
$$|x| = \begin{cases} x \text{ if } x \geq 0 \\ -x \text{ if } x < 0 \end{cases}$$

For example, $|5| = 5$, $|-5| = 5$, and $-|-3| = -3$.

Since absolute value represents distance on the number line, to represent the distance between the numbers x and b, we can write $|x - b|$, where $|x - b|$ is defined as follows.

▶
$$|x - b| = \begin{cases} x - b \text{ if } x - b \geq 0 \\ -(x - b) \text{ if } x - b < 0 \end{cases}$$

The conditions $x - b \geq 0$ and $x - b < 0$ are equivalent to the simpler statements $x \geq b$ and $x < b$. When $b = 0$, this more general definition becomes our earlier definition of absolute value.

Example 2 Solve $|x| = 3$.

Since either $|3| = 3$ or $|-3| = 3$, the solution set is $\{3, -3\}$.

Example 3 Solve $|x - 4| = 2$.

Again we have two cases.

(1) If $x \geq 4$, $|x - 4| = x - 4$, so that
$$x - 4 = 2$$
$$x = 6.$$

(2) If $x < 4$, $|x - 4| = -(x - 4)$, and
$$-(x - 4) = 2$$
$$x = 2.$$

Thus, the solution set is $\{2, 6\}$.

In Section 1.4, we defined inequality relations. We can use these relations to describe intervals on the number line. For example, the set of numbers $\{x \mid 1 < x < 5\}$ corresponds to the set of points on the number line between 1 and 5 as shown in Figure 1.2. The fact that 1 and 5 are not included in the set is shown by the open circles at those points. When we write an inequality such as $1 < x < 5$ we must keep in mind that this represents the compound sentence $1 < x$ *and* $x < 5$, where a **compound sentence** refers to two simple sentences joined with a connective, such as *and* or *or*.

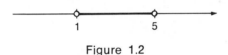

Figure 1.2

We can use the same method to describe other intervals on the number line, as shown in Figure 1.3, where a solid circle indicates that the point belongs to the set.

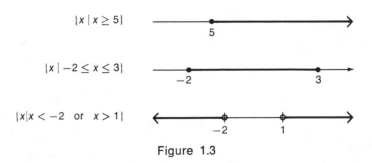

$\{x \mid x \geq 5\}$

$\{x \mid -2 \leq x \leq 3\}$

$\{x \mid x < -2 \quad \text{or} \quad x > 1\}$

Figure 1.3

Now, let us consider absolute value inequalities. To solve the inequality $|x| < 5$, we use the definition of $|x|$. There are two cases:

(1) $x < 0$ and $|x| < 5$

If $x < 0$, $|x| = -x$, and so $|x| < 5$ becomes $-x < 5$, or $x > -5$. Thus, $x < 0$ and $|x| < 5$ becomes $x < 0$ and $x > -5$, or $-5 < x < 0$.

(2) $x \geq 0$ and $|x| < 5$

If $x \geq 0$, $|x| = x$, and so $|x| < 5$ becomes $x < 5$. Thus, $x \geq 0$ and $|x| < 5$ becomes $x \geq 0$ and $x < 5$, or $0 \leq x < 5$.

From the first case, we know every number x satisfying $-5 < x < 0$ is a solution, and from the second case all numbers x such that $0 \leq x < 5$ are solutions. The final solution set is the *union* of the solution sets from the two cases. It can be written $\{x \mid -5 < x < 0 \text{ or } 0 \leq x < 5\}$, or simply

$$\{x \mid -5 < x < 5\}.$$

The solution of the inequality $|x| > 5$ again requires two cases:

(1) $x < 0$ and $|x| > 5$

Here $|x| = -x$ so $|x| > 5$ becomes $-x > 5$, or $x < -5$. Now $x < 0$ and $|x| > 5$ becomes $x < 0$ and $x < -5$, or simply $x < -5$.

(2) $x \geq 0$ and $|x| > 5$

Here $|x| = x$ so $|x| > 5$ becomes $x > 5$. Now $x \geq 0$ and $|x| > 5$ becomes $x \geq 0$ and $x > 5$, or simply $x > 5$.

Again the final solution set is the union of the solution sets from case (1) and case (2). It can be written $\{x \mid x < -5 \text{ or } x > 5\}$. Generalizing, we have the following.

THEOREM 1.9   For all real numbers a and b, $b \geq 0$:
(a) $|a| \leq b$ if and only if $-b \leq a \leq b$
(b) $|a| \geq b$ if and only if $a \leq -b$ or $a \geq b$.

Example 4   Solve $|x| \geq 3$.
  From Theorem 1.9(b), if $|x| \geq 3$, then $x \leq -3$ or $x \geq 3$, as shown on the number line of Figure 1.4. The solution set here is written $\{x \mid x \leq -3 \text{ or } x \geq 3\}$.

Figure 1.4

Example 5   Solve $|x - 2| < 5$.
  From Theorem 1.9(a), letting $a = x - 2$ and $b = 5$, we have

$$|x - 2| < 5$$
$$-5 < x - 2 < 5$$
$$-3 < x < 7.$$

The solution set is written $\{x \mid -3 < x < 7\}$.

Properties of operations with absolute value are given in the following theorem.

THEOREM 1.10   For all real numbers a and b:

(a) $|ab| = |a| \cdot |b|$

(b) $|a + b| \le |a| + |b|$

(c) $|a - b| \ge |a| - |b|$

(d) $\left|\dfrac{a}{b}\right| = \dfrac{|a|}{|b|}$, $(b \ne 0)$.

We omit the proof of these theorems. The proofs depend on the definitions of addition and multiplication for real numbers, and the definition of absolute value. Some examples of such proofs are included in the exercises.

1.5 Exercises

Complete the following statements by choosing the correct relation: $=$, $\le$, *or* $\ge$.

1. $|5|$ _____ $|-5|$
2. $|x|$ _____ $|-x|$
3. $-|x|$ _____ $|x|$
4. $-|-x|$ _____ $|x|$
5. $|x - 4|$ _____ $|4 - x|$
6. $|x - 8|$ _____ $|8 - x|$
7. If $x < y$, then $|x - y|$ _____ $y - x$.
8. If $x < y$, then $|x - y|$ _____ $x - y$.
9. If $|x| \le 5$, then -5 _____ x _____ 5.
10. If $|3x| \le 1$, then -1 _____ $3x$ _____ 1.
11. If $|x| \ge 4$, then x _____ 4 or x _____ -4.
12. If $|x + 2| \ge 3$, then $x + 2$ _____ 3 or $x + 2$ _____ -3.
13. $|3| \cdot |-5|$ _____ $|3(-5)|$
14. $|3| \cdot |x|$ _____ $|3x|$
15. $|4 + 3|$ _____ $|4| + |3|$
16. $|5 - 1|$ _____ $|5| + |-1|$
17. $|-2 + 3|$ _____ $|-2| + |3|$
18. $|5 - 1|$ _____ $|5| - |1|$
19. $|3 - 2|$ _____ $|3| - |2|$
20. $|3 - 4|$ _____ $|3| - |4|$
21. $\dfrac{|7|}{|-1|}$ _____ $\left|\dfrac{7}{-1}\right|$
22. $\dfrac{|-1|}{|-3|}$ _____ $\left|\dfrac{-1}{-3}\right|$

Graph the following intervals on the number line.

23. $x \geq 5$
24. $-2 < x < 4$
25. $x + 2 \leq 8$ and $x \geq 3$
26. $x - 1 > 0$ and $x < 0$
27. $x > 3$ or $x < 1$
28. $2x > -2$ or $x + 2 < 4$
29. $x \geq 2$ or $x > 5$

Solve.

30. $|x| = 4$
31. $|x| = -2$
32. $|-x| = -1$
33. $|-x| = 1$
34. $|x - 2| = 5$
35. $|2x + 5| = 3$
36. $\left| x - \dfrac{1}{2} \right| = 3$
37. $\left| \dfrac{x}{3} + 1 \right| = 2$

Chapter Two

Algebraic Expressions and Exponents

An **algebraic expression** is any expression of variables or constants obtained by performing a finite number of the basic operations of addition, subtraction, multiplication, division (except by zero), or extraction of roots. The simplest such algebraic expression is a polynomial, which we shall discuss in the first section.

2.1 Polynomials

If n is any natural number, and a is any real number, then we define the symbol a^n by writing

▶ $$a^n = a \cdot a \cdot a \cdots a,$$

where a appears as a factor n times. Here n is called the **exponent,** or **power,** a is called the **base,** and the expression a^n is called an **exponential.** Thus

$$4^3 = 4 \cdot 4 \cdot 4 = 64, \quad \text{and} \quad (-6)^2 = 36.$$

Any product of a constant (called the **coefficient**) and variables raised to powers is called a **term.** A finite sum of terms with whole number exponents on the variables, such as $-2x^3 + 4x^2 - 5$, or $2m^3n^2y - 3$, or $3p$, is called a **polynomial.**

If every term in a polynomial contains only one variable, such as x, the polynomial is called a **polynomial in x.** If every coefficient in a polynomial is a real number, the polynomial is called a **polynomial over the reals; a polynomial over the rationals** and a **polynomial over the integers** are defined in a similar way. Thus, $6x^3 - 4x^2 + 3x - 5$ is a polynomial in x over the integers, and $\sqrt{3}m^4 - 5m^3y + 3y^4$ is a polynomial over the reals.

The highest exponent in a polynomial in one variable is called the **degree** of the polynomial. A constant is said to have degree zero. By the definition above, $3x^4 - 5x^2 + 2x + 3$ is a polynomial of degree 4. A term containing more than one variable is said to have degree equal to the sum of all exponents appearing on the variables in the term. For example, $-3x^4y^3z^5$ is of degree $4 + 3 + 5 = 12$. The degree of a polynomial in more than one variable is equal to the highest degree of any term appearing in the polynomial. Thus, $2x^4y^3 - 3x^5y + x^6y^2$ is of degree 8. A polynomial containing exactly three terms is called a **trinomial,** one containing exactly two terms is a **binomial,** while a single term polynomial is called a **monomial.**

Capital letters are often used to name polynomials; we shall use the notation $P(x)$ to represent a polynomial P containing x as the only variable. If $P(x) = 3x^2 - 4x + 5$, we shall use $P(2)$ to represent the value of the polynomial P when $x = 2$. Here,

$$P(x) = 3x^2 - 4x + 5,$$

and so
$$\begin{aligned} P(2) &= 3 \cdot 2^2 - 4 \cdot 2 + 5 \\ &= 3 \cdot 4 - 4 \cdot 2 + 5 \\ &= 12 - 8 + 5 \\ &= 9. \end{aligned}$$

In the same way, $P(-1) = 12$ and $P(\frac{1}{2}) = \frac{15}{4}$.

Sums and differences of polynomials can be found by using the distributive property. Thus

$$\begin{aligned} 3m^5 - 8m^5 &= (3 - 8)m^5 \\ &= -5m^5, \end{aligned}$$

and

$$\begin{aligned} (2y^4 - 3y^2 + y) + (4y^4 + 7y^2 + 6y) &= (2 + 4)y^4 + (-3 + 7)y^2 + (1 + 6)y \\ &= 6y^4 + 4y^2 + 7y. \end{aligned}$$

Before we can find products of polynomials we need to develop a few properties of exponents. By definition, a^m (m a natural number, a any real number) means $a \cdot a \cdots a$, where a appears as a factor m times. In the same way, a^n means $a \cdot a \cdots a$, where a appears as a factor n times. Thus in the product $a^m \cdot a^n$, a appears as a factor a total of $m + n$ times. Hence, for any natural numbers m and n and any real number a,

$$a^m \cdot a^n = a^{m+n}.$$

Similar proofs can be given for the other properties of exponents summarized in the following theorem.

THEOREM 2.1   For any natural numbers m and n,
and for any real numbers a and b:

(a) $a^m \cdot a^n = a^{m+n}$

(b) $\dfrac{a^m}{a^n} = \begin{cases} a^{m-n} \text{ if } m > n \\ 1 \text{ if } m = n \\ \dfrac{1}{a^{n-m}} \text{ if } m < n \end{cases}$ $(a \neq 0)$

(c) $(a^m)^n = a^{mn}$

(d) $(ab)^m = a^m b^m$

(e) $\left(\dfrac{a}{b}\right)^m = \dfrac{a^m}{b^m}$, $(b \neq 0)$.

Example 1   Each of the following examples is justified by one or more parts of Theorem 2.1, and the properties of multiplication.

(a) $(4x^3)(-3x^5) = (4)(-3)x^3 \cdot x^5 = -12x^8$

(b) $-3^4(-3)^5 = -(3)^4 \cdot [-(3)^5] = 3^9$

(c) $-3^5(-3)^4 = -3^9$

(d) $\dfrac{k^{r+1}}{k^r} = k^{(r+1)-r} = k$, for any natural number r, $k \neq 0$

We can also use the distributive property to find the product of two polynomials when one of them is not a monomial. For example, if we treat $3x - 4$ as a single quantity, we can use the distributive property to write

$$(3x - 4)(2x^2 - 3x + 5) = (3x - 4)(2x^2) + (3x - 4)(-3x) + (3x - 4)(5).$$

If we use the distributive property again, we have

$(3x - 4)(2x^2 - 3x + 5)$
$$\begin{aligned}
&= (3x)(2x^2) - 4(2x^2) + (3x)(-3x) - 4(-3x) + (3x)5 - 4(5) \\
&= 6x^3 - 8x^2 - 9x^2 + 12x + 15x - 20 \\
&= 6x^3 - 17x^2 + 27x - 20.
\end{aligned}$$

It is sometimes more convenient to write such a product as shown below.

$$\begin{array}{r}
2x^2 - 3x + 5 \\
3x - 4 \\
\hline
- 8x^2 + 12x - 20 \\
6x^3 - 9x^2 + 15x \\
\hline
6x^3 - 17x^2 + 27x - 20
\end{array}$$

Example 2

$$\begin{aligned}
(6m + 1)(4m - 3) &= (6m)(4m) + (6m)(-3) + 1(4m) + 1(-3) \\
&= 24m^2 - 14m - 3
\end{aligned}$$

By Theorem 2.1 (b), we know that if a is any nonzero real number, and m and n are natural numbers, then

$$\frac{a^m}{a^n} = \begin{cases} a^{m-n} \text{ if } m > n \\ 1 \text{ if } m = n \\ \dfrac{1}{a^{n-m}} \text{ if } m < n. \end{cases}$$

What happens if we try to generalize this a little? For example,

$$1 = \frac{6^4}{6^4},$$

but if we try to subtract exponents, as in the theorem, we have

$$1 = \frac{6^4}{6^4} = 6^{4-4} = 6^0.$$

We shall use this example to make the following definition plausible. If a is any nonzero real number,

▶ $$a^0 = 1.$$

The symbol 0^0 is meaningless.

Example 3

(a) $3^0 = 1$ (b) $(-4)^0 = 1$
(c) $-4^0 = -1$ (d) $-(-4)^0 = -1$

It can be shown that the conditions of Theorem 2.1 can be broadened to include the cases when m or n equal 0; parts of the proof of this statement are included in the exercises.

2.1 Exercises

Let $P(x) = 2x^2 - 4x$ and $Q(x) = 2x + 1$. Find each of the following.

1. $P(-3)$
2. $P(4)$
3. $Q(1)$
4. $Q(-3)$
5. $P(\frac{1}{2})$
6. $Q(\frac{1}{4})$
7. $Q(m)$

8. $P(r)$
9. $P(-m)$
10. $Q(1 - r)$
11. $P(7 + h)$
12. $P(-3 + 2r)$
13. $P(-2x + 3)$
14. $Q\left(\dfrac{x-1}{2}\right)$

15. $P(x + r)$
16. $Q(x - r)$
17. $Q[P(2)]$
18. $P[Q(-1)]$
19. $P[P(-2)]$
20. $Q[Q(0)]$

Perform each of the following indicated operations.

21. $(3x^2 - 4x + 5) + (-2x^2 + 3x - 2)$
22. $(4m^3 - 3m^2 + 5) + (-3m^3 - m^2 + 5)$

23. $(12y^2 - 8y + 6) - (3y^2 - 4y + 2)$
24. $(8p^2 - 5p) - (3p^2 - 2p + 4)$
25. $-3(4q^2 - 3q + 2) + 2(-q^2 + q - 4)$
26. $2(3r^2 - 4r + 2) - 3(-r^2 + 4r - 5)$
27. $(4r - 1)(7r + 2)$
28. $(5m - 6)(3m + 4)$
29. $(6p + 5)(3p - 7)$
30. $(2z + 1)(3z - 4)$
31. $(5r + 2)(5r - 2)$
32. $(6z + 5)(6z - 5)$
33. $(2k - 3)^2$
34. $(3p + 5)^2$
35. $(k + 2)(12k^3 - 3k^2 + k + 1)$
36. $(2p - 1)(3p^2 - 4p + 5)$
37. $(2z - 1)^3$
38. $(3p - 1)(9p^2 + 3p + 1)$
39. $(2m + 1)(4m^2 - 2m + 1)$
40. $(x - 1)(x^2 - 1)$

Simplify each of the following. Assume all variables appearing as exponents represent natural numbers.

41. $(3m)^2(-2m)^3(-3m^4)$

42. $(-p^3)(2p^2)^3(-2p^4)$

43. $\dfrac{(8m^2)^3(12m^4)^2(-6m^3)}{(-4m^3)^4(3m^5)^2(2m^2)}$

44. $\dfrac{(3^3)^3(-2z^4)^5(32z^3)}{(2z^2)^3(-3z^5)^2(-2z^2)^3}$

45. $y^m \cdot y^{1+m}$

46. $z^{2r}z^{r+3}$

47. $\dfrac{(2m^n)^2(-4m^{2+n})}{8m^{4n}}$, $(n > 2)$

48. $\dfrac{(3 - y)^k(3 - y)^{2k}}{(3 - y)^2}$, $(k > 1)$

Prove each of the following statements for n = 0 and any natural number m.

49. $a^m a^n = a^{m+n}$
50. $(a^m)^n = a^{mn}$

2.2 Quotients of Polynomials

We shall define the **quotient** of the polynomials $P(x)$ and $Q(x)$ as

$$\frac{P(x)}{Q(x)} = K \quad \text{if and only if} \quad P(x) = K \cdot Q(x),$$

for some algebraic expression K. A quotient of polynomials is not defined if any of the variables assume a value for which the denominator is zero. If the expression K is a polynomial, then $P(x)$ is said to be exactly (or evenly) divisible by $Q(x)$; if K is not a polynomial, then $P(x)$ is not exactly divisible by $Q(x)$. To find the quotient of a polynomial $P(x)$ and a monomial $Q(x)$, we use the definition of division and find the product

$$P(x) \cdot \frac{1}{Q(x)}, \qquad (Q(x) \neq 0).$$

Example 4

(a) $\dfrac{2m^5 - 6m^3}{2m^3} = (2m^5 - 6m^3) \cdot \dfrac{1}{2m^3} = m^2 - 3$

This is because $2m^5 - 6m^3 = 2m^3(m^2 - 3)$. Since the quotient $m^2 - 3$ is a polynomial, we say that $2m^5 - 6m^3$ is exactly divisible by $2m^3$.

(b) $\dfrac{3y^6x^3 - 6y^3x^6 + 8y^5x}{3y^3x^3} = \dfrac{3y^6x^3}{3y^3x^3} - \dfrac{6y^3x^6}{3y^3x^3} + \dfrac{8y^5x}{3y^3x^3} = y^3 - 2x^3 + \dfrac{8y^2}{3x^2}$

Here the quotient is not a polynomial, and so $3y^6x^3 - 6y^3x^6 + 8y^5x$ is not exactly divisible by $3y^3x^3$.

To divide one polynomial by another, it is often convenient to use a **division algorithm.** (An algorithm is an orderly procedure for performing an operation.) For example, we can use a division algorithm here very similar to that used for whole numbers.

$$
\begin{array}{r}
2m^2 - 3m\ + \tfrac{1}{2} \\
2m - 1\,\overline{\big)\,4m^3 - 8m^2 + 4m + 6} \\
\underline{4m^3 - 2m^2} \\
-6m^2 + 4m \\
\underline{-6m^2 + 3m} \\
m + 6 \\
\underline{m - \tfrac{1}{2}} \\
1\tfrac{3}{2}
\end{array}
$$

Hence, $\dfrac{4m^3 - 8m^2 + 4m + 6}{2m - 1} = 2m^2 - 3m + \tfrac{1}{2} + \dfrac{1\tfrac{3}{2}}{2m - 1}.$

We shall often need to divide a polynomial by a first degree binomial of the form $x + k$, where the coefficient of x is 1. Thus it will prove beneficial to develop a shortcut for this process. To begin, let us consider an example.

$$
\begin{array}{r}
3x^2 + 10x\ + 40 \\
x - 4\,\overline{\big)\,3x^3 -\ \ 2x^2 \qquad\quad - 150} \\
\underline{3x^3 - 12x^2} \\
10x^2 \\
\underline{10x^2 - 40x} \\
40x - 150 \\
\underline{40x - 160} \\
10
\end{array}
$$

To simplify this process, we can omit all variables and write only coefficients, using 0 to represent the coefficient of any missing variables. Since the coefficient of x in the division is always 1 in problems of this type, we can omit it too. These omissions simplify the problem as follows.

$$
\begin{array}{r}
3 \quad 10 \quad 40 \\
-4\,\overline{\smash{\big)}\,3 - 2 \quad\;\; 0 - 150} \\
\mathbf{3} - 12 \\
\hline
10 \\
\mathbf{10} - 40 \\
\hline
40 - 150 \\
\mathbf{40} - 160 \\
\hline
10
\end{array}
$$

The boldface numbers are repetitions of the numbers directly above, and can be omitted.

$$
\begin{array}{r}
3 \quad 10 \quad 40 \\
-4\,\overline{\smash{\big)}\,3 - 2 \quad\;\; 0 - 150} \\
-12 \\
\hline
10 \\
-40 \\
\hline
40 - 150 \\
-160 \\
\hline
10
\end{array}
$$

The entire problem can now be compacted, with the top row of numbers omitted, since it merely duplicates the bottom row.

$$
\begin{array}{r}
-4\,\overline{\smash{\big)}\,3 - \;\,\cdot 2 \quad\;\; 0 - 150} \\
-12 - 40 - 160 \\
\hline
3 \quad 10 \quad 40 \quad\;\; 10
\end{array}
$$

We obtained the bottom row by subtracting -12, -40, and -160 from the corresponding terms above. For reasons that will become clear later, we change the -4 at the left to 4, which also changes the sign of the numbers in the second row, and then add. Doing this, we have

$$
\begin{array}{r}
4\,\overline{\smash{\big)}\,3 - 2 \quad\;\; 0 - 150} \\
12 \quad 40 \quad 160 \\
\hline
3 \quad 10 \quad 40 \quad\;\; 10
\end{array}
$$

This abbreviated process is called **synthetic division.**

Example 5 Divide $5m^3 - 6m^2 - 28m - 2$ by $m + 2$.

Using synthetic division, we begin by writing

$$
-2\,\overline{\smash{\big)}\,5 - 6 - 28 - 2}
$$

First, we bring down the 5.

$$-2\overline{\smash{\big)}\ 5 - 6 - 28 - 2}$$

$$5$$

Now, multiply (-2) by (5) to get -10, and add. The result is -16.

$$
\begin{array}{r}
-2\overline{\smash{\big)}\ 5 - 6 - 28 - 2} \\
\underline{-\ 10 } \\
5 - 16
\end{array}
$$

Here, $(-2)(-16) = 32$. Again we add.

$$
\begin{array}{r}
-2\overline{\smash{\big)}\ 5 - 6 - 28 - 2} \\
\underline{-\ 10 32 } \\
5 - 16 4
\end{array}
$$

Since $(-2)(4) = -8$, we have

$$
\begin{array}{r}
-2\overline{\smash{\big)}\ 5 - 6 - 28 - 2} \\
\underline{-\ 10 32 - 8} \\
5 - 16 4 - 10
\end{array}
$$

The quotient is read directly from the bottom row.

$$\frac{5m^3 - 6m^2 - 28m - 2}{m + 2} = 5m^2 - 16m + 4 + \frac{-10}{m + 2}$$

The degree of the quotient will always be one less than the degree of the polynomial to be divided.

2.2 Exercises

Perform each of the following divisions.

1. $\dfrac{4m^3 - 8m^2 + 16m}{2m}$

2. $\dfrac{30k^5 - 12k^3 + 18k^2}{6k^2}$

3. $\dfrac{6y^2 + y - 2}{2y - 1}$

4. $\dfrac{16r^2 + 2r - 3}{2r + 1}$

5. $\dfrac{20p^2 + 11p - 3}{4p + 3}$

6. $\dfrac{32m^2 + 20m - 3}{8m - 1}$

7. $\dfrac{12x^2 - 7x - 12}{4x - 5}$

8. $\dfrac{12p^2 - 13p - 16}{3p - 2}$

9. $\dfrac{8z^2 + 14z - 20}{2z + 5}$

10. $\dfrac{6a^2 - 13a - 18}{2a + 1}$

11. $\dfrac{2x^3 - 11x^2 + 19x - 10}{2x - 5}$

12. $\dfrac{3p^3 - 11p^2 + 5p + 3}{3p + 1}$

13. $\dfrac{15x^3 + 11x^2 + 20}{3x + 4}$

14. $\dfrac{4r^3 + 2r^2 - 14r + 15}{2r + 5}$

15. $\dfrac{4z^5 - 4z^2 - 5z + 3}{2z^2 + z + 1}$

16. $\dfrac{12z^4 - 25z^3 + 35z^2 - 26z + 10}{4z^2 - 3z + 5}$

Use synthetic division for each of the following.

17. $\dfrac{m^4 - 3m^3 - 4m^2 + 12m}{m - 2}$

18. $\dfrac{p^4 - 3p^3 - 5p^2 + 2p - 16}{p + 2}$

19. $\dfrac{3x^3 - 11x^2 - 20x + 3}{x - 5}$

20. $\dfrac{4p^3 + 8p^2 - 16p - 9}{p + 3}$

21. $\dfrac{4m^3 - 3m - 2}{m + 1}$

22. $\dfrac{3q^3 - 4q + 2}{q - 1}$

23. $\dfrac{x^5 + 3x^4 + 2x^3 + 2x^2 + 3x + 1}{x + 2}$

24. $\dfrac{m^6 - 3m^4 + 2m^3 - 6m^2 - 5m + 3}{m + 2}$

25. $\dfrac{\dfrac{1}{3}x^3 - \dfrac{2}{9}x^2 + \dfrac{1}{27}x + 1}{x - \dfrac{1}{3}}$

26. $\dfrac{x^3 + x^2 + \dfrac{1}{2}x + \dfrac{1}{8}}{x + \dfrac{1}{2}}$

27. $\dfrac{x^4 - 1}{x - 1}$

28. $\dfrac{x^7 + 1}{x + 1}$

29. $\dfrac{x^4 + 1}{x - 1}$

30. $\dfrac{x^7 - 1}{x + 1}$

2.3 Special Products and Factoring

The products mentioned in the following theorem occur so often in practice that they are referred to as special products.

THEOREM 2.2   For all real numbers a, b, c, d,
and for all variables x and y:

(a) $x^2 + 2xy + y^2 = (x + y)^2$
(perfect square trinomial)

(b) $x^3 + 3x^2y + 3xy^2 + y^3 = (x + y)^3$

(c) $x^2 - y^2 = (x + y)(x - y)$
(difference of two squares)

(d) $x^2 + (a + b)x + ab = (x + a)(x + b)$

(e) $acx^2 + (ad + bc)x + bd = (ax + b)(cx + d)$

(f) $x^3 + y^3 = (x + y)(x^2 - xy + y^2)$
(sum of two cubes)

(g) $x^3 - y^3 = (x - y)(x^2 + xy + y^2)$
(difference of two cubes).

This result can be proved by the methods of multiplication of polynomials presented in Section 2.2.

The process of finding polynomials whose product equals a given polynomial is called **factoring** the given polynomial. For example, by (c) above,

$$4x^2 - 9 = (2x + 3)(2x - 3).$$

Example 6   Factor $m^2 + 2m - 24$.

We need to find two real numbers a and b such that $(m + a)(m + b) = m^2 + 2m - 24$. By Theorem 2.2(d) we see that the coefficient for m must be $(a + b)$ and the constant term must equal ab. Thus, $a + b = 2$ and $ab = -24$. That is, we need two numbers whose sum is 2 and whose product is -24. Since -4 and 6 satisfy these requirements, we can write

$$m^2 + 2m - 24 = (m - 4)(m + 6).$$

Example 7   Factor $6p^2 - 7pq - 5q^2$.

We use Theorem 2.5(e). We need real numbers a, b, c, and d such that $ac = 6$, $ad + bc = -7$, and $bd = -5$. In general, we use the method of trial and error to find these numbers. By inspection, we can find that

$$6p^2 - 7pq - 5q^2 = (3p - 5q)(2p + q).$$

Although it is true that $x^2 - 5 = (x + \sqrt{5})(x - \sqrt{5})$, we shall in general be interested only in factors having integer coefficients. For this reason, we say that $x^2 - 5$ cannot be factored over the integers. Note that $x^2 - 5$ can be factored over the reals, but not over the integers. Unless otherwise specified, factor shall mean to factor over the integers.

Example 8 Factor $m^3 - 64n^3$.

In Theorem 2.2(g), if we replace x with m and y with $4n$, we have

$$m^3 - 64n^3 = m^3 - (4n)^3$$
$$= (m - 4n)[m^2 + m(4n) + (4n)^2]$$
$$= (m - 4n)(m^2 + 4mn + 16n^2).$$

Example 9 Factor $(a - 1)^3 + 1$.

We can use Theorem 2.2(f) with $x = a - 1$ and $y = 1$. We have

$$(a - 1)^3 + 1 = [(a - 1) + 1][(a - 1)^2 - (a - 1) + 1]$$
$$= a(a^2 - 2a + 1 - a + 1 + 1)$$
$$= a(a^2 - 3a + 3).$$

Example 10 Factor $p^6 - q^6$.

Use Theorem 2.2(g) with $x = p^2$ and $y = q^2$.

$$p^6 - q^6 = (p^2 - q^2)(p^4 + p^2q^2 + q^4)$$

Since $p^2 - q^2$ can be further factored, as $(p + q)(p - q)$, the result is

$$p^6 - q^6 = (p + q)(p - q)(p^4 + p^2q^2 + q^4).$$

Example 11 Factor $p^6 - q^6$ using Theorem 2.2(c).

We have $p^6 - q^6 = (p^3)^2 - (q^3)^2$
$$= (p^3 - q^3)(p^3 + q^3)$$
$$= (p - q)(p^2 + pq + q^2)(p + q)(p^2 - pq + q^2).$$

From this result, we see that we did not factor completely in Example 10 above. For this reason, the approach of Example 11 is better for this type of problem.

2.3 Exercises

Factor the following. Assume all variables used as exponents are real numbers.

1. $12mn - 8m$
2. $3pq - 18pqr$
3. $6px^2 - 8px^3 - 12px$
4. $9m^2n^3 - 18m^3n^2 + 27m^2n^4$
5. $4p^2 + 3p - 1$
6. $6x^2 + 7x - 3$
7. $2z^2 + 7z - 30$
8. $4m^2 + m - 3$
9. $6q^2 - q - 12$
10. $10b^2 - 19b - 15$
11. $12r^2 + 24r - 15$
12. $12p^2 + p - 20$
13. $18r^2 - 3rs - 10s^2$
14. $12m^2 + 16mn - 35n^2$
15. $64p^2q^2 - 121m^2n^2$
16. $169r^2 - 81s^2$
17. $81m^4 - 16n^8$
18. $256p^4 - 81q^4$
19. $9x^2 - 6x^3 + x^4$
20. $m^{16} - 1$
21. $m^6 - 1$
22. $a^3 + b^6$

23. $8m^6 - 27n^3$

24. $p^6 - q^6$

25. $x^2 - (x - y)^2$

26. $16m^2 - 25(m - n)^2$

27. $8y^3 + 27(x - y)^3$

28. $64p^6 - 81(p^2 - q)^3$

29. $(x + y)^2 + 2(x + y)z - 15z^2$

30. $(m + n)^2 + 3(m + n)p - 10p^2$

31. $6(a + b)^2 + (a + b)c - 40c^2$

32. $12(g + h)^2 - 5(g + h)j - 25j^2$

33. $(p + q)^2 - (p - q)^2$

34. $(p - q)^2 - (p + q)^2$

35. $m^{2n} - 16$

36. $p^{4n} - 49$

37. $x^{3n} - y^{6n}$

38. $a^9 - b^9$

39. $2x^{2n} - 23x^n y^n - 39y^{2n}$

2.4 Rational Expressions

Any expression which can be stated as the quotient of two polynomials (with nonzero denominator) is called a **rational expression.** Since a rational expression is analogous to a rational number (which is a quotient of integers), many of the results we proved in Chapter 1 for rational numbers generalize quickly to rational expressions. In particular, based on Theorem 1.6, we have the following result.

THEOREM 2.3   For all polynomials $P(x)$, $Q(x) \neq 0$, $R(x)$, and $S(x) \neq 0$, we have:

(a) $\dfrac{P(x)}{Q(x)} = \dfrac{R(x)}{S(x)}$ if and only if $P(x) \cdot S(x) = Q(x) \cdot R(x)$.

(b) $\dfrac{P(x)}{Q(x)} = \dfrac{P(x) \cdot S(x)}{Q(x) \cdot S(x)}$.

(fundamental theorem of rational expressions)

(c) $\dfrac{P(x)}{Q(x)} \cdot \dfrac{R(x)}{S(x)} = \dfrac{P(x) \cdot R(x)}{Q(x) \cdot S(x)}$.

(multiplication of rational expressions)

(d) $\dfrac{P(x)}{Q(x)} + \dfrac{R(x)}{Q(x)} = \dfrac{P(x) + R(x)}{Q(x)}$.

(addition of rational expressions)

(e) $\dfrac{P(x)}{Q(x)} - \dfrac{R(x)}{Q(x)} = \dfrac{P(x) - R(x)}{Q(x)}$.

(subtraction of rational expressions)

(f) $\dfrac{P(x)}{Q(x)} \div \dfrac{R(x)}{S(x)} = \dfrac{P(x) \cdot S(x)}{Q(x) \cdot R(x)}$, $(R(x) \neq 0)$.

(division of rational expressions)

We must be particularly careful to indicate the domain of the variable when using part (b) of Theorem 2.3. For example, -2 is not in the domain of the rational expression

$$\frac{x+6}{x+2},$$

since $x = -2$ makes the denominator zero. In the same way, -2 and -4 are not in the domain of

$$\frac{(x+6)(x+4)}{(x+2)(x+4)}.$$

Although

$$\frac{x+6}{x+2} = \frac{(x+6)(x+4)}{(x+2)(x+4)}$$

by part (b) of the theorem, we should remember that this holds only when

$$x \neq -2,\ x \neq -4.$$

Example 12

$$\frac{3m^2 - 2m - 8}{3m^2 + 14m + 8} \cdot \frac{3m+2}{3m+4} = \frac{(m-2)(3m+4)}{(m+4)(3m+2)} \cdot \frac{3m+2}{3m+4}$$

$$= \frac{(m-2)(3m+4)(3m+2)}{(m+4)(3m+2)(3m+4)}$$

$$= \frac{m-2}{m+4} \qquad \text{for all } m \neq -4,$$

$$m \neq -\tfrac{2}{3},\ m \neq -\tfrac{4}{3}$$

From now on, we shall not state (unless specifically required) all restrictions on the variable. But keep in mind that no statement involving rational expressions is meaningful for values of the variable which make any denominator zero.

Example 13

$$\frac{6p^2 - 7pq - 20q^2}{6p^2 - 25pq + 25q^2} \div \frac{6p^2 + 11pq + 4q^2}{2p^2 + 9pq + 4q^2}$$

$$= \frac{(2p - 5q)(3p + 4q)(2p + q)(p + 4q)}{(2p - 5q)(3p - 5q)(2p + q)(3p + 4q)}$$

$$= \frac{p + 4q}{3p - 5q}$$

Example 14 $\quad \dfrac{y+2}{y-1} + \dfrac{y}{y+1} = \dfrac{(y+2)(y+1)}{(y-1)(y+1)} + \dfrac{y(y-1)}{(y-1)(y+1)}$

(Here we use the fact that $(y-1)(y+1)$ is common denominator.)

$$= \frac{y^2 + 3y + 2}{(y-1)(y+1)} + \frac{y^2 - y}{(y-1)(y+1)}$$

$$= \frac{y^2 + 3y + 2 + y^2 - y}{(y-1)(y+1)}$$

$$= \frac{2y^2 + 2y + 2}{(y-1)(y+1)}$$

$$= \frac{2(y^2 + y + 1)}{(y-1)(y+1)}$$

We usually leave the denominator in factored form in a problem of this type.

Any quotient of two rational expressions in fractional form is called a **complex fraction.** Complex fractions can often be simplified by using Theorem 2.3, as shown in the following example.

Example 15 Simplify $\dfrac{\dfrac{a}{a+1}+\dfrac{1}{a}}{\dfrac{1}{a}+\dfrac{1}{a+1}}$.

We can first simplify numerator and denominator respectively.

$$\frac{\dfrac{a}{a+1}+\dfrac{1}{a}}{\dfrac{1}{a}+\dfrac{1}{a+1}}=\frac{\dfrac{a^2+1(a+1)}{a(a+1)}}{\dfrac{1(a+1)+1(a)}{a(a+1)}}$$

$$=\frac{\dfrac{a^2+a+1}{a(a+1)}}{\dfrac{2a+1}{a(a+1)}}$$

$$=\frac{a^2+a+1}{a(a+1)}\cdot\frac{a(a+1)}{2a+1}$$

$$=\frac{a^2+a+1}{2a+1}$$

We can also simplify the complex fraction by multiplying both numerator and denominator by the common denominator of all the denominators, in this case $a(a+1)$. We have

$$\frac{\dfrac{a}{a+1}+\dfrac{1}{a}}{\dfrac{1}{a}+\dfrac{1}{a+1}}=\frac{\left(\dfrac{a}{a+1}+\dfrac{1}{a}\right)a(a+1)}{\left(\dfrac{1}{a}+\dfrac{1}{a+1}\right)a(a+1)}$$

$$=\frac{a^2+(a+1)}{(a+1)+a}$$

$$=\frac{a^2+a+1}{2a+1}.$$

2.4 Exercises

Perform each of the following operations.

1. $\dfrac{2x-2}{3}\cdot\dfrac{6x-6}{(x-1)^3}$

2. $\dfrac{3m-15}{4m-20}\cdot\dfrac{m^2-10m+25}{12m-60}$

3. $\dfrac{a^2-a-6}{a+2}\div\dfrac{a-3}{a+2}$

4. $\dfrac{m^2+11m+30}{m+6}\div\dfrac{m+5}{m+6}$

5. $\dfrac{p^2 - p - 12}{p^2 - 2p - 15} \cdot \dfrac{p^2 - 9p + 20}{p^2 - 8p + 16}$

6. $\dfrac{x^2 + 2x - 15}{x^2 + 11x + 30} \cdot \dfrac{x^2 + 2x - 24}{x^2 - 8x + 15}$

7. $\dfrac{2m^2 - 5m - 12}{2m^2 + 5m - 12} \div \dfrac{2m^2 + 9m + 9}{2m^2 + 3m - 9}$

8. $\dfrac{3z^2 + z - 2}{4z^2 - z - 5} \div \dfrac{3z^2 + 11z + 6}{4z^2 + 7z - 15}$

9. $\left(1 + \dfrac{1}{x}\right)\left(1 - \dfrac{1}{x}\right)$

10. $\left(3 + \dfrac{2}{y}\right)\left(3 - \dfrac{2}{y}\right)$

11. $\dfrac{x^3 + y^3}{x^2 - y^2} \cdot \dfrac{x + y}{x^2 - xy + y^2}$

12. $\dfrac{4y^3 - 125}{4y^2 - 20y + 25} \cdot \dfrac{2y - 5}{y}$

13. $\dfrac{x^3 + y^3}{x^3 - y^3} \cdot \dfrac{x^2 - y^2}{x^2 + 2xy + y^2}$

14. $\dfrac{x^2 - y^2}{(x - y)^2} \cdot \dfrac{x^2 - xy + y^2}{x^2 - 2xy + y^2} \div \dfrac{x^3 + y^3}{(x - y)^4}$

15. $\dfrac{m}{m + n} + \dfrac{n}{m + n}$

16. $\dfrac{3}{y} + \dfrac{4}{y}$

17. $\dfrac{1}{y} + \dfrac{1}{y + 1}$

18. $\dfrac{2}{3(x - 1)} + \dfrac{1}{4(x - 1)}$

19. $\dfrac{2}{a + b} - \dfrac{1}{2(a + b)}$

20. $\dfrac{3}{m} - \dfrac{1}{m - 1}$

21. $\dfrac{1}{a + 1} - \dfrac{1}{a - 1}$

22. $\dfrac{1}{x + z} + \dfrac{1}{x - z}$

23. $\dfrac{m + 1}{m - 1} + \dfrac{m - 1}{m + 1}$

24. $\dfrac{2}{x - 1} + \dfrac{1}{1 - x}$

25. $\dfrac{3}{a - 2} - \dfrac{1}{2 - a}$

26. $\dfrac{3}{m-n} - \dfrac{m}{m+n}$

27. $\dfrac{1}{a^2 - 5a + 6} - \dfrac{1}{a^2 - 4}$

28. $\dfrac{-3}{m^2 - m - 2} - \dfrac{1}{m^2 + 3m + 2}$

29. $\dfrac{1}{x^2 + x - 12} - \dfrac{1}{x^2 - 7x + 12} + \dfrac{1}{x^2 - 16}$

30. $\dfrac{2}{2p^2 - 9p - 5} + \dfrac{p}{3p^2 - 17p + 10} - \dfrac{2p}{6p^2 - p - 2}$

31. $\left(\dfrac{3}{p-1} - \dfrac{2}{p+1}\right)\left(\dfrac{p-1}{p}\right)$

32. $\left(\dfrac{y}{y^2 - 1} - \dfrac{y}{y^2 - 2y + 1}\right)\left(\dfrac{y-1}{y+1}\right)$

33. $\dfrac{\dfrac{1}{x+h} - \dfrac{1}{x}}{h}$

34. $\dfrac{1}{h}\left(\dfrac{1}{(x+h)^2 + 9} - \dfrac{1}{x^2 + 9}\right)$

35. $\dfrac{1 + \dfrac{1}{x}}{1 - \dfrac{1}{x}}$

36. $\dfrac{2 - \dfrac{2}{y}}{2 + \dfrac{2}{y}}$

37. $\dfrac{1 + \dfrac{1}{1-b}}{1 - \dfrac{1}{1+b}}$

38. $m - \dfrac{m}{m + \dfrac{1}{2}}$

39. $1 + \dfrac{1}{1 + \dfrac{1}{x}}$

40. $1 - \dfrac{1}{1 - \dfrac{1}{x}}$

41. $1 - \dfrac{1}{1 + \dfrac{1}{1 + \dfrac{1}{1 + \dfrac{1}{1+1}}}}$

42. $1 - \dfrac{1}{1 + \dfrac{1}{1 - \dfrac{1}{1 + \dfrac{1}{1 - \dfrac{1}{1+1}}}}}$

2.5 Integer Exponents

We have discussed natural number and zero exponents, and now we need to generalize this work to include all rational numbers as exponents. In this section we shall begin by discussing integer exponents (negative as well as nonnegative.) It would certainly be desirable to define integer exponents in such a way that all past results are still valid. For example, we know that for any nonnegative integers m and n, and any nonzero real number a,

$$\frac{a^m}{a^n} = a^{m-n}, \qquad \text{if } m > n.$$

To generalize this result, we let $m = 0$, and then we have

$$\frac{a^0}{a^n} = a^{0-n} = a^{-n}.$$

Since $a^0 = 1$, we have

$$a^{-n} = \frac{a^0}{a^n} = \frac{1}{a^n}.$$

Hence if negative integer exponents are to satisfy the same rules as nonnegative integer exponents, we must define a^{-n} as follows: for all integers n and all real numbers $a \neq 0$,

▶
$$a^{-n} = \frac{1}{a^n}.$$

For example, $3^{-1} = \dfrac{1}{3}$, $2^{-3} = \dfrac{1}{2^3} = \dfrac{1}{8}$, and $-4^{-2} = \dfrac{-1}{4^2} = \dfrac{-1}{16}$.

It can be shown that Theorem 2.1 can be generalized to include all integer exponents, and not just nonnegative integers. The result is given below.

THEOREM 2.4   For all integers m and n, and all real numbers $a \neq 0$ and $b \neq 0$, we have:

(a) $a^m \cdot a^n = a^{m+n}$

(b) $\dfrac{a^m}{a^n} = a^{m-n}$

(c) $(a^m)^n = a^{mn}$

(d) $(ab)^m = a^m b^m$

(e) $\left(\dfrac{a}{b}\right)^n = \dfrac{a^n}{b^n}.$

Note that our definition of integer exponents permits us to simplify part (b) of Theorem 2.1. Thus, instead of writing as we did before

$$\frac{m^5}{m^7} = \frac{1}{m^2}, \qquad (\text{since } 5 < 7)$$

we can now simply write

$$\frac{m^5}{m^7} = m^{5-7} = m^{-2}.$$

Example 16

(a) $2^{-4} \cdot 2^5 = 2^{-4+5} = 2^1 = 2$ Theorem 2.4 (a)

(b) $3x^{-2}(4^{-1}x^{-5})^2 = 3x^{-2}(4^{-2}x^{-10})$ Theorem 2.4 (d) and (c)

$\qquad\qquad\quad = 3 \cdot 4^{-2}x^{-12}$ Theorem 2.4 (a)

$\qquad\qquad\quad = \dfrac{3}{16x^{12}}$ definition of a negative exponent

Example 17 Simplify $\dfrac{(x+y)^{-1}}{x^{-1}+y^{-1}}$.

Using the definition of negative integer exponent, we can write

$$\frac{(x+y)^{-1}}{x^{-1}+y^{-1}} = \frac{\dfrac{1}{x+y}}{\dfrac{1}{x}+\dfrac{1}{y}}$$

$$= \frac{\dfrac{1}{x+y}}{\dfrac{x+y}{xy}}$$

$$= \frac{xy}{(x+y)^2}.$$

Example 18 Simplify $\dfrac{x^{-1}}{x^{-1}-y^{-1}}$.

Here we have

$$\frac{x^{-1}}{x^{-1}-y^{-1}} = \frac{\dfrac{1}{x}}{\dfrac{1}{x}-\dfrac{1}{y}}$$

$$= \frac{\dfrac{1}{x}}{\dfrac{y-x}{xy}}$$

$$= \frac{y}{y-x}.$$

2.5 Exercises

Simplify each of the following.

1. 2^{-3}
2. 3^{-2}
3. $(-4)^{-3}$
4. $(-5)^{-2}$
5. $(\frac{1}{2})^{-3}$
6. $(\frac{2}{3})^{-2}$
7. $2^{-3} - 2^{-2}$
8. $3^{-3} + 3^{-1}$
9. $(2^{-3})^{-2}$
10. $[(3^{-2})^2]^{-1}$

Write each of the following with no negative exponents. Assume all variables are nonzero real numbers.

11. $\dfrac{2^{-1}x^3y^{-3}}{xy^{-2}}$

12. $\dfrac{5^{-2}m^2y^{-1}}{5^2m^{-1}y^{-2}}$

13. $\dfrac{p^3r^2}{p^{-1}r^{-3}}$

14. $\dfrac{a^3b^{-2}}{a^{-1}b^4}$

15. $(-2x^{-3}y^2)^{-2}$

16. $(-3p^2q^{-2}s^{-1})^{12}$

17. $\dfrac{(3^{-1}m^{-2}n^2)^{-2}}{(mn)^{-1}}$

18. $\dfrac{(-4x^3y^{-2})^{-2}}{(4x^5y^4)^{-1}}$

19. $\dfrac{a^{-1}+b^{-1}}{(ab)^{-1}}$

20. $\dfrac{p^{-1}-q^{-1}}{(pq)^{-1}}$

21. $\dfrac{r^{-1}+q^{-1}}{r^{-1}-q^{-1}}\cdot\dfrac{r-q}{r+q}$

22. $\dfrac{xy^{-1}+yx^{-1}}{x^2+y^2}$

23. $(a+b)^{-1}(a^{-1}+b^{-1})$

24. $(m^{-1}+n^{-1})^{-1}$

25. $\dfrac{x-9y^{-1}}{(x-3y^{-1})(x+3y^{-1})}$

26. $\dfrac{(m+n)^{-1}}{m^{-2}-n^{-2}}$

2.6 Rational Exponents

We have discussed natural number exponents, and in the previous section we found that all results for natural number exponents apply also to integer exponents, provided we define integer exponents as we did in the last section. In this section we shall discuss rational number exponents, and find that rational exponents properly defined also obey the same rules as natural number exponents.

To begin, let us define the symbol $\sqrt[n]{a}$ as follows. If a is a nonnegative real number, and n is any natural number, or if a is negative, and n is an odd natural number, then

$$\sqrt[n]{a}=b \qquad \text{if and only if} \qquad b^n=a.$$

We call $\sqrt[n]{a}$ an **nth root of a.** When $a\ge 0$ and n is an even natural number, then $\sqrt[n]{a}$ represents the *positive* nth root of a. The general proof that such nth roots exist depends upon the completeness axiom (see Chapter 1) and is left for the reader to complete in Exercise 41. It is customary to write $\sqrt[2]{a}$ as $\sqrt{a}$ and to read both symbols as "square root of a", while $\sqrt[3]{a}$ is read "cube root of a." Note that $\sqrt[5]{32}=2$ since $2^5=32$. It is important to note the restrictions stated in the definition. As long as a is a nonnegative real number, and n is any natural number, then $\sqrt[n]{a}$ is always a real number. However, if a is negative, then $\sqrt[n]{a}$ is a real number only for odd values of the natural number n. Thus, $\sqrt[3]{-27}=-3$, since $(-3)^3=-27$, but $\sqrt[4]{-16}$ is not a real number, since there is no real number whose fourth power is -16.

The symbol $\sqrt[n]{a}$ is called a **radical,** while a itself is called a **radicand,** and n is called the **index** of the radical. Note that if a is positive and n is even, there are two different nth roots of a. One of these roots is positive, written $\sqrt[n]{a}$, while the other is negative and is written $-\sqrt[n]{a}$. Thus both -3 and 3 are 4th roots of 81, since $(-3)^4 = 81$ and also $3^4 = 81$. If n is odd, then for any real value of a there is exactly one nth root. The only cube root of 125 is 5, since 5 is the only real number with the property that $5^3 = 125$. If a is negative, and n is even, there are no real nth roots of a.

The basic properties of nth roots are summarized in the following results.

THEOREM 2.5   For any real numbers a and b and any natural number n for which the following roots are real, we have:

(a) $(\sqrt[n]{a})^n = a$

(b) $\sqrt[n]{a^n} = a$ if either $\begin{cases} a > 0 \\ a < 0 \text{ and } n \text{ is odd} \end{cases}$

(c) $\sqrt[n]{a^n} = |a|$ for $a < 0$, n even

(d) $\sqrt[n]{a} \cdot \sqrt[n]{b} = \sqrt[n]{ab}$

(e) $\dfrac{\sqrt[n]{a}}{\sqrt[n]{b}} = \sqrt[n]{\dfrac{a}{b}}$

(f) $\sqrt[m]{\sqrt[n]{a}} = \sqrt[mn]{a}.$

We shall prove only part of this result, with the remainder left as exercises. To prove (a), let $x = \sqrt[n]{a}$. Then, by the definition of nth root, $x^n = a$. By substitution, $(\sqrt[n]{a})^n = x^n = a$, which is what we wanted to show.

To verify (d), let $x = \sqrt[n]{a}$, and $y = \sqrt[n]{b}$. Then, by the definition of nth root, $x^n = a$ and $y^n = b$. Hence, $x^n \cdot y^n = ab$. However, by our previous work with exponents, we can write $x^n \cdot y^n = ab$ as

$$(xy)^n = ab,$$

which means xy is an nth root of ab, or

$$xy = \sqrt[n]{ab}.$$

Substituting the original values of x and y gives

$$\sqrt[n]{a} \cdot \sqrt[n]{b} = \sqrt[n]{ab}.$$

Let us now define the symbol $a^{1/n}$. We want to do this so as to be consistent with all earlier work on exponents. Thus, we must define $a^{1/n}$ so as to preserve the rule $(b^m)^n = b^{mn}$. If we replace b^m with $a^{1/n}$, we have

$$(a^{1/n})^n = a^{n/n} = a^1 = a.$$

Thus, to be consistent with past work we must define $a^{1/n}$ so that it is an nth root of a. That is, if n is a natural number and if $\sqrt[n]{a}$ exists, then

▶
$$a^{1/n} = \sqrt[n]{a}.$$

What about rational exponents in general? Before we can define a symbol such as $a^{m/n}$, we might first note that in order for past results to be valid, we will need to have

$$(a^{1/n})^m = a^{m/n}$$

and
$$(a^m)^{1/n} = a^{m/n}.$$

Therefore, before we define $a^{m/n}$ we must first prove the equality of $(a^{1/n})^m$ and $(a^m)^{1/n}$.

THEOREM 2.6   For all integers m and n and all real numbers a for which all the indicated roots exist,

$$(a^m)^{1/n} = (a^{1/n})^m.$$

To prove this, we can use some of the previous work of this chapter.

$$[(a^{1/n})^m]^n = (a^{1/n})^{mn} = [(a^{1/n})^n]^m = a^m$$

Hence, $(a^{1/n})^m$ is an nth root of a^m, or $(a^{1/n})^m = (a^m)^{1/n}$.

Since $(a^{1/n})^m = (a^m)^{1/n}$ we shall define $a^{m/n}$ as follows. For all integers m and n and all real numbers a for which all the following roots exist,

▶
$$a^{m/n} = (a^m)^{1/n} = (a^{1/n})^m$$

or, equivalently, $a^{m/n} = \sqrt[n]{a^m} = (\sqrt[n]{a})^m$.

It can now be shown that all the results of Theorem 2.4 concerning integer exponents also apply to rational exponents. We shall not prove this, but we shall assume it from now on.

Example 19

(a) $(-27)^{4/3} = [(-27)^{1/3}]^4 = (-3)^4 = 81$
     Also, $(-27)^{4/3} = [(-27)^4]^{1/3} = 81$. (Note that $(-27)^4$ is positive.)

(b) $8^{4/3} = (2^3)^{4/3} = 2^4 = 16$

(c) $(-81)^{5/4} = \sqrt[4]{-81^5}$ which is not a real number.

(d) $81^{5/4} \cdot 4^{-3/2} = (3^4)^{5/4} \cdot (2^2)^{-3/2} = 3^5 \cdot 2^{-3} = \dfrac{3^5}{8}$

(e) $\left(\dfrac{3m^{5/6}}{y^{3/4}}\right)^2 \cdot \left(\dfrac{8y^3}{m^6}\right)^{2/3} = \dfrac{9m^{5/3}}{y^{3/2}} \cdot \dfrac{4y^2}{m^4} = \dfrac{36y^{1/2}}{m^{7/3}}$

2.6 Exercises

Simplify each of the following that exist. Assume all variables represent nonnegative real numbers.

1. $4^{1/2}$
2. $25^{1/2}$
3. $8^{2/3}$
4. $81^{3/4}$
5. $27^{-2/3}$

6. $32^{-4/5}$
7. $(\frac{4}{9})^{-3/2}$
8. $(\frac{1}{8})^{-5/3}$
9. $125^{-2/3}$
10. $64^{5/6}$

11. $(16p^4)^{1/2}$
12. $(36r^6)^{1/2}$
13. $(27x^6)^{2/3}$
14. $(64a^{12})^{5/6}$

Write each of the following with only positive exponents. Assume all variables represent positive real numbers, and variables used as exponents represent rational numbers.

15. $\left(\dfrac{x^4y^3z}{16x^{-6}yz^5}\right)^{-1/2}$

16. $\left(\dfrac{p^3r^9}{27p^{-3}r^{-6}}\right)^{-1/3}$

17. $(r^{1/2} - r^{-1/2})^2$
18. $(p^{1/2} - p^{-1/2})(p^{1/2} + p^{-1/2})$

19. $\dfrac{k^{3/2} \cdot k^{-1/2}}{k^{1/4} \cdot k^{3/4}}$

20. $\dfrac{m^{2/5} \cdot m^{3/5} \cdot m^{-4/5}}{m^{1/5} \cdot m^{-6/5}}$

21. $\dfrac{x^{-2/3} \cdot x^{4/3}}{x^{1/2} \cdot x^{-3/4}}$

22. $\dfrac{-4a^{1/2} \cdot a^{2/3}}{a^{-5/6}}$

23. $\dfrac{8y^{2/3}y^{-1}}{2^{-1}y^{3/4} \cdot y^{-1/6}}$

24. $\dfrac{9 \cdot k^{1/3} \cdot k^{-1/2} \cdot k^{-1/6}}{k^{-2/3}}$

25. $\dfrac{m^{1-a} \cdot m^a}{m^{-1/2}}$

26. $\dfrac{y^{3-b}y^{2b-1}}{y^{1/2}}$

27. $(r^{3/p})^{2p} \cdot (r^{1/p})^{p^2}$
28. $(m^{2/x})^{x/3} \cdot (m^{x/4})^{2/x}$
29. $[(x^{1/2} - x^{-1/2})^2 + 4]^{1/2}$
30. $[(x^{1/2} + x^{-1/2})^2 - 4]^{1/2}$

Write each of the following using rational exponents instead of radical signs. Assume all variables represent positive real numbers.

31. $\sqrt{12}$
32. $\sqrt{40}$
33. $\sqrt[3]{x^2}$
34. $p\sqrt[3]{p^4}$
35. $\sqrt[3]{y^4}$

36. $\sqrt[3]{z^5}$
37. $y^3\sqrt[4]{y^6}$
38. $p^2\sqrt[3]{\sqrt[4]{p^8}}$
39. $m^{-2/3}\sqrt[4]{m^6}$
40. $y^{-3/2}\sqrt[4]{\sqrt{y^6}}$

41. We can use the completeness axiom to verify that a number such as $\sqrt{2}$ actually exists. (Recall the completeness axiom: a nonempty set of real numbers with an upper bound has a least upper bound.) Let

$$A = \{x \,|\, x^2 < 2\}.$$

(a) Verify that A is not empty.

(b) Verify that $x = 2$ is an upper bound for set A.

(c) We showed in (b) that set A has an upper bound. By the completeness axiom, what conclusion can we draw about set A?

(d) What is a good name for the least upper bound of set A?

2.7 Simplifying Radicals

We shall consider an expression containing radicals **simplified** when all the following conditions are satisfied.

▶ (1) All radicals are removed from any denominators (a process called **rationalizing the denominator**).

▶ (2) All possible factors have been removed from under radical signs.

▶ (3) The index on the radical is as small as possible.

Using this definition, we write, for example,

$$\sqrt[3]{81x^5y^7z^6} = \sqrt[3]{27 \cdot 3x^3 \cdot x^2y^6 \cdot y \cdot z^6} = 3xy^2z^2\sqrt[3]{3x^2y}.$$

Also, $\sqrt[3]{\dfrac{3}{4}} = \sqrt[3]{\dfrac{3 \cdot 4 \cdot 4}{4 \cdot 4 \cdot 4}} = \sqrt[3]{\dfrac{48}{64}} = \dfrac{\sqrt[3]{48}}{\sqrt[3]{64}} = \dfrac{\sqrt[3]{48}}{4} = \dfrac{2\sqrt[3]{6}}{4} = \dfrac{\sqrt[3]{6}}{2}.$

Although we can write $\sqrt[6]{3^2}$ as $3^{2/6} = 3^{1/3} = \sqrt[3]{3}$, extreme care must be used in simplifying radicals such as $\sqrt[6]{x^2}$, since x can be positive or negative. Note that

$$\sqrt[6]{(-8)^2} = (-8)^{2/6} = [(-8)^2]^{1/6} = 64^{1/6} = 2;$$

while, even though $2/6 = 1/3$, $(-8)^{1/3} = -2$. Hence, in this case

$$\sqrt[6]{x^2} \neq \sqrt[3]{x}.$$

If x is nonnegative, then it is always true that $x^{m/n} = x^{mp/np}$; however it is not necessarily true if x is negative (although, it may be since $(-8)^{1/3} = (-8)^{3/9} = -2$). Reducing rational exponents on negative bases must be considered case by case.

The process of rationalizing the denominator, mentioned in (1) above, is perhaps best illustrated with a few examples.

Example 20 Rationalize the denominator of $\dfrac{1}{1 - \sqrt{2}}$.

The best approach to a problem such as this is to multiply both numerator and denominator by the conjugate of the denominator, in this case $1 + \sqrt{2}$. Doing this, we obtain

$$\frac{1}{1 - \sqrt{2}} = \frac{1 \cdot (1 + \sqrt{2})}{(1 - \sqrt{2})(1 + \sqrt{2})} = \frac{1 + \sqrt{2}}{1 - 2} = -1 - \sqrt{2}.$$

Example 21 Find a decimal approximation to $\dfrac{3}{2 + \sqrt{5}}$.

We could find a decimal approximation for $\sqrt{5}$ in the Appendix:

$$\sqrt{5} \approx 2.236.$$

Then we could write

$$\frac{3}{2 + \sqrt{5}} \approx \frac{3}{2 + 2.236} = \frac{3}{4.236}.$$

which requires a lengthy calculation. It would be better to first rationalize the denominator.

$$\frac{3}{2 + \sqrt{5}} = \frac{3(2 - \sqrt{5})}{(2 + \sqrt{5})(2 - \sqrt{5})} = \frac{6 - 3\sqrt{5}}{4 - 5} = -6 + 3\sqrt{5}$$

Now we have

$$\frac{3}{2 + \sqrt{5}} = -6 + 3\sqrt{5} \approx -6 + 3(2.236) = -6 + 6.708 = .708.$$

Example 22   Rationalize the denominator of $\dfrac{1}{x^{1/3} + y}$.

We use a process similar to, but more complicated than, the process used above for square roots. Recall that $a^3 + b^3 = (a + b)(a^2 - ab + b^2)$. Here we can let $a = x^{1/3}$ and $b = y$, and then multiply top and bottom by

$$a^2 - ab + b^2 = x^{2/3} - x^{1/3}y + y^2.$$

Doing this, we have

$$\frac{1}{x^{1/3} + y} = \frac{1(x^{2/3} - x^{1/3}y + y^2)}{(x^{1/3} + y)(x^{2/3} - x^{1/3}y + y^2)} = \frac{x^{2/3} - x^{1/3}y + y^2}{x + y^3}.$$

Example 23   Simplify $\sqrt[3]{64m^4n^5} - \sqrt[3]{-27m^{10}n^{14}}$.

We have

$$\sqrt[3]{64m^4n^5} - \sqrt[3]{-27m^{10}n^{14}} = 4mn\sqrt[3]{mn^2} - (-3)m^3n^4\sqrt[3]{mn^2}$$
$$= 4mn\sqrt[3]{mn^2} + 3m^3n^4\sqrt[3]{mn^2}$$
$$= (4 + 3m^2n^3)\, mn\sqrt[3]{mn^2}.$$

Example 24   Write $\sqrt[3]{\sqrt{6}}$ using rational exponents.

Here $\sqrt[3]{\sqrt{6}} = \sqrt[3]{6^{1/2}} = (6^{1/2})^{1/3} = 6^{1/6}$.

2.7 Exercises

Simplify each of the following. Assume all variables represent positive real numbers.

1. $\sqrt[3]{81}$

2. $\sqrt{50}$

3. $\sqrt[4]{\dfrac{32}{81}}$

4. $\sqrt[3]{\dfrac{81}{16}}$

5. $\sqrt{24 \cdot 3^3 \cdot 2^4}$

6. $\sqrt[3]{16 \cdot (-2)^4 \cdot 2^8}$

7. $\sqrt{5} + \sqrt{20}$

8. $\sqrt{6} + \sqrt{54}$

9. $4\sqrt{3} - 5\sqrt{12} + 3\sqrt{75}$

10. $2\sqrt{5} - 3\sqrt{20} + 2\sqrt{45}$

11. $\sqrt{50} - 8\sqrt{8} + 4\sqrt{18}$

12. $6\sqrt{27} - 3\sqrt{12} + 5\sqrt{48}$

13. $3\sqrt[3]{16} - 4\sqrt[3]{2}$

14. $\sqrt[3]{2} - \sqrt[3]{16} + 2\sqrt[3]{54}$

15. $\dfrac{1}{\sqrt{3}} - \dfrac{2}{\sqrt{12}} + 2\sqrt{3}$

16. $\dfrac{1}{\sqrt{2}} + \dfrac{3}{\sqrt{8}} + \dfrac{1}{\sqrt{32}}$

17. $\sqrt{98x^5z^8}$

18. $\sqrt[3]{16z^5x^8y^4}$

19. $\sqrt[3]{81p^5x^2y^6}$

20. $\sqrt{242a^4b^3}$

21. $\sqrt[3]{x^{2/3}y^3}$

22. $\sqrt[4]{m^5 \cdot n^{3/2}}$

23. $\sqrt{\dfrac{2}{3x}}$

24. $\sqrt{\dfrac{5}{3p}}$

25. $\sqrt{\dfrac{9x}{2xy}}$

26. $\sqrt{\dfrac{8y}{9yz}}$

27. $\sqrt[3]{\dfrac{8}{x^2}}$

28. $\sqrt[3]{\dfrac{9}{16p^4}}$

29. $\sqrt{\dfrac{x^5y^3}{z^2}}$

30. $\sqrt[3]{\dfrac{k^5m^3 \cdot r^2}{r^8}}$

31. $\sqrt[4]{\dfrac{a^8b^5}{c^3}}$

32. $\sqrt{\dfrac{g^3h^5}{r^3}}$

33. $\sqrt[3]{\dfrac{9x^5y^6}{z^5w^2}}$

34. $\sqrt[4]{\dfrac{32m^5p^8}{q^3}}$

35. $\sqrt[4]{\dfrac{r^5 \cdot s^3}{t^2}}$

36. $\sqrt[3]{\dfrac{a^8 \cdot b^5}{c^2 \cdot d^5}}$

37. $\sqrt[4]{\dfrac{x^2}{y^5}}$

38. $\sqrt[4]{\dfrac{32}{8y^3}}$

39. $\dfrac{\sqrt[3]{mn} \cdot \sqrt[3]{m^2}}{\sqrt[3]{n^2}}$

40. $\dfrac{\sqrt[3]{8m^2n^3} \cdot \sqrt[3]{2m^2}}{\sqrt[3]{32m^4n^3}}$

41. $\sqrt{a^3b^5} - 2\sqrt{a^7b^3} + \sqrt{a^3b^9}$

42. $\sqrt{p^7q^3} - \sqrt{p^5q^9} + \sqrt{p^9q}$

43. $\dfrac{x}{\sqrt{x+1}}$

44. $\dfrac{\sqrt{p}}{1 - \sqrt{p}}$

45. $\dfrac{2}{\sqrt{1+r}}$

46. $\dfrac{y}{\sqrt{y+4}}$

47. $\dfrac{\sqrt{x} + \sqrt{x+1}}{\sqrt{x} - \sqrt{x+1}}$

48. $\dfrac{\sqrt{p} + \sqrt{p^2-1}}{\sqrt{p} - \sqrt{p^2-1}}$

Rationalize the numerator of each of the following.

49. $\dfrac{\sqrt{3}}{2}$ 50. $\dfrac{\sqrt{6}}{5}$ 51. $\dfrac{1 + \sqrt{2}}{2}$ 52. $\dfrac{1 - \sqrt{3}}{3}$

53. $\dfrac{\sqrt{x}}{1 + \sqrt{x}}$

54. $\dfrac{\sqrt{p}}{1 - \sqrt{p}}$

55. $\dfrac{\sqrt{x} + \sqrt{x+1}}{\sqrt{x} - \sqrt{x+1}}$

56. $\dfrac{\sqrt{p} + \sqrt{p^2 - 1}}{\sqrt{p} - \sqrt{p^2 - 1}}$

Chapter Three

First Degree Relations
and Functions

Relations and functions are a central idea of modern mathematics. In fact, much of the rest of this text is devoted to a discussion of some of the many types of relations and functions: quadratic, such as $y = x^2$; cubic, such as $y = (x + 1)^3$; exponential, such as $y = 2^x$; and so on. In this chapter, we discuss first degree relations and functions, such as $y = 2x - 4$ or $y = x - 2$. To begin, we will first define the term relation, and then the term function.

3.1 Relations and Functions

An **ordered pair** of numbers consists of two numbers written in parentheses, in which the sequence of numbers is important, such as (2, 4). Here 2 is called the **first component** and 4 the **second component.** A **relation** is any set of ordered pairs. Thus, $\{(2, -1), (4, -2)\}$ is a relation, as is

$$R = \{(x, y) \mid x < 4 \text{ and } y > 0\}.$$

(This is read "the set of all ordered pairs (x, y) such that $x < 4$ and $y > 0$.) The set of all first components of the ordered pairs of a relation is called the **domain** of the relation, while the set of all second components is called the **range** of the relation. In relation R above, the domain can be expressed as $\{x \mid x < 4\}$, while the range is $\{y \mid y > 0\}$.

Example 1 Find the domain and range of each of the following.

(a) $\{(x, y) \mid y = x^2\}$

Here x can take on any value, while y is always nonnegative, since $y = x^2$. The domain can be written $\{x \mid x \text{ is a real number}\}$, while the range is written $\{y \mid y \geq 0\}$.

(b) $\{(x, y) \mid y = \sqrt{x}\}$

Since $\sqrt{x}$ is defined only for nonnegative x, the domain here is $\{x \mid x \geq 0\}$. The symbol $\sqrt{x}$ indicates the nonnegative square root of x, and so the range is $\{y \mid y \geq 0\}$.

A **function** is a special type of relation with the property that for each domain element (x-value) there is exactly one range element (y-value). By this definition, $\{(x, y) \mid x = y^2\}$ is not a function (although it is a relation) since the domain element $x = 16$ leads to two range numbers, $y = 4$ and $y = -4$. On the other hand, the relations $\{(x, y) \mid y = x^2\}$ and $\{(x, y) \mid y = 2x - 5\}$ are functions since both have exactly one value of y for each value of x.

Letters such as f, g, and h are often used to name functions. Thus, we can write

$$f = \{(x, y) \mid y = 3x - 4\}.$$

For this function f, if $x = 5$, then $y = 3 \cdot 5 - 4 = 11$. Hence, $(5, 11) \in f$. This is often abbreviated by saying $f(5) = 11$ (read "f of 5 equals 11.") In the same way, $f(-3) = 3(-3) - 4 = -13$, $f(0) = -4$, and $f(x) = 3x - 4$. Using this new notation we shall often write

$$f(x) = 3x - 4$$

or

$$y = 3x - 4$$

instead of

$$f = \{(x, y) \mid y = 3x - 4\}.$$

Example 2   Let $f(x) = 3x - 4$ and $g(x) = 5x + 6$.

(a)   Find $g(-6)$.
$g(-6) = 5(-6) + 6 = -30 + 6 = -24$.

(b)   Find $f(4) + g(4)$.
$f(4) + g(4) = (3 \cdot 4 - 4) + (5 \cdot 4 + 6) = 8 + 26 = 34$.

(c)   Find $f[g(2)]$.
Since $g(2) = 5 \cdot 2 + 6 = 16$, we have $f[g(2)] = f(16) = 44$.

(d)   Find $g[g(-1)]$.
Since $g(-1) = 1$, we have $g[g(-1)] = g(1) = 11$.

An expression such as $f[g(x)]$ is called a **composite function.**

Example 3   Let $f(x) = 5x - 1$ and $g(x) = x^2$. Find $f[g(x)]$ and $g[f(x)]$.

We have $f[g(x)] = f(x^2) = 5x^2 - 1$,

and $g[f(x)] = g(5x - 1) = (5x - 1)^2$.

It is often helpful to draw a graph of the ordered pairs belonging to a relation. To do this we normally use the crossed number lines of a Cartesian coordinate system (named for its inventor, the philosopher-mathematician René Descartes). We shall let the number lines cross at right angles at their respective zero points, with positive numbers measured to the right on the horizontal number line, called the **x-axis,** and up along the vertical number line, called the **y-axis.** Thus in Figure 3.1, A represents the point whose coordinates are given by the ordered pair (3, 4), while B represents (−5, 6), C represents (−2, −4), D represents (4, −3) and E represents (−3, 0). Point A is said to be in the first **quadrant,** B is in the second quadrant, C is in the third quadrant, and D is in the fourth quadrant. The points of the axes themselves, such as E, belong to no quadrant.

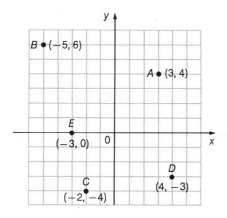

Figure 3.1

We can derive a useful graphical test for determining when a relation is also a function. By the definition of function, for each value of x in the domain there must be exactly one value of y in the range. Thus for a relation to be a function, as shown in Figure 3.2, a vertical line must never cut the graph of

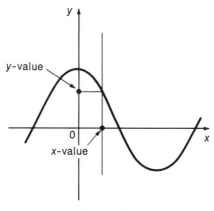

Figure 3.2

the relation in more than one point. Hence, Figure 3.3 represents a function, while Figure 3.4 and Figure 3.5 do not.

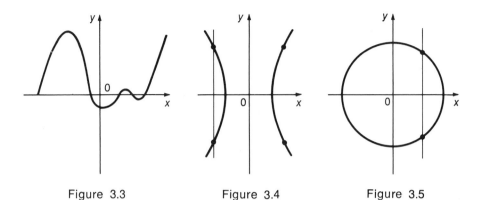

Figure 3.3   Figure 3.4   Figure 3.5

3.1 Exercises

Give the domain and range of each of the following relations. Identify those which are functions.

1. $\{(x,y)\,|\,y = 2x\}$
2. $\{(x,y)\,|\,y - 2 = x\}$
3. $\{(x,y)\,|\,y < x + 1\}$
4. $\{(x,y)\,|\,3x - 1 < y\}$
5. $\{(x,y)\,|\,y = x^2\}$
6. $\{(x,y)\,|\,y = (x - 2)^2\}$
7. $\{(x,y)\,|\,y = \sqrt{16 - x^2}\}$

8. $\{(x,y)\,|\,y = |x|\}$
9. $\{(x,y)\,|\,y = -2\}$
10. $\{(x,y)\,|\,x = 3\}$
11. $\left\{(x,\,y)\,|\,y = \dfrac{1}{x}\right\}$
12. $\left\{(x,\,y)\,|\,y = \dfrac{2}{x - 1}\right\}$

Let $f(x) = 3x - 1$ and $g(x) = x^2$. Find each of the following.

13. $f(0)$
14. $f(-1)$
15. $f(-3)$
16. $f(4)$

17. $g(2)$
18. $g(0)$
19. $f(a)$
20. $g(b)$

21. $f[g(2)]$
22. $g[f(2)]$
23. $f[f(1)]$
24. $g[g(-2)]$

For all positive integers x, let $f(x) = 2^x$. Find each of the following.

25. $f(1)$
26. $f(2)$

27. $f(4)$
28. $f(m)$

Let $g = \{(x,\,y)\,|\,y = 3^{-x}$ for all integers $x\}$. Find each of the following.

29. $g(0)$
30. $g(2)$

31. $g(-1)$
32. $g(-2)$

A function f with the property that $f(-x) = f(x)$ for all values of x is called an

even function. A function f such that $f(-x) = -f(x)$ *is called an odd function.
Identify each of the following as odd, even, or neither.*

33. $f(x) = x^2$
34. $f(x) = x^3$
35. $f(x) = x^4 + x^2 + 5$

36. $f(x) = x^3 - x + 1$
37. $f(x) = 2x + 3$
38. $f(x) = |x|$

3.2 Linear Functions

As mentioned in the introduction to this chapter, there are many special kinds
of functions. In this section we discuss an important first degree function (all
terms having degree no greater than 1), the linear function. First, we define a
linear relation as any relation of the form

$$\{(x, y) \,|\, ax + by = c; \, a, b \text{ not both } 0\}.$$

Here a, b, and c are real numbers, a restriction which will be assumed in the
future. Any linear relation which is also a function is called a **linear function.**
Hence, $\{(x, y) \,|\, 2x = y\}$ is a linear function since it is equivalent to

$$\{(x, y) \,|\, 2x - y = 0\},$$

where $a = 2$, $b = -1$, and $c = 0$. From now on, we shall abbreviate relations
such as $\{(x, y) \,|\, 2x = y\}$ by merely writing the associated equation, in this case
$2x = y$.

Which (if any) linear relations are not also linear functions? A relation is
not a function if for some value of x, there is more than one value of y. If we
let x equal some given real number, say $x = x_1$, then the definition of linear
relation yields the equation $ax_1 + by = c$. For fixed real number values of a,
x_1, and c, the only way there can be more than one value of y which satisfies
$ax_1 + by = c$ is if $b = 0$. Hence, we have the following result.

THEOREM 3.1   If $b \neq 0$, then $ax + by = c$ is a linear function.

By Theorem 3.1, the only equations leading to linear relations that are not also
functions are those of the form $x = k$, for some real number k.

Every linear relation has a graph which is a straight line. To find this
straight line graph, it is often useful to find the x-intercept and y-intercept.
An **x-intercept** is an x-value (if any) at which the graph of the relation crosses
the x-axis, while a **y-intercept** is a y-value (if any) at which the graph crosses
the y-axis. Note that the graph of Figure 3.6 has x-intercept x_1 and y-intercept
y_1.

To find the x-intercept of a line $ax + by = c$, we can let $y = 0$ (see Figure
3.6; note that when the graph crosses the x-axis we have $y = 0$). Doing this,

we have

$$ax + b \cdot 0 = c$$
$$ax = c$$
$$x = c/a,$$

where $a \neq 0$. In the same way, to find the y-intercept we let $x = 0$. Doing so, we have

$$a \cdot 0 + by = c$$
$$by = c$$
$$y = c/b,$$

where $b \neq 0$. Hence, we have proved the following result.

THEOREM 3.2   For any linear function $ax + by = c$, with $a \neq 0$ and $b \neq 0$, the x-intercept is c/a and the y-intercept is c/b.

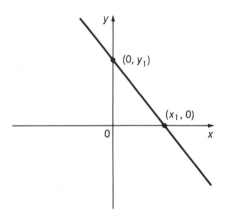

Figure 3.6

Example 4   Graph $3x + 2y = 6$.

Here $a = 3$, $b = 2$ and $c = 6$. Hence, the x-intercept is given by $c/a = 6/3 = 2$ and the y-intercept by $c/b = 6/2 = 3$. The graph is shown in Figure 3.7 on page 56.

Example 5   Graph $y = -3$.

To have an x-intercept there must be a value of x to make $y = 0$. However, here $y = -3 \neq 0$. Hence the line has no x-intercept. This happens only if the line is parallel to the x-axis, as shown in Figure 3.8 on page 56.

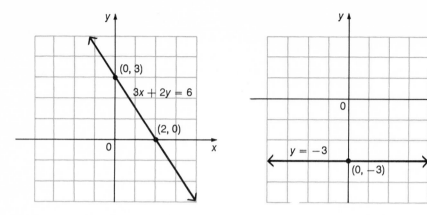

Figure 3.7   Figure 3.8

Example 6 Find the simultaneous solution set (the set of all ordered pairs satisfying both equations) of $2x - y = 7$ and $x - 2y = 8$.

One method we can use to find the solution set is to graph both lines and from the graph estimate the point (or points) of intersection (if any), as was done for this example in Figure 3.9. From Figure 3.9 we see that $\{(2, -3)\}$ is the simultaneous solution set.

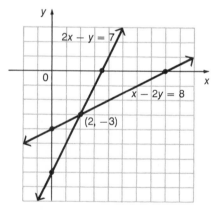

Figure 3.9

3.2 Exercises

Graph each of the following.

1. $y = 3x$	5. $2x + 5y = 10$	9. $x = 2$
2. $x + y = 4$	6. $3x + 4y = 12$	10. $y = -3$
3. $3x - y = 6$	7. $4x - 3y = 9$	11. $y = -5$
4. $2x + 3y = 6$	8. $5x + 3y = 15$	12. $x = 6$

Find the point of intersection (if any) of the following pairs of lines by graphing.

13. $x + y = 6$
    $x - y = 2$

14. $x + 2y = 9$
    $x - y = 0$

15. $3x + 4y = 7$
    $2x + 5y = 0$

16. $4x + y + 1 = 0$
    $2x + 3y = 7$

17. $3x + 2y = -2$
    $3x = 6 - 2y$

18. $4x + 5y = 10$
    $10y = 5 - 8x$

3.3 Straight Lines

As mentioned in the previous section, linear relations of the form $ax + by = c$ have straight lines as graphs. An important characteristic of a straight line is its slope, a numerical measure of the steepness of the line. To find this measure, let us consider the line going through the two distinct points (x_1, y_1) and (x_2, y_2), as shown in Figure 3.10. The difference

$$x_2 - x_1$$

is called the **change in x,** and is denoted Δx (read "delta x",) where Δ is the Greek letter delta. In the same way, the **change in y** can be written

$$\Delta y = y_2 - y_1.$$

The **slope** of a line (usually denoted m) is defined to be the change in y divided by the change in x, or

$$\blacktriangleright \qquad m = \frac{\Delta y}{\Delta x} = \frac{y_2 - y_1}{x_2 - x_1} \qquad (\Delta x \neq 0).$$

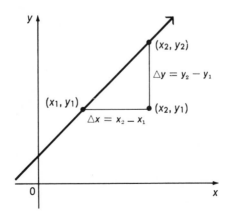

Figure 3.10

It can be shown (using theorems for similar triangles) that the slope is independent of the choice of points on the line. That is, the slope of a line remains the same no matter which pair of distinct points of the line determines the slope.

Example 7 Find the slope of the line through $(-4, 8)$ and $(2, -3)$.

We can choose $x_1 = -4$, $y_1 = 8$, $x_2 = 2$ and $y_2 = -3$. Hence, $\Delta y = -3 - 8 = -11$ and $\Delta x = 2 - (-4) = 6$. Thus the slope is given by

$$m = \frac{\Delta y}{\Delta x} = \frac{-11}{6}.$$

Example 8 Find the slope of the line $3x - 4y = 12$.

We first find any two points that lie on the line, for example, $(0, -3)$ and $(4, 0)$. Using these points and the definition of slope, we have

$$m = \frac{\Delta y}{\Delta x} = \frac{0 - (-3)}{4 - 0} = \frac{3}{4}.$$

Example 9 Find the slope of the line $y = -3$.

Again, we select any two points that lie on the line. Here we might select $(5, -3)$ and $(-2, -3)$. We can then use the definition of slope.

$$m = \frac{\Delta y}{\Delta x} = \frac{-3 - (-3)}{-2 - 5} = 0$$

Recall that the graph of $y = -3$ is a horizontal line. In general, as suggested by this example, the slope of a horizontal line is 0.

Example 10 Find the slope of $x = 2$.

Here $(2, 1)$ and $(2, -4)$ lie on the line. The slope is given by

$$m = \frac{\Delta y}{\Delta x} = \frac{-4 - 1}{2 - 2} = -\frac{5}{0},$$

which is not a real number. The graph of $x = 2$ is a vertical line; by generalizing, a vertical line has no slope.

The next theorem provides a convenient method to determine the slope and y-intercept of a line. We shall then use these numbers as an aid in drawing the graph.

THEOREM 3.3 SLOPE-INTERCEPT FORM

If a linear function is given by $y = mx + b,$
then m is the slope of the line and b is the y-intercept.

The proof that b is the y-intercept will be left as an exercise. To prove that m is the slope we must consider several cases; the proof of the case where $m \neq 0$, $b \neq 0$ is as follows.

As above, to find the slope we find two distinct points that lie on the line. Here (as can be seen by substitution) the points $(0, b)$ and $(-b/m, 0)$ lie on the line. Since $b \neq 0$, these points are distinct. Using these two points and the definition of slope we can find the slope of the line.

$$\text{slope} = \frac{\Delta y}{\Delta x} = \frac{0 - b}{-\dfrac{b}{m} - 0} = \frac{-b}{-\dfrac{b}{m}} = m$$

Hence, the slope is given by m, which is what we wanted to show. To complete the proof, we need to consider the case where $m \neq 0$, and $b = 0$, and the case where $m = 0$ and $b \neq 0$.

Example 11   Find the slope and y-intercept of $3x - y = 2$. Graph the line.

We first write $3x - y = 2$ in the form $y = mx + b$. Here, $y = 3x - 2$. From this we see that the slope is $m = 3$ and the y-intercept is $b = -2$. To draw the graph we first locate the y-intercept. (See Figure 3.11.) By the definition of slope, $m = \dfrac{\Delta y}{\Delta x}$, if $\Delta x = 1$, then $\Delta y = m$. Hence, if x changes by 1 unit, y changes by m units. Here we know $m = 3 = 3/1$. Thus, if x changes 1 unit, y changes 3 units. We can use this fact to find another point of the graph, as shown in Figure 3.11.

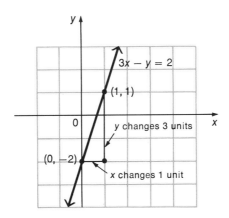

Figure 3.11

Sometimes we know the slope of a line together with one point that the line passes through. In such a case it is often helpful to use the point-slope form of the equation of a line given in the next theorem.

THEOREM 3.4   POINT-SLOPE FORM

If a line has slope m and passes through (x_1, y_1), then the equation of the line is given by
$$y - y_1 = m(x - x_1).$$

We leave it to the reader to show that the point (x_1, y_1) satisfies the given equation and that m is the slope of the line $y - y_1 = m(x - x_1)$. (Hint: Solve the equation for y and then use Theorem 3.3.)

Example 12   Write and simplify the equation of the line going through $(-4, 1)$ with slope $m = -3$.

Here, $x_1 = -4$, $y_1 = 1$ and $m = -3$. Using the point-slope form of the equation of a line, we have

$$y - 1 = -3[x - (-4)]$$
$$y - 1 = -3(x + 4)$$
$$y - 1 = -3x - 12$$
$$y = -3x - 11.$$

Example 13   Find and simplify the equation of the line through $(-3, 2)$ and $(2, -4)$.

Here $m = \dfrac{-4 - 2}{2 - (-3)} = \dfrac{-6}{5}$. If we let $x_1 = -3$ and $y_1 = 2$, we have

$$y - 2 = -\frac{6}{5}[x - (-3)]$$
$$5(y - 2) = -6(x + 3)$$
$$5y - 10 = -6x - 18$$
$$5y + 6x + 8 = 0.$$

Verify that we get the same equation if we use $(2, -4)$ instead of $(-3, 2)$ in the point-slope form.

3.3 Exercises

Find the slope of each of the following lines.

1. $y = 2x$
2. $y = 3x - 2$
3. $y = 4x - 5$
4. $y = -2x + 1$
5. $3x + 4y = 6$
6. $2x + y = 8$
7. through $(-2, 1)$ and $(3, 2)$

8. through $(-2, 3)$ and $(-1, 2)$
9. through $(0, 4)$ and $(-1, -3)$
10. through $(-4, -3)$ and $(5, 0)$
11. $y = 4$
12. $x = -6$
13. the x-axis
14. the y-axis

Find an equation for each of the following lines.

15. through $(-1, 3)$, $m = 2$
16. through $(2, 4)$, $m = -1$
17. through $(-5, 4)$, $m = -\frac{3}{2}$
18. through $(-4, 3)$, $m = \frac{3}{4}$
19. through $(2, 0)$, $m = -\frac{3}{4}$
20. through $(0, -1)$, $m = \frac{3}{5}$
21. through $(-3, 2)$, $m = 0$
22. through $(6, 1)$, $m = 0$
23. through $(-8, 1)$, no slope
24. through $(0, 4)$, no slope

25. through $(4, 2)$ and $(1, 3)$
26. through $(8, -1)$ and $(4, 3)$
27. through $(-1, 3)$ and $(3, 4)$
28. through $(6, 0)$ and $(3, 2)$
29. through $(0, 3)$ and $(4, 0)$
30. through $(-3, 0)$ and $(0, -5)$
31. x-intercept 3, y-intercept -2
32. x-intercept -2, y-intercept 4
33. x-intercept -5, no y-intercept
34. y-intercept 3, no x-intercept

Graph each of the following lines.

35. through $(-1, 2)$, $m = \frac{3}{2}$
36. through $(-2, 8)$, $m = -1$
37. through $(3, -4)$, $m = -\frac{1}{3}$

38. through $(-2, -3)$, $m = -\frac{3}{4}$
39. through $(-3, -4)$, $m = -\frac{3}{5}$
40. through $(3, 4)$, $m = \frac{2}{3}$

Two lines having the same slope are parallel, while two lines whose slopes have a product of -1 are perpendicular. Write equations for each of the following lines.

41. through $(-1, 4)$, parallel to $x + 3y = 5$
42. through $(2, -5)$, parallel to $y - 4 = 2x$
43. through $(3, -4)$, perpendicular to $x + y = 4$
44. through $(-2, 6)$, perpendicular to $2x - 3y = 5$
45. Do the points $(4, 3)$, $(2, 0)$, and $(-18, -12)$ lie on the same line? (Hint: find the equation of the line through two of the points.)
46. Do the points $(4, -5)$, $(3, -5/2)$, and $(-6, 18)$ lie on a line?
47. Prove that b is the y-intercept in Theorem 3.3.

3.4 Solving Equations and Inequalities

In this section we shall discuss methods of finding the solution sets of linear equations and inequalities. First, we shall say that any two equations having the same domain and the same solution set are **equivalent.** Hence, $2x = 16$ and $3x - 5 = 19$ are equivalent, since they both have the same solution set, $\{8\}$. The next theorem, which follows from the addition and multiplication properties of equality, is useful in finding the solution set of linear equations.

THEOREM 3.5   If $P(x)$, $Q(x)$, and $R(x)$ are algebraic expressions, then:

(a) $P(x) = Q(x)$ and $P(x) + R(x) = Q(x) + R(x)$ are equivalent;

(b) $P(x) = Q(x)$ and $P(x) \cdot R(x) = Q(x) \cdot R(x)$ are equivalent, if $R(x) \neq 0$.

To solve an equation such as

$$\frac{3p - 1}{3} - \frac{2p}{p - 1} = p,$$

we first obtain a (simpler) equivalent equation by multiplying both sides by $3(p - 1)$, where we must assume $p \neq 1$, (Why?) From this, we have,

$$(3p - 1)(p - 1) - 3(2p) = 3p(p - 1)$$
$$3p^2 - 4p + 1 - 6p = 3p^2 - 3p.$$

We can obtain an even simpler equivalent equation by combining terms and adding $-3p^2$ to both sides.

$$-10p + 1 = -3p$$

If we now add $10p$ to both sides, we obtain the equivalent equation

$$1 = 7p.$$

Multiplying both sides by $\frac{1}{7}$ gives the equivalent equation

$$\tfrac{1}{7} = p,$$

so that the solution set of the given equation is $\{\tfrac{1}{7}\}$.

To solve

$$\frac{x}{x - 2} = \frac{2}{x - 2} + 2,$$

we multiply both sides of the equation by $x - 2$, assuming that $x - 2 \neq 0$. This gives

$$x = 2 + 2x - 4$$
$$x = 2.$$

But we had to assume that $x - 2 \neq 0$ in order to use Theorem 3.5(b). Since $x = 2$, $x - 2 = 0$, and the theorem does not apply. The solution set is $\varnothing$.

Example 14   Solve $J\left(\dfrac{x}{k} + a\right) = x$ for x.

If we first multiply both sides by k, we have

$$kJ\left(\frac{x}{k} + a\right) = kx$$
$$kJ\left(\frac{x}{k}\right) + kJa = kx$$
$$Jx + kJa = kx.$$

If we add $-Jx$ to both sides, we have

$$kJa = kx - Jx$$
$$kJa = x(k - J).$$

If we assume $k \neq J$, we can multiply both sides by $\dfrac{1}{k-J}$ to find the solution

$$\frac{kJa}{k-J} = x.$$

Example 15   Melinda and Hortense win equal amounts of cash in a contest. Melinda will spend her money in 8 years, while Hortense will require 10. If they both work together on only one of the prizes, how long will it take them to spend it?

Let x represent the number of years it will take the two women working together to spend a single prize. In one year, Melinda will spend $\frac{1}{8}$ of the amount while Hortense will spend $\frac{1}{10}$. Hence,

$$\frac{1}{8} + \frac{1}{10} = \frac{1}{x}$$
$$5x + 4x = 40$$
$$x = \frac{40}{9} \text{ years.}$$

Working together, the women can spend one prize in $\frac{40}{9}$ years, or about 4 years, 5 months.

To solve a linear inequality, such as $3x - 5 < 7$, we use the following properties of inequalities which are consequences of Theorem 1.8.

THEOREM 3.6   If $P(x)$, $Q(x)$, and $R(x)$ are algebraic expressions, then:
(a) $P(x) < Q(x)$ and $P(x) + R(x) < Q(x) + R(x)$ are equivalent (have the same solution set);
(b) $P(x) < Q(x)$ and $P(x) \cdot R(x) < Q(x) \cdot R(x)$ are equivalent, if $R(x) > 0$;
(c) $P(x) < Q(x)$ and $P(x) \cdot R(x) > Q(x) \cdot R(x)$ are equivalent, if $R(x) < 0$.

To use the theorem to solve the inequality $3x - 5 < 7$, we write the following series of equivalent inequalities.

$$3x - 5 + 5 < 7 + 5$$
$$3x < 12$$
$$x < 4$$

Thus, the solution set of $3x - 5 < 7$ is $\{x \mid x < 4\}$.

Example 16 Solve $4 - 3y \leq 7 + 2y$.

We can write the following series of equivalent inequalities.

$$4 - 3y \leq 7 + 2y$$
$$-5y \leq 3$$
$$y \geq -\frac{3}{5}$$

From this last inequality, we see that the solution set is $\{y \mid y \geq -\frac{3}{5}\}$.

To solve an equation involving absolute values, such as $|2x + 5| = 9$, we use the definition of absolute value to write

$$2x + 5 = 9 \quad \text{or} \quad 2x + 5 = -9.$$

The solution to this compound sentence is

$$x = 2 \quad \text{or} \quad x = -7.$$

A similar technique is used for inequalities, as shown in the next two examples.

Example 17 Solve $|3p - 1| \leq 5$.

By Theorem 1.9, $|x| \leq b$ for nonnegative b implies $-b \leq x \leq b$. Hence, for $|3p - 1| \leq 5$ we may write

$$-5 \leq 3p - 1 \leq 5$$
$$-4 \leq 3p \leq 6$$
$$-\frac{4}{3} \leq p \leq 2.$$

The solution set is $\{p \mid -\frac{4}{3} \leq p \leq 2\}$.

Example 18 Solve $|4m - 3| \geq 13$.

Again by Theorem 1.9, for all nonnegative b, $|x| \geq b$ implies $x \geq b$ or $x \leq -b$. Hence, we have

$$4m - 3 \geq 13 \quad \text{or} \quad 4m - 3 \leq -13$$
$$4m \geq 16 \quad \text{or} \quad 4m \leq -10$$
$$m \geq 4 \quad \text{or} \quad m \leq -\frac{5}{2}.$$

The solution set can be written $\{m \mid m \geq 4 \text{ or } m \leq -\frac{5}{2}\}$.

3.4 Exercises

Solve each of the following.

1. $4x - 1 = 15$
2. $-3y + 2 = 5$

3. $-2[m - (4 + 2m) + 3] = 2m + 2$

4. $4[2p - (3 - p) + 5] = -7p - 2$

5. $\dfrac{3x - 2}{7} = \dfrac{x + 2}{5}$

6. $\dfrac{2p + 5}{5} = \dfrac{p + 2}{3}$

7. $\dfrac{x}{3} - 7 = 6 - \dfrac{3x}{4}$

8. $\dfrac{4p - 2}{7} = \dfrac{3p - 2}{4}$

9. $\dfrac{m}{2} - \dfrac{1}{m} = \dfrac{6m + 5}{12}$

10. $\dfrac{6z + 2}{5} - \dfrac{3z + 5}{3} = \dfrac{z + 1}{15}$

11. $\dfrac{2r}{r - 1} = 5 + \dfrac{2}{r - 1}$

12. $\dfrac{5}{2a + 3} + \dfrac{1}{a - 6} = 0$

13. $\dfrac{x + 4}{x - 3} - \dfrac{8x}{2x + 5} + \dfrac{3x}{x - 3} = 0$

14. $\dfrac{5}{2p + 3} - \dfrac{3}{p - 2} = \dfrac{4}{2p + 3}$

15. $2x + 1 \le 9$

16. $3y - 2 < 10$

17. $-3p - 2 \le 1$

18. $-5r + 3 \le -2$

19. $|a - 2| = 1$

20. $|3 - x| = 2$

21. $|3m - 1| = 2$

22. $|4p + 2| = 5$

23. $|5 - 3x| = 3$

24. $|-2 + 5a| = 3$

25. $|a - 4| < 6$

26. $|z + 5| \le 2$

27. $|3m - 2| > 1$

28. $|4z + 6| \ge 6$

29. $|8b + 5| \ge 7$

30. $|2m - 5| > 3$

Solve each of the following for the letter indicated.

31. $PV = k$ for V

32. $s = \frac{1}{2}gt^2$ for g

33. $V = V_0 + gt$ for g

34. $C = \frac{5}{9}(F - 32)$ for F

35. $\dfrac{1}{R} = \dfrac{1}{r_1} + \dfrac{1}{r_2}$ for R, (all denominators not equal to 0)

36. $A = \frac{1}{2}h(b + B)$ for B

37. Harry can do a job in 9 hours, and Mert in 6. How long will it take them if they work together on the same job?

38. An inlet pipe can fill Tom's pool in 5 hours, while an outlet pipe can empty it in 8 hours. In his haste to watch television, Tom left both pipes open. How long would it then take to fill the pool?

39. Suppose Tom discovered his error (see Exercise 38) after an hour long program. If he then closed the outlet pipe, how much longer would be needed to fill the pool?

3.5 Graphing Linear Inequalities

A line divides a plane into three sets of points–the points of the line itself, and the points belonging to the two regions determined by the line. Each of these

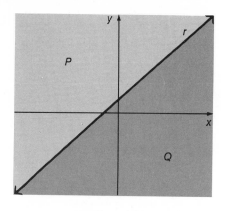

Figure 3.12

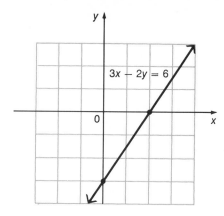

Figure 3.13

two regions is called a **half-plane.** In Figure 3.12, line r divides the plane into three different sets of points — line r, half-plane P, and half-plane Q. Note that r belongs neither to P nor to Q.

We shall use the idea of a half-plane as an aid in graphing linear inequalities, as described in the next theorem which we state without proof.

THEOREM 3.7   The graph of the linear inequality $ax + by < c$ is the half-plane, determined by the line $ax + by = c$, whose points satisfy the inequality $ax + by < c$. Also, the graph of the linear inequality $ax + by \leq c$ is the union of the appropriate half-plane determined by the line $ax + by = c$ and the points of the line itself.

By Theorem 3.7, the graph of $3x - 2y \leq 6$ is a half-plane and its boundary line; to find the graph we first draw the line $3x - 2y = 6$, as shown in Figure 3.13. To determine which half-plane belongs to the graph, we select any point not on the line. If and only if this point satisfies the inequality, then the half-plane from which the point was chosen belongs to the graph. Often, $(0, 0)$ is a useful point for this test. If we select $(0, 0)$ here, we have $3(0) - 2(0) \leq 6$. Since this statement is true, the half-plane including $(0, 0)$ belongs to the graph of $3x - 2y \leq 6$. The final graph is shown in Figure 3.14.

Example 19   Graph $x + 4y > 4$.

Here the graph consists only of a half-plane. It is customary to indicate this by making the boundary line dashed, as was done in Figure 3.15.

Example 20   Graph $y < -1$.

Recall that $y = -1$ is a horizontal line. Using this fact we can complete the graph, as shown in Figure 3.16.

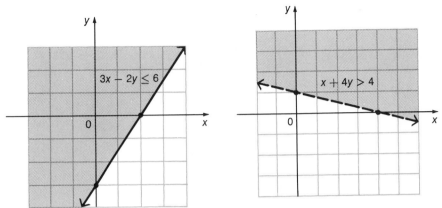

Figure 3.14

Figure 3.15

Example 21 Graph the intersection of $2x - y \le 4$ and $x < 3$.

Figure 3.17 shows the graphs of both relations. The heavily shaded area represents the intersection of the two regions. Note that the point of intersection of the boundary lines, (3, 2), belongs to the first relation, but not the second, and hence not to the intersection. To show this, we use an open dot on the graph at (3, 2).

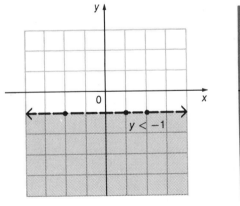

Figure 3.16

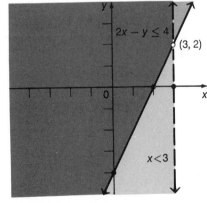

Figure 3.17

3.5 Exercises

Graph each of the following.

1. $x \le 3$
2. $y \le -2$
3. $x + 2y \le 6$
4. $x - y \ge 2$
5. $2x + 3y \ge 4$
6. $4y - 3x < 5$

7. $3x - 5y > 6$

8. $x < 3 + 2y$

9. $5x \le 4y - 2$

10. $2x > 3 - 4y$

Graph the intersection of the following relations.

11. $x - 3y < 4$ and $x \le 0$

12. $2x + y > 5$, and $y \ge 0$

13. $3x + 2y \ge 6$ and $y \le 2$

14. $4x + 3y \le 6$ and $x \le -2$

15. $x + 2y < 4$ and $3x - y > 5$

16. $4x + 1 < 2y$ and $-x > y - 3$

17. $2x + 3y < 6$ and $x - 5y \ge 10$

18. $3x - 4y > 6$ and $2x + 3y > 4$

3.6 Graphing Nonlinear First Degree Relations

Many first degree relations which we shall use in later work are **nonlinear**. Thus, the equations defining these relations *cannot* be written in the form $ax + by = c$. In this section we shall graph several such nonlinear relations. For example, to graph $y = |x|$, let us recall that $|x| \ge 0$ for all values of x. As long as $x \ge 0$, then $|x| = x$. On the other hand, if $x < 0$ then $|x| = -x$. Using these facts, and selecting some ordered pairs belonging to the relation, we obtain the graph shown in Figure 3.18. Note that the graph is the graph of a function.

Example 22 Graph $y < |x - 2|$.

As in earlier work with linear inequalities, we begin by graphing the boundary, $y = |x - 2|$. We select some points that satisfy the boundary relation, and then sketch the graph, as shown in Figure 3.19. To determine which region belongs to the relation we select any point off the boundary line, such as $(5, -3)$. Since $-3 < |5 - 2|$, we shade the side of the graph including $(5, -3)$, as shown in Figure 3.19.

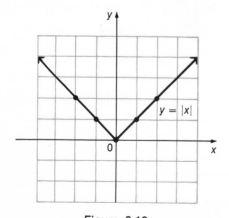

Figure 3.18

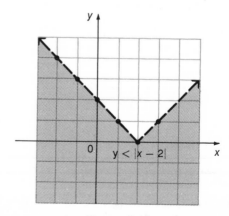

Figure 3.19

Example 23 Graph $y \geq |3x - 4|$.

Again, we graph the boundary $y = |3x - 4|$. Since $(0, 0)$ does not satisfy the relation, we shade the region that does not include $(0, 0)$. The graph is shown in Figure 3.20.

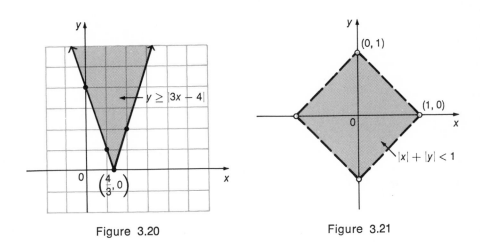

Figure 3.20   Figure 3.21

Example 24 Graph $|x| + |y| < 1$.

First, we graph the boundary, $|x| + |y| = 1$. We can do this by considering the following cases. First, if $x \geq 0$ and $y \geq 0$, then $|x| + |y| = 1$ becomes $x + y = 1$. Hence, in quadrant I we graph $x + y = 1$. Next, in quadrant II, we have $x < 0$ and $y > 0$. Hence, $|x| + |y| = 1$ becomes $-x + y = 1$. In quadrant III we have $-x - y = 1$, while we have $x - y = 1$ in quadrant IV. We now graph these dashed lines in the appropriate quadrants. Note that the four points $(0, 1)$, $(1, 0)$, $(-1, 0)$, $(0, -1)$ are not part of the graph. Since $(0, 0)$ satisfies $|x| + |y| < 1$, we have the graph of Figure 3.21.

The **greatest integer function,** written $y = [x]$ is defined by saying that $[x]$ represents the greatest integer less than or equal to x. Hence, $[8] = 8$, $[-5] = -5$, $[\pi] = 3$, $[-3.4] = -4$, and so on. To graph this function, we might note that for $0 \leq x < 1$, we have $[x] = 0$. Also, for $1 \leq x < 2$, we have $[x] = 1$. Thus, the graph, as shown in Figure 3.22, is composed of a series of line segments. Note that in each case the left endpoint of the segment is included, and the right endpoint is excluded. The greatest integer function is an example of a **step function.**

Example 25 Graph the **nearest integer function,** $y = \langle x \rangle$, defined by saying $\langle x \rangle$ is the distance from x to the nearest integer.

Note here that $\langle \frac{3}{4} \rangle = \frac{1}{4}$, since the distance from $\frac{3}{4}$ to the nearest integer, 1, is $\frac{1}{4}$. Also, $\langle 12.3 \rangle = .3$, $\langle -4.1 \rangle = .1$, and $\langle \frac{3}{2} \rangle = \frac{1}{2}$. Using these points we can sketch the graph, as shown in Figure 3.23. The nearest integer function is an example of a **periodic function**.

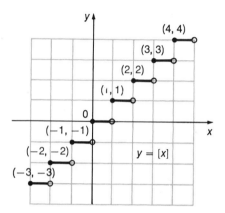

Figure 3.22

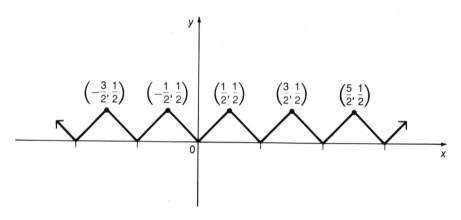

Figure 3.23

Example 26 Graph the relation

$$y = \begin{cases} x + 1 \text{ if } x \leq 2 \\ -2x + 7 \text{ if } x > 2. \end{cases}$$

For $x \leq 2$, we graph $y = x + 1$. For $x > 2$, we graph $y = -2x + 7$, as shown in Figure 3.24.

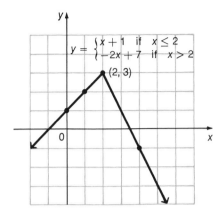

Figure 3.24

3.6 Exercises

Graph each of the following relations.

1. $y \leq |x|$
2. $x \geq |y + 1|$
3. $x \leq |y - 3|$
4. $y \geq |x + 2|$
5. $2x > |3y - 1|$

6. $x - 1 < |3y + 2|$
7. $y = |x| - x$
8. $x = |y| + y$
9. $y < |x| - x$
10. $x < |y| + y$

11. $y = [x + \frac{1}{2}]$ (Hint: this is similar to the greatest integer function.)

12. $y = [2x - 1]$
13. $y = [3x - 2]$
14. $y = [3x] + 1$
15. $y = \langle 2x \rangle$
16. $y = \langle 3x \rangle - 1$
17. $x + |y| = 1$
18. $|x| - y = 1$
19. $|x + y| = 1$
20. $|x + 3y| = 2$

21. $y = \begin{cases} x - 1 \text{ if } x \leq 3 \\ 2 \text{ if } x > 3 \end{cases}$

22. $y = \begin{cases} 6 - x \text{ if } x \leq 3 \\ 3x - 6 \text{ if } x > 3 \end{cases}$

23. $y = \begin{cases} 4 - x \text{ if } x < 2 \\ 1 + 2x \text{ if } x \geq 2 \end{cases}$

24. $y = \begin{cases} -2 \text{ if } x \geq 1 \\ 2 \text{ if } x \leq 1 \end{cases}$

3.7 Arithmetic Progressions

In this section we shall investigate a special type of first degree function–an arithmetic sequence, or arithmetic progression. First, let us say that a **sequence** is a function whose domain is a subset of the set of positive integers. The range elements of this function are called the **terms** of the sequence. Since the domain of every sequence is the same, we can exhibit a given sequence merely by listing its terms. In fact, we often refer to the list of terms as the sequence itself. For example, in using the terminology "the sequence 2, 4, 6, 8, ...," we actually mean "the sequence whose terms (range elements), listed in order, are 2, 4, 6, 8, ..." The terms of the sequence

$$a = \{(x, a(x)) \,|\, a(x) = 2x + 1, x \text{ is a positive integer}\}$$

are $a(1) = 3$, $a(2) = 5$, $a(3) = 7$, ..., $a(n) = 2n + 1$, ... Note that the terms of a sequence have an order automatically determined by the order of the positive integers. Hence it makes sense to speak of the first term of a sequence, the second term, the third term, and so on.

It is customary to write the term $a(1)$ as a_1, $a(2)$ as a_2, and in general, $a(n)$ as a_n. Using this notation, we see that for the sequence given above, $a_5 = 2(5) + 1 = 11$, $a_9 = 19$, and $a_k = 2k + 1$.

A sequence in which each term after the first is obtained by adding a fixed number (called the **common difference**) to the previous term is called an **arithmetic sequence** or **arithmetic progression.** Thus, the sequence

$$5, 9, 13, 17, 21, \ldots$$

is an arithmetic progression since each term after the first is obtained from the previous one by adding 4. That is,

$$9 = 5 + 4$$
$$13 = 9 + 4$$
$$17 = 13 + 4$$
$$21 = 17 + 4,$$

and so on. In general, if a_1 is the first term of an arithmetic sequence and d is the common difference, the sequence is given by

$$a_1, a_1 + d, a_1 + 2d, a_1 + 3d, \ldots, a_1 + (n - 1)d, \ldots.$$

As shown above, the **nth term of an arithmetic sequence** is given by

▶ $$a_n = a_1 + (n - 1)d.$$

Example 27 Find the 13th term of the sequence $-3, 1, 5, 9, \ldots$.
Here $a_1 = -3$ and $d = 4$. Hence, $a_{13} = -3 + (13 - 1)4 = 45$.

Example 28 Suppose $a_8 = -16$ and $a_{16} = -40$. Find a_1.
We know $a_8 = a_1 + (8 - 1)d$. Hence, $-16 = a_1 + 7d$ or $a_1 = -16 - 7d$. Also, $-40 = a_1 + 15d$ and $a_1 = -40 - 15d$. From these two equations, we have $-16 - 7d = -40 - 15d$, or $d = -3$. Since $-16 = a_1 + 7d$, we have $a_1 = 5$.

Given a_1 and d we can find the sum of the first n terms of an arithmetic progression. To begin, we write the sum of the first n terms of an arithmetic progression.

$$S_n = a_1 + [a_1 + d] + [a_1 + 2d] + \cdots + [a_1 + (n - 1)d]$$

Next, we write this same sum in reversed order.

$$S_n = [a_1 + (n-1)d] + [a_1 + (n-2)d] + \cdots + [a_1 + d] + a_1$$

Now we add the respective sides of these two equations.

$$S_n + S_n = (a_1 + [a_1 + (n-1)d]) + ([a_1 + d] + [a_1 + (n-2)d])$$
$$+ \cdots + ([a_1 + (n-1)d] + a_1)$$

From this, we have

$$2S_n = [2a_1 + (n-1)d] + [2a_1 + (n-1)d] + \cdots + [2a_1 + (n-1)d].$$

Since there are n of the $[2a_1 + (n-1)d]$ terms on the right, we have

$$2S_n = n[2a_1 + (n-1)d]$$

▶
$$S_n = \frac{n}{2}[2a_1 + (n-1)d].$$

Since $a_n = a_1 + (n-1)d$, we can also write

▶
$$S_n = \frac{n}{2}[a_1 + a_n].$$

Example 29 Find the sum of the first 60 positive integers.

Here $n = 60$, $a_1 = 1$, and $a_{60} = 60$. We have

▶
$$S_{60} = \frac{60}{2}(1 + 60) = 30 \cdot 61 = 1830.$$

Example 30 The sum of the first 17 terms of an arithmetic progression is 187. If $a_{17} = -13$, find a_1 and d.

We can use the formula developed above. We have

$$187 = \frac{17}{2}(a_1 - 13)$$
$$374 = 17(a_1 - 13)$$
$$22 = a_1 - 13$$
$$a_1 = 35.$$

We know $a_{17} = a_1 + (17 - 1)d$. Hence,

$$-13 = 35 + 16d$$
$$-48 = 16d$$
$$d = -3.$$

3.7 Exercises

Find the eighth term of each of the following arithmetic progressions.

1. $a_1 = 5, d = 2$
2. $a_1 = -3, d = -4$
3. $a_1 = -4, d = 2$
4. $a_1 = 0, d = 5$
5. $a_3 = 2, d = 1$
6. $a_4 = 5, d = -2$
7. $a_1 = 8, a_2 = 6$
8. $a_1 = 6, a_2 = 3$

9. $a_1 = 12, a_3 = 6$
10. $a_2 = 5, a_4 = 1$
11. $a_{10} = 6, a_{12} = 15$
12. $a_{15} = 8, a_{17} = 2$
13. $a_1 = x, a_2 = x + 3$
14. $a_2 = y + 1, d = -3$
15. $a_6 = 2m, a_7 = 3m$
16. $a_5 = 4p + 1, a_7 = 6p + 7$

Find the sum of the first 10 terms for each of the following arithmetic progressions.

17. $a_1 = 8, d = 3$
18. $a_1 = 2, d = 6$
19. $a_1 = 12, d = -2$
20. $a_1 = -9, d = 4$
21. $a_3 = 5, a_4 = 8$
22. $a_2 = 9, a_4 = 13$

23. $5, 9, 13, \ldots$
24. $8, 6, 4, \ldots$
25. $3\frac{1}{2}, 5, 6\frac{1}{2}, \ldots$
26. $2\frac{1}{2}, 3\frac{3}{4}, 5, \ldots$
27. $a_1 = 10, a_{10} = 5\frac{1}{2}$
28. $a_1 = -8, a_{10} = -\frac{5}{4}$

We may write $f(1) + f(2) + f(3) + \cdots + f(n)$ in abbreviated form as

$$\sum_{i=1}^{n} f(i).$$

($\sum$ is the Greek letter sigma which is used to represent sum.) For instance, if $f(i) = 5i - 3$ and $n = 6$, then

$$\sum_{i=1}^{6} (5i - 3) = (5 \cdot 1 - 3) + (5 \cdot 2 - 3)$$
$$+ (5 \cdot 3 - 3) + (5 \cdot 4 - 3) + (5 \cdot 5 - 3) + (5 \cdot 6 - 3)$$
$$= 2 + 7 + 12 + 17 + 22 + 27$$
$$= 87.$$

Evaluate each of the following sums.

29. $\displaystyle\sum_{i=1}^{3} i$

30. $\displaystyle\sum_{i=1}^{5} (2i + 1)$

31. $\displaystyle\sum_{i=1}^{6} (3i + 2)$

32. $\displaystyle\sum_{i=1}^{7} (2 - 4i)$

33. $\displaystyle\sum_{i=4}^{9} (5i + 6)$

34. $\displaystyle\sum_{i=4}^{12} (7i + 3)$

35. Find the sum of all integers between 50 and 72.
36. Find the sum of all integers between -9 and 31.
37. If a clock strikes the proper number of bongs each hour on the hour, how many bongs will it bong in a month of 30 days?
38. A stack of telephone poles has 30 in the first row, 29 in the second, and so on, with one pole in the last row. How many poles are in the stack?

Chapter Four

Second Degree Relations and Functions

Many important applications of mathematics require a knowledge of second degree relations. In this chapter, we discuss methods of graphing such relations, as well as methods of finding real number solutions of second degree equations.

4.1 The Quadratic Function

A function of the form $y = ax^2 + bx + c$, $(a \neq 0)$ a, b, and c real numbers, is called a **quadratic function.** (Without the restriction $a \neq 0$ we could have $y = 0x^2 + bx + c = bx + c$, which is a linear function.) Perhaps the simplest quadratic function is $y = x^2$, where $a = 1$, $b = 0$, and $c = 0$. To find some ordered pairs belonging to the graph of this function we can choose values for x and then find the corresponding values for y, as shown in the chart accompanying Figure 4.1. These pairs can then be graphed, and a smooth curve

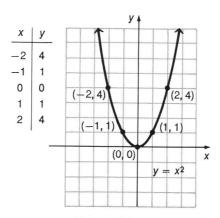

Figure 4.1

drawn through them, as in Figure 4.1. This curve is called a **parabola.** Since the coefficient of x^2 in the equation $y = x^2$ is positive, the parabola opens upward. On the other hand, the coefficient of x^2 in the equation $y = -x^2$ is negative, and the graph of this parabola would open downward.

Parabolas have many practical applications. For example, cross sections of spotlight reflectors, or radar dishes, form parabolas. Every quadratic function has a graph which is a parabola.

Example 1 Graph $y = 2x^2 + 8x + 5$.

To graph such a parabola, it is often convenient to rewrite the equation in a special form by **completing the square.** We want to write $y = 2x^2 + 8x + 5$ in the form $y = m(x + n)^2 + p$. To do so, we write

$$y = 2(x^2 + 4x \qquad) + 5.$$

The process of completing the square requires that we now convert the expression in parentheses, $x^2 + 4x$, into a perfect square by adding an appropriate number. To find this number, take half the coefficient of x ($\frac{1}{2} \cdot 4 = 2$) and square the result ($2^2 = 4$). Add the square, 4, to the quantity inside the parentheses, and subtract 8 outside the parentheses to compensate for adding $2 \cdot 4 = 8$. The result is

$$y = 2(x^2 + 4x + 4) + 5 - 8.$$

We can now write the expression as

$$y = 2(x + 2)^2 - 3.$$

Since $2(x + 2)^2 \geq 0$ for all real values of x, we see that the lowest possible value of y is $y = -3$, and that y has this value when $x = -2$. Thus, the lowest point on this graph, called the **vertex,** is $(-2, -3)$. By finding other ordered pairs belonging to the function, we get the graph shown in Figure 4.2.

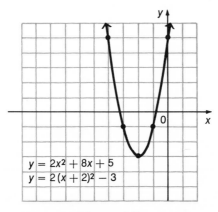

Figure 4.2

Example 2 Graph $y = -3x^2 - 2x + 1$.

Before completing the square here, we first rewrite the expression as

$$y = -3(x^2 + \tfrac{2}{3}x \qquad) + 1.$$

We then add $(\tfrac{1}{2} \cdot \tfrac{2}{3})^2 = \tfrac{1}{9}$ to the quantity inside the parentheses, and correspondingly, subtract $(-3)(\tfrac{1}{9}) = -\tfrac{1}{3}$ outside the parentheses.

$$y = -3(x^2 + \tfrac{2}{3}x + \tfrac{1}{9}) + 1 - (-\tfrac{1}{3})$$
$$y = -3(x + \tfrac{1}{3})^2 + \tfrac{4}{3}$$

Since $-3(x + \tfrac{1}{3})^2 \leq 0$ for all real values of x, the parabola will open downward, and hence the vertex, $(-\tfrac{1}{3}, \tfrac{4}{3})$, will be the *highest* point on this graph. By selecting some ordered pairs of the graph, we get the result of Figure 4.3.

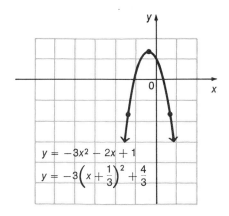

Figure 4.3

In the same way, we can complete the square on the general quadratic function

$$y = ax^2 + bx + c \qquad (a \neq 0).$$

Following the steps of the previous examples, we have

$$y = a\left(x^2 + \frac{b}{a}x \qquad\right) + c.$$

Now we take half of $\dfrac{b}{a}\left(\dfrac{1}{2} \cdot \dfrac{b}{a} = \dfrac{b}{2a}\right)$ and square this result.

This gives

$$y = a\left(x^2 + \frac{b}{a}x + \frac{b^2}{4a^2}\right) + c - \frac{b^2}{4a}$$
$$= a\left(x + \frac{b}{2a}\right)^2 + \frac{4ac - b^2}{4a}.$$

We use this last statement to prove the following result.

THEOREM 4.1   The vertex of the graph of the quadratic function $y = ax^2 + bx + c$, $(a \neq 0)$ is given by

$$\left(-\frac{b}{2a}, \frac{4ac - b^2}{4a}\right).$$

This vertex is the highest point on the graph if $a < 0$, and the lowest point on the graph if $a > 0$.

Example 3   Find the vertex of $y = -2x^2 + 3x - 5$.

Here $a = -2$, $b = 3$, and $c = -5$. Theorem 4.1 is a convenient way to find the vertex. Using the result of this theorem, we have

$$\left(-\frac{b}{2a}, \frac{4ac - b^2}{4a}\right) = \left(\frac{-3}{2(-2)}, \frac{4(-2)(-5) - 3^2}{4(-2)}\right)$$

$$= \left(\frac{3}{4}, -\frac{31}{8}\right).$$

Here the vertex is the highest point on the graph. The graph of this parabola could be obtained by plotting this vertex and some other ordered pairs that satisfy the equation.

Example 4   Graph $y < 2x^2 - 3$.

Using a result analagous to Theorem 3.7 for graphing linear inequalities, we first graph the boundary parabola, $y = 2x^2 - 3$, as shown in Figure 4.4. Then, we select any point not on the parabola, such as (0, 0). Since (0, 0) does not satisfy the original inequality, we shade the portion of the graph not including (0, 0), as in Figure 4.4.

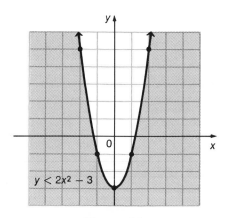

Figure 4.4

4.1 Exercises

Graph each of the following parabolas and identify the vertex.

1. $y = x^2 + 1$
2. $y = 2x^2$
3. $y = (x - 2)^2$
4. $y = (x + 4)^2$
5. $y = (x + 3)^2 - 4$
6. $y = (x - 5)^2 - 4$
7. $y = -2(x + 3)^2 + 2$
8. $y = -3(x - 2)^2 + 1$
9. $y = -\frac{1}{2}(x + 1)^2 - 3$

10. $y = \frac{2}{3}(x - 2)^2 - 1$
11. $y = x^2 - 2x + 3$
12. $y = x^2 + 6x + 5$
13. $y = x^2 + 8x + 13$
14. $y = x^2 - 12x + 30$
15. $y = -x^2 - 4x + 2$
16. $y = -x^2 + 6x - 6$
17. $y = 2x^2 - 4x + 5$
18. $y = -3x^2 + 24x - 46$

Graph each of the following relations.

19. $y < 3x^2 + 2$
20. $y \le x^2 - 4$
21. $y > (x - 1)^2 + 2$
22. $y > 2(x + 3)^2 - 1$
23. Farmer Frieda wishes to fence a rectangular area which lies against a river, so that only three sides need to be fenced. She has 320 feet of fencing. What should the dimensions of the field be if the enclosed area is to be a maximum? (Hint: let x represent one side of the area, with $320 - 2x$ the other side. Graph the area parabola, $A = x(320 - 2x)$ and investigate the vertex.)
24. What would be the maximum area that could be enclosed by Farmer Frieda's 320 feet of fencing if she had to enclose all 4 sides of her field?
25. Find two numbers whose sum is 20 and whose product is a maximum.

4.2 Zeros of Quadratic Functions

A **zero** of a function $y = f(x)$ is a value of x that makes $f(x) = 0$. Thus, -2 is a zero of $y = 2x^2 + 3x - 2$, since $y = 2(-2)^2 + 3(-2) - 2 = 0$. In this section we shall discuss methods of finding real zeros of quadratic functions. The simplest method of finding such zeros, but one that is not always applicable, is by factoring. This method depends on the following result.

THEOREM 4.2   ZERO-FACTOR THEOREM
If a and b are real numbers with $ab = 0$,
then $a = 0$ or $b = 0$.

To prove this, let us assume that $ab = 0$ and $a \ne 0$. We must then show that $b = 0$. We have

$$a \neq 0$$

$$ab = 0$$

$$\frac{1}{a}(ab) = \frac{1}{a}(0)$$

(Why do we know that $\frac{1}{a}$ is a real number?)

$$\left(\frac{1}{a} \cdot a\right)b = 0$$

$$b = 0.$$

In a similar way, we can show that if $b \neq 0$, then $a = 0$.

Example 5 Find the zeros of $y = 6x^2 + 7x - 3$.

We want to find all the values of x that make $6x^2 + 7x - 3 = 0$. To use the zero-factor theorem, we factor $6x^2 + 7x - 3 = 0$ as $(3x - 1)(2x + 3) = 0$. Thus, we have

$$(3x - 1)(2x + 3) = 0,$$

and by the zero-factor theorem, $(3x - 1)(2x + 3) = 0$ only if

$$3x - 1 = 0 \quad \text{or} \quad 2x + 3 = 0.$$

The solution set of this compound sentence, $\{\frac{1}{3}, -\frac{3}{2}\}$, gives the zeros of the function.

Not all quadratic functions have zeros which can be found by factoring ($y = x^2 + x + 6$ is an example). Hence, we need a more general method for finding zeros. To find this more general method, we complete the square on the general quadratic equation

$$ax^2 + bx + c = 0 \quad (a \neq 0).$$

We begin by writing

$$a\left(x^2 + \frac{b}{a}x \quad\right) + c = 0.$$

We can now complete the square in a manner similar to that used when looking for the vertex of a parabola. First, we take half of b/a, and square the result, obtaining $b^2/4a^2$. We add this quantity inside the parentheses, and subtract $b^2/4a$, which compensates for adding $a(b^2/4a^2) = b^2/4a$.

$$a\left(x^2 + \frac{b}{a}x + \frac{b^2}{4a^2}\right) + c - \frac{b^2}{4a} = 0$$

Now we add $\dfrac{b^2}{4a} - c$ $\left(\text{which equals } \dfrac{b^2 - 4ac}{4a}\right)$ to both sides, and then simplify.

$$a\left(x^2 + \frac{b}{a}x + \frac{b^2}{4a^2}\right) = \frac{b^2 - 4ac}{4a}$$

We can now write the expression in parentheses on the left-hand side as the square of a binomial.

$$a\left(x + \frac{b}{2a}\right)^2 = \frac{b^2 - 4ac}{4a}$$

If we now divide both sides of the equation by a, we get

$$\left(x + \frac{b}{2a}\right)^2 = \frac{b^2 - 4ac}{4a^2}.$$

This last statement leads to the compound sentence

$$x + \frac{b}{2a} = \sqrt{\frac{b^2 - 4ac}{4a^2}} \quad \text{or} \quad x + \frac{b}{2a} = -\sqrt{\frac{b^2 - 4ac}{4a^2}},$$

which we can also write as

$$x = \frac{-b + \sqrt{b^2 - 4ac}}{2a} \quad \text{or} \quad x = \frac{-b - \sqrt{b^2 - 4ac}}{2a}.$$

A simplified form of this result is given in Theorem 4.3.

THEOREM 4.3   THE QUADRATIC FORMULA
The zeros of the quadratic function $y = ax^2 + bx + c$, $(a \neq 0)$ are given by

$$x = \frac{-b \pm \sqrt{b^2 - 4ac}}{2a}.$$

Example 6   Find the zeros of $y = x^2 - 4x + 1$.
    Here $a = 1$, $b = -4$, and $c = 1$. Using the quadratic formula, we have

$$x = \frac{-b \pm \sqrt{b^2 - 4ac}}{2a}$$

$$x = \frac{4 \pm \sqrt{16 - 4}}{2}$$

$$x = \frac{4 \pm 2\sqrt{3}}{2}$$

$$x = 2 \pm \sqrt{3}.$$

The zeros are $2 + \sqrt{3}$ and $2 - \sqrt{3}$.

    The expression $b^2 - 4ac$ appearing in the quadratic formula is called the **discriminant**.

▶ If $b^2 - 4ac > 0$, the function has **two** distinct real zeros.
▶ If $b^2 - 4ac = 0$, the function has exactly **one** real zero.
▶ If $b^2 - 4ac < 0$, the function has **no** real zeros.

We shall discuss nonreal zeros of quadratic functions in a later chapter.

4.2 Exercises

Find all real solutions of the following equations.

1. $x(x - 2)(x + 3) = 0$
2. $(2m - 1)(3m + 2)(4m + 5) = 0$
3. $p^2 - 5p + 6 = 0$
4. $q^2 + 2q - 8 = 0$
5. $6z^2 - 5z - 50 = 0$
6. $6r^2 + 7r = 3$
7. $8k^2 + 14k + 3 = 0$
8. $18r^2 - 9r - 2 = 0$
9. $-m^2 + m + 1 = 0$
10. $y^2 - 3y + 2 = 0$
11. $2s^2 + 2s = 3$
12. $t^2 - t = 3$
13. $3a^2 + 7a - 2 = 0$
14. $4z^2 - 4z = 3$
15. $x^2 - 6x + 7 = 0$
16. $11p^2 - 7p + 1 = 0$
17. $12r^2 + 5r = 3$

18. $m^2 - 4m + 7 = 0$
19. $9p^2 - 30p = 25$
20. $4n^2 + 12n + 9 = 0$
21. $4 - \dfrac{11}{x} - \dfrac{3}{x^2} = 0$
22. $3 - \dfrac{4}{p} = \dfrac{2}{p^2}$
23. $2 - \dfrac{5}{r} + \dfrac{3}{r^2} = 0$
24. $2 - \dfrac{5}{k} + \dfrac{2}{k^2} = 0$
25. $m^2 - \sqrt{2}m - 1 = 0$
26. $z^2 - \sqrt{3}z - 2 = 0$
27. $\sqrt{2}p^2 - 3p + \sqrt{2} = 0$
28. $-\sqrt{6}k^2 - 2k + \sqrt{6} = 0$
29. $mx^2 + nx + 2 = 0$
30. $(p + q)r^2 - 2pr + p = 0$

Evaluate the discriminant $b^2 - 4ac$ for each of the following, and use it to predict the type of zeros of the function.

31. $y = x^2 + 8x + 16$
32. $y = x^2 - 5x + 4$
33. $y = 3x^2 - 5x + 2$
34. $y = 8x^2 - 14x + 3$

35. $y = 4x^2 - 6x - 3$
36. $y = 2x^2 - 4x + 1$
37. $y = 9x^2 + 11x + 4$
38. $y = 3x^2 - 8x + 5$

4.3 Equations Quadratic in Form

An equation that can be made into a quadratic equation, either by substituting variables or by raising both sides to a power, is described as **quadratic in form.** For example,

$$12m^4 - 11m^2 + 2 = 0$$

can be converted into a quadratic by letting $x = m^2$. Doing this, we have

$$12x^2 - 11x + 2 = 0,$$

which can be solved by factoring.

$$(3x - 2)(4x - 1) = 0$$
$$x = \tfrac{2}{3} \quad \text{or} \quad x = \tfrac{1}{4}.$$

Since $x = m^2$, we have

$$m^2 = \tfrac{2}{3} \quad \text{or} \quad m^2 = \tfrac{1}{4}$$
$$m = \pm\sqrt{\tfrac{2}{3}} \quad \text{or} \quad m = \pm\sqrt{\tfrac{1}{4}}$$
$$m = \pm\frac{\sqrt{6}}{3} \quad \text{or} \quad m = \pm\frac{1}{2}.$$

Example 7   Solve $(a - 2)^{2/3} + (a - 2)^{1/3} - 2 = 0$.
Here we can let $x = (a - 2)^{1/3}$. By substitution, we have

$$x^2 + x - 2 = 0,$$

which can be factored as follows.

$$(x + 2)(x - 1) = 0$$
$$x = -2 \quad \text{or} \quad x = 1.$$

Since $x = (a - 2)^{1/3}$, we have

$$(a - 2)^{1/3} = -2 \quad \text{or} \quad (a - 2)^{1/3} = 1$$
$$a - 2 = -8 \quad \text{or} \quad a - 2 = 1$$
$$a = -6 \quad \text{or} \quad a = 3.$$

These solutions can be verified by substituting in the original equation.

Example 8   Solve $\left(m - \dfrac{1}{m}\right)^2 + 2\left(m - \dfrac{1}{m}\right) = 8$.

Let $x = m - \dfrac{1}{m}$. Then we have

$$x^2 + 2x - 8 = 0,$$
$$(x + 4)(x - 2) = 0$$
$$x = -4 \quad \text{or} \quad x = 2.$$

Since $x = m - \dfrac{1}{m}$, we have

$$m - \frac{1}{m} = -4 \quad \text{or} \quad m - \frac{1}{m} = 2$$
$$m^2 - 1 = -4m \quad \text{or} \quad m^2 - 1 = 2m$$
$$m^2 + 4m - 1 = 0 \quad \text{or} \quad m^2 - 2m - 1 = 0.$$

Using the quadratic formula, we can write

$$m = \frac{-4 \pm \sqrt{16 + 4}}{2} \quad \text{or} \quad m = \frac{2 \pm \sqrt{4 + 4}}{2}$$

$$m = -2 \pm \sqrt{5} \quad \text{or} \quad m = 1 \pm \sqrt{2}.$$

To solve an equation such as $x = \sqrt{15 - 2x}$ we can square both sides. That this procedure is valid is shown by the next theorem.

THEOREM 4.4   The solution set of an equation $P(x) = Q(x)$ is a subset of the solution set of the equation $[P(x)]^n = [Q(x)]^n$, for any positive integer value of n.

Note that the theorem only asserts that the solution set of the original equation is a *subset* of the solution set of the new equation. The new equation may have more solutions or **roots** than the original equation. Any such extra roots are called **extraneous roots.** To identify extraneous roots, all proposed roots must be checked in the *original* equation.

By Theorem 4.4, the equation $x = \sqrt{15 - 2x}$ can be solved by squaring both sides.

$$x^2 = (\sqrt{15 - 2x})^2$$
$$x^2 = 15 - 2x$$
$$x^2 + 2x - 15 = 0$$
$$(x + 5)(x - 3) = 0$$
$$x = -5 \quad \text{or} \quad x = 3.$$

We must now check the proposed roots in the original equation,

$$x = \sqrt{15 - 2x}.$$

If $x = -5$, does $x = \sqrt{15 - 2x}$?   If $x = 3$, does $x = \sqrt{15 - 2x}$?
$$-5 \neq \sqrt{15 + 10} \qquad\qquad\qquad 3 = \sqrt{15 - 6}$$
$$-5 \neq 5. \qquad\qquad\qquad\qquad\quad 3 = 3.$$

By this check, only $x = 3$ is a root. The proposed root $x = -5$, which does not work in the original equation, is an extraneous root.

Example 9   Solve $\sqrt{2x + 3} - \sqrt{x + 1} = 1$.
We can write the equation as

$$\sqrt{2x + 3} = 1 + \sqrt{x + 1}.$$

If we now square both sides, we get

$$2x + 3 = 1 + 2\sqrt{x + 1} + x + 1$$
$$x + 1 = 2\sqrt{x + 1}.$$

One side of the equation still contains a radical, and to eliminate it, we need to square both sides again.

$$x^2 + 2x + 1 = 4(x + 1)$$
$$x^2 - 2x - 3 = 0$$
$$(x - 3)(x + 1) = 0$$
$$x = 3 \quad \text{or} \quad x = -1.$$

We must check these proposed roots in the original equation.

$$\text{If } x = 3, \text{ does } \sqrt{2x + 3} - \sqrt{x + 1} = 1?$$
$$\sqrt{9} - \sqrt{4} = 1$$
$$3 - 2 = 1.$$

$$\text{If } x = -1, \text{ does } \sqrt{2x + 3} - \sqrt{x + 1} = 1?$$
$$\sqrt{1} - \sqrt{0} = 1$$
$$1 - 0 = 1.$$

Here both proposed solutions, $x = 3$ and $x = -1$, belong to the solution set, $\{3, -1\}$.

4.3 Exercises

Solve each of the following equations.

1. $\sqrt{x} + 1 = 2$
2. $\sqrt{x} - 3 = 5$
3. $m^4 + 2m^2 - 15 = 0$
4. $p^4 - p^2 - 6 = 0$
5. $2z^4 + 5z^2 - 3 = 0$
6. $3k^4 + 7k^2 - 6 = 0$
7. $2r^4 - 7r^2 + 5 = 0$
8. $4x^4 - 8x^2 + 3 = 0$
9. $(g - 2)^2 - 6(g - 2) + 8 = 0$
10. $(p + 2)^2 - 2(p + 2) - 15 = 0$
11. $-(r + 1)^2 - 3(r + 1) + 3 = 0$
12. $2(z - 4)^2 - 2(z - 4) - 3 = 0$
13. $6(k + 2)^4 - 11(k + 2)^2 + 4 = 0$
14. $8(m - 4)^4 - 10(m - 4)^2 + 3 = 0$
15. $\left(m + \dfrac{1}{m} \right)^2 + 3\left(m + \dfrac{1}{m} \right) - 10 = 0$
16. $\left(p - \dfrac{1}{p} \right)^2 + \left(p - \dfrac{1}{p} \right) = 30$
17. $(r - 1)^{2/3} + (r - 1)^{1/3} = 12$
18. $(y + 3)^{2/3} - 2(y + 3)^{1/3} - 3 = 0$

19. $p = \sqrt{5p - 6}$

20. $m = \sqrt{8 - 2m}$

21. $x = -\sqrt{\dfrac{50 + 5x}{6}}$

22. $s = \sqrt{\dfrac{3 - 2s}{2}}$

23. $\sqrt{18 + \sqrt{x}} = \sqrt{2x + 3}$

24. $\sqrt{9 + \sqrt{x}} = \sqrt{x - 3}$

25. $\sqrt{2x + 5} + \sqrt{5x - 1} = 6$

26. $\sqrt{2x + 1} - \sqrt{x - 3} = 2$

4.4 Quadratic Inequalities

For what values of x is $x^2 - x - 12 < 0$? To find out, we might begin by graphing the parabola $y = x^2 - x - 12$. This can be done by completing the square

$$y = (x^2 - x + \tfrac{1}{4}) - 12 - \tfrac{1}{4}$$
$$y = (x - \tfrac{1}{2})^2 - \tfrac{49}{4}.$$

The graph of this parabola is shown in Figure 4.5. From the graph of the para-

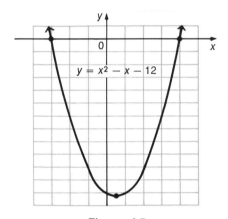

Figure 4.5

bola we can see that $y = x^2 - x - 12$ is negative for all values of x between -3 and 4. That is, the solution set of $x^2 - x - 12 < 0$ is $\{x \mid -3 < x < 4\}$. This solution set is graphed in Figure 4.6.

Figure 4.6

Since we want only the solution set, we can develop a shortcut. To solve $x^2 - x - 12 < 0$, we begin by finding the values of x which satisfy $x^2 - x - 12 = 0$.

$$x^2 - x - 12 = 0$$
$$(x + 3)(x - 4) = 0$$
$$x = -3 \quad \text{or} \quad x = 4.$$

Note that these roots of the equation are the points where the graph in Figure 4.5 crosses the x-axis. For all x-values between these points, the y-values will be negative since the graph there is below the x-axis. In the same way, the graph is above the x-axis when $x > 4$ or $x < -3$, so that the corresponding y-values are positive. Hence, all we need to do is determine one y-value in each of the three regions indicated in Figure 4.6. If one y-value is negative in a particular region, all y-values in that region will be negative and $x^2 - x - 12$ will be negative. By testing three points, using one x-value from each of the three regions, we can determine the solution.

In summary, to solve a quadratic inequality:

▶ (1) Find the roots of the corresponding quadratic equation.

▶ (2) Test one point in each of the regions determined by the roots to find the solution.

For example, to solve the inequality $2x^2 + 5x - 12 \geq 0$, we first find the roots of $2x^2 + 5x - 12 = 0$, namely, $x = \frac{3}{2}$ and $x = -4$. These two points, $x = \frac{3}{2}$ and $x = -4$, divide the number line into three regions, as shown in Figure 4.7. We select a point in each region and determine whether or not it

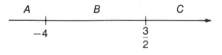

Figure 4.7

satisfies $2x^2 + 5x - 12 \geq 0$. If we try $x = -5$ from region A, we have

$$2(-5)^2 + 5(-5) - 12 = 13 \geq 0.$$

Since -5 satisfies the inequality, all the points of region A belong to the solution set. If we try $x = 0$, from region B, we have

$$2(0)^2 + 5(0) - 12 = -12 \ngeq 0.$$

This one point from region B did not satisfy the inequality, and thus no point from region B belongs to the solution set. For $x = 2$, from region C, we have

$$2(2)^2 + 5(2) - 12 = 6 \geq 0,$$

so that all points of C belong to the solution set. Hence, the solution set is $\{x \mid x \leq -4 \text{ or } x \geq \frac{3}{2}\}$, the graph of which is shown in Figure 4.8.

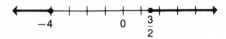

Figure 4.8

Example 10 Solve the inequality $1 \leq \dfrac{5}{x+4}$.

We could begin by multiplying both sides of the inequality by $x + 4$, but we would then have to consider whether $x + 4$ is positive or negative. To avoid this, we multiply both sides by $(x + 4)^2$, which is always nonnegative. Doing this, we have

$$(x + 4)^2 \leq 5(x + 4)$$
$$x^2 + 8x + 16 \leq 5x + 20$$
$$x^2 + 3x - 4 \leq 0.$$

The roots of $x^2 + 3x - 4 = 0$ are $x = -4$ and $x = 1$. By testing a point from each of the three regions, we see that the solution set of $x^2 + 3x - 4 \leq 0$ is $\{x \mid -4 \leq x \leq 1\}$. However, the solution set of $1 \leq \dfrac{5}{x+4}$, is

$$\{x \mid -4 < x \leq 1\},$$

since $x = -4$ is meaningless in the original inequality.

We can also work this example in another way. We know that if we multiply both sides of

$$1 \leq \frac{5}{x+4}$$

by a positive number we do not change the direction of the inequality symbol. Thus, let us assume $x + 4 > 0$ and write:

CASE 1
$$x + 4 > 0$$
$$1(x + 4) \leq \frac{5}{x+4}(x + 4)$$
$$x + 4 \leq 5$$
$$x \leq 1.$$

Here we have $x + 4 > 0$, or $x > -4$, and $x \leq 1$. The solution set of $x > -4$ and $x \leq 1$ is given by

$$\{x \mid x > -4 \text{ and } x \leq 1\} = \{x \mid x > -4\} \cap \{x \mid x \leq 1\}$$
$$= \{x \mid -4 < x \leq 1\}.$$

On the other hand, if we multiply both sides of the inequality by a negative number, we must change the direction of the inequality. Thus we have:

CASE 2
$$x + 4 < 0$$
$$1(x + 4) \geq \frac{5}{x + 4}(x + 4)$$
$$x + 4 \geq 5$$
$$x \geq 1.$$

The solution set here is given by

$$\{x \mid x < -4\} \cap \{x \mid x \geq 1\} = \emptyset.$$

The solution set of the original inequality is the *union* of the solution sets from case 1 and case 2, or

$$\{x \mid -4 < x \leq 1\} \cup \emptyset = \{x \mid -4 < x \leq 1\}.$$

4.4 Exercises

Solve each of the following inequalities.

1. $x^2 \leq 9$
2. $p^2 > 16$
3. $y^2 - 10y + 25 < 25$
4. $m^2 + 6m + 9 < 9$
5. $r^2 + 4r + 4 \geq 3$
6. $z^2 + 6z + 9 < 8$
7. $x^2 - x \leq 6$
8. $r^2 + r < 12$

9. $2k^2 - 9k > -4$
10. $3n^2 < 10 - 13n$
11. $x^2 > 0$
12. $p^2 < -1$
13. $\dfrac{3}{x - 6} \leq 2$
14. $\dfrac{x + 1}{x + 2} \leq 3$

15. $\dfrac{1}{k - 2} < \dfrac{1}{3}$
16. $\dfrac{1}{m - 1} < 1$
17. $x^3 - 4x \leq 0$
18. $r^3 - 9r \geq 0$

4.5 Circles

In this section we shall discuss the circle. A **circle** is the set of points in a plane which lie a given distance from a given point. The given distance is called the **radius** and the given point is called the **center.** As an aid in developing the equation of a circle, we state the distance formula (see Exercise 27).

THEOREM 4.5   DISTANCE FORMULA
The distance between the points (x_1, y_1) and (x_2, y_2) is given by

$$\sqrt{(x_2 - x_1)^2 + (y_2 - y_1)^2}.$$

For example, the distance between $(-5, 3)$ and $(3, 18)$ is

$$\sqrt{[3 - (-5)]^2 + (18 - 3)^2} = \sqrt{8^2 + 15^2} = 17.$$

Figure 4.9 shows a circle of radius 3 with center at the origin. If we select any point (x, y) on the circle, and use the distance formula to find the distance from (x, y) to the center, we have

$$\sqrt{(x - 0)^2 + (y - 0)^2} = 3$$
$$\sqrt{x^2 + y^2} = 3$$
$$x^2 + y^2 = 9.$$

Thus, $x^2 + y^2 = 9$ is the equation of a circle with center at the origin and radius 3. Note the characteristics of this equation: it contains both x^2 and y^2 and both have equal coefficients.

Example 11 Find the equation of a circle with center at $(-2, 4)$ and radius 4 (see Figure 4.10).

We again use the distance formula.

$$\sqrt{(x + 2)^2 + (y - 4)^2} = 4$$
$$(x + 2)^2 + (y - 4)^2 = 16$$

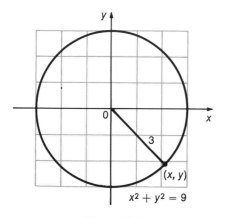

Figure 4.9

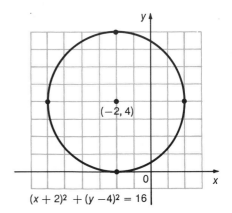

Figure 4.10

Generalizing from these examples, we see that the distance formula can be used to find the equation of a circle, as shown in the next theorem.

THEOREM 4.6 The equation of a circle having center (h, k) and radius r is

$$(x - h)^2 + (y - k)^2 = r^2.$$

Example 12 Graph $x^2 - 6x + y^2 + 10y = -25$.

We suspect that this equation represents a circle since the second degree terms have equal coefficients. To see if this equation really represents a circle, we complete the square on both x and y.

$$(x^2 - 6x + 9) + (y^2 + 10y + 25) = -25 + 9 + 25$$
$$(x - 3)^2 + (y + 5)^2 = 9$$

The equation is that of a circle having center at $(3, -5)$ and radius 3, with the graph as shown in Figure 4.11.

Example 13 Graph $x^2 + 10x + y^2 - 4y + 33 = 0$.

Completing the square gives

$$(x^2 + 10x + 25) + (y^2 - 4y + 4) = -33 + 25 + 4$$
$$(x + 5)^2 + (y - 2)^2 = -4.$$

There are no ordered pairs (x, y) satisfying this condition (why?).

Example 14 Graph $y = \sqrt{16 - x^2}$.

If we square both sides, we have $y^2 = 16 - x^2$, or $x^2 + y^2 = 16$, which is a circle with center at the origin and radius 4. However, in the original equation $y = \sqrt{16 - x^2}$, so that y can take on only nonnegative values. Hence, the graph is a semicircle, as shown in Figure 4.12. The domain of the function is $\{x \mid -4 \leq x \leq 4\}$, and the range is $\{y \mid 0 \leq y \leq 4\}$.

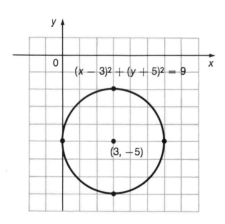

Figure 4.11

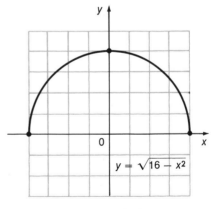

Figure 4.12

4.5 Exercises

Find the distance between each of the following pairs of points.

1. $(2, 4)$, $(3, -1)$
2. $(0, 4)$, $(-2, 3)$
3. $(-5, 1)$, $(2, 3)$
4. $(-2, 1)$, $(4, 2)$

5. Use the distance formula to verify that $(1, 1)$, $(2, -2)$, and $(3, -5)$ lie on the same line. (Hint: if the three points do not lie on the same line, they will be the vertices of a triangle.)
6. Verify that $(1, 1)$, $(5, -11)$, and $(4, 2)$ are the vertices of a right triangle. (Hint: if the sum of the squares of the length of two sides of a triangle equals the square of the third side, then the triangle is a right triangle.)

Graph each of the following.

7. $x^2 + y^2 = 36$
8. $x^2 + y^2 = 81$
9. $(x - 4)^2 + (y + 3)^2 = 4$
10. $(x + 3)^2 + (y - 2)^2 = 16$
11. $(x + 2)^2 + (y - 5)^2 = 12$
12. $(x - 4)^2 + (y - 3)^2 = 8$
13. $y = \sqrt{36 - x^2}$

14. $x = \sqrt{121 - y^2}$
15. $x = -\sqrt{8 - y^2}$
16. $y = -\sqrt{12 - x^2}$
17. $x^2 + (y + 3)^2 \leq 16$
18. $(x - 4)^2 + (y + 3)^2 \leq 9$
19. $y \leq \sqrt{49 - x^2}, y \geq 0$
20. $x > \sqrt{16 - y^2}, x \geq 0$

Find the center and radius of each of the following.

21. $x^2 + 6x + y^2 + 8y = -9$
22. $x^2 - 4x + y^2 + 12y = -4$
23. $x^2 - 12x + y^2 + 10y = -25$

24. $x^2 + 8x + y^2 - 6y = -16$
25. $x^2 + 8x + y^2 - 14y = -65$
26. $x^2 - 2x + y^2 = -1$

27. Prove the distance formula (Theorem 4.5).

4.6 Ellipses and Hyperbolas

An **ellipse** is the set of points in a plane the sum of whose distances from two distinct fixed points is constant. The two fixed points are called the **foci** (singular: **focus**) of the ellipse. See Figure 4.13. Using the distance formula and the definition of ellipse we have the following theorem (see Exercise 23).

THEOREM 4.7   The ellipse with x-intercepts $x = a$ and $x = -a$ and y-intercepts $y = b$ and $y = -b$ has the equation

$$\frac{x^2}{a^2} + \frac{y^2}{b^2} = 1.$$

Note that the coefficients of x^2 and y^2 are different, but both positive.

To graph an equation such as $4x^2 + 9y^2 = 36$, we first divide through by 36, obtaining

$$\frac{x^2}{9} + \frac{y^2}{4} = 1.$$

This equation is of the form shown in Theorem 4.7. Hence, the graph is an ellipse having x-intercepts $x = 3$ and $x = -3$, with y-intercepts given by

$y = 2$ and $y = -2$. Additiònal ordered pairs which satisfy the equation of the ellipse may be found to complete the graph. The graph of this ellipse is shown in Figure 4.14.

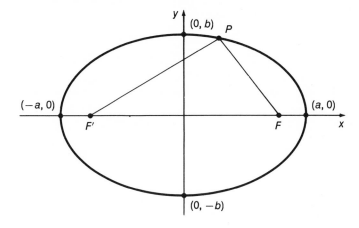

If F and F' are the foci of an ellipse, then for any point P on the ellipse, $PF' + PF = 2a$.

Figure 4.13

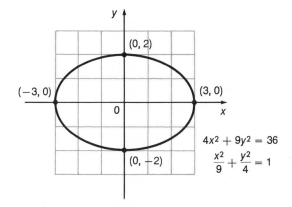

$$4x^2 + 9y^2 = 36$$
$$\frac{x^2}{9} + \frac{y^2}{4} = 1$$

Figure 4.14

Example 15   Graph $16x^2 + y^2 \leq 16$.

Dividing through by 16 gives the equation of the boundary, $x^2 + \dfrac{y^2}{16} = 1$.

This equation represents an ellipse having x-intercepts of $x = 1$ and $x = -1$, with y-intercepts $y = 4$ and $y = -4$. Since the point $(0, 0)$ satisfies $16x^2 + y^2 \leq 16$, the graph also includes the interior of the ellipse, as shown in Figure 4.15.

The graph in Figure 4.16 illustrates the idea of symmetry. For each point A of the graph, there is a corresponding point A', such that $AB = A'B$,

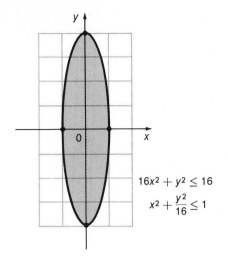

$16x^2 + y^2 \leq 16$

$x^2 + \dfrac{y^2}{16} \leq 1$

Figure 4.15

Figure 4.16

where AA' is perpendicular to the x-axis, and B is on the x-axis. Thus the graph is said to be **symmetric with respect to the x-axis.** Verify that the figure is also symmetric with respect to the y-axis.

A **hyperbola** is the set of points in a plane the difference of whose distances from two fixed points (called foci) is constant. The following theorem depends on the distance formula and the idea of symmetry.

THEOREM 4.8   A hyperbola having x-intercepts $x = a$ and $x = -a$, centered about the origin and symmetric to the x-axis, has an equation of the form

$$\frac{x^2}{a^2} - \frac{y^2}{b^2} = 1,$$

while a hyperbola having y-intercepts $y = a$ and $y = -a$, centered about the origin and symmetric to the y-axis, has an equation of the form

$$\frac{y^2}{a^2} - \frac{x^2}{b^2} = 1.$$

As an example, let us graph the hyperbola

$$\frac{x^2}{16} - \frac{y^2}{9} = 1.$$

By Theorem 4.8, the x-intercepts are $x = 4$ and $x = -4$. However, if $x = 0$, we have

$$-\frac{y^2}{9} = 1,$$

or $$y^2 = -9,$$

which has no real solutions. Hence, the graph has no y-intercepts. To complete the graph we can find some other ordered pairs that belong to it. For example, if $x = 6$, we have

$$\frac{6^2}{16} - \frac{y^2}{9} = 1$$

$$-\frac{y^2}{9} = 1 - \frac{36}{16}$$

$$\frac{y^2}{9} = \frac{20}{16}$$

$$y^2 = \frac{180}{16} = \frac{45}{4}$$

$$y = \pm\frac{3\sqrt{5}}{2} \approx \pm3.4.$$

Thus the graph includes the points $(6, 3.4)$ and $(6, -3.4)$. Also, if $x = -6$, we have $y \approx \pm3.4$. Hence, the graph also includes $(-6, 3.4)$ and $(-6, -3.4)$, as shown in Figure 4.17.

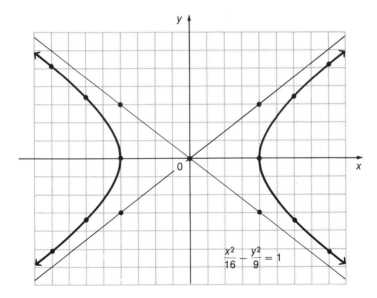

Figure 4.17

The straight lines approached by the graph of the hyperbola of Figure 4.17 are called the **asymptotes** of the hyperbola. A hyperbola of the form

$$\frac{x^2}{a^2} - \frac{y^2}{b^2} = 1$$

will have asymptotes whose equations are given by

$$\frac{x^2}{a^2} - \frac{y^2}{b^2} = 0.$$

This can be simplified by writing

$$\frac{x^2}{a^2} = \frac{y^2}{b^2},$$

from which

$$\frac{x}{a} = \frac{y}{b} \qquad \text{or} \qquad \frac{x}{a} = -\frac{y}{b},$$

so that

$$y = \frac{b}{a}x \qquad \text{or} \qquad y = -\frac{b}{a}x$$

are the equations of the asymptotes. In the same way,

$$\frac{y^2}{a^2} - \frac{x^2}{b^2} = 1$$

has asymptotes given by

$$y = \frac{a}{b}x \qquad \text{or} \qquad y = -\frac{a}{b}x$$

just as above. In general then, we have the following result.

THEOREM 4.9   The asymptotes of the hyperbola $\dfrac{x^2}{a^2} - \dfrac{y^2}{b^2} = 1$ are given by

$$\boldsymbol{y = \frac{b}{a}x} \qquad \text{or} \qquad \boldsymbol{y = -\frac{b}{a}x,}$$

while the asymptotes of the hyperbola $\dfrac{y^2}{a^2} - \dfrac{x^2}{b^2} = 1$ are given by

$$\boldsymbol{y = \frac{a}{b}x} \qquad \text{or} \qquad \boldsymbol{y = -\frac{a}{b}x.}$$

Example 16   Graph $4y^2 - x^2 = 16$.

If we divide both sides by 16, we have

$$\frac{y^2}{4} - \frac{x^2}{16} = 1.$$

Note that the y-intercepts are $y = 2$ and $y = -2$, and that there are no x-intercepts. We can use Theorem 4.9 to find the asymptotes. Here the asymptotes are given by

$$y = \frac{1}{2}x \qquad \text{or} \qquad y = -\frac{1}{2}x.$$

Using the intercepts and asymptotes, we get the graph shown in Figure 4.18.

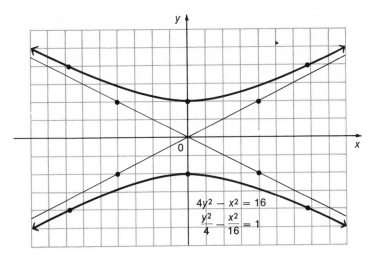

Figure 4.18

Example 17 Graph $y \le \sqrt{1 + 4x^2}$, $y \ge 0$.

If we square both sides of $y = \sqrt{1 + 4x^2}$, we get

$$y^2 = 1 + 4x^2$$

or
$$y^2 - 4x^2 = 1,$$

which is a hyperbola with y-intercepts $y = 1$ and $y = -1$. Since $\sqrt{1 + 4x^2}$ is always nonnegative, we have only half a hyperbola, as shown in Figure 4.19. We then select any point not on the hyperbola and try it in the original inequality. Since $(0, 0)$ satisfies this inequality we shade the side of the graph containing $(0, 0)$, as shown in Figure 4.19.

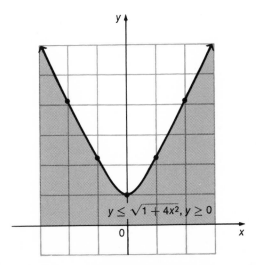

Figure 4.19

The graphs of the second degree relations we have studied, *parabolas*, *hyperbolas*, *ellipses* and *circles*, are called **conic sections** since each can be obtained by cutting a cone with a plane, as shown in Figure 4.20.

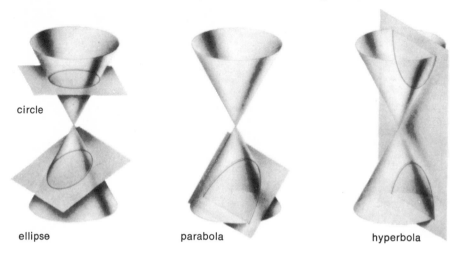

circle

ellipse   parabola   hyperbola

Figure 4.20

Note that the conic sections have second degree equations of the form

$$Ax^2 + Cy^2 + Dx + Ey + F = 0,$$

where either A or C must be nonzero. (The letter B is reserved for an xy term, which we have not discussed here.) The special characteristics of each of the relations we have discussed are summarized below.

▶ *parabola*   either A or C equals 0, but not both
▶ *circle* $A = C \neq 0$
▶ *ellipse* $A \neq C,\ AC > 0$
▶ *hyperbola* $AC < 0$

4.6 Exercises

Sketch the graph of each of the following.

1. $\dfrac{x^2}{9} + \dfrac{y^2}{4} = 1$

2. $\dfrac{x^2}{16} + \dfrac{y^2}{36} = 1$

3. $\dfrac{x^2}{9} + y^2 = 1$

4. $\dfrac{y^2}{16} - \dfrac{x^2}{9} = 1$

5. $\dfrac{x^2}{6} + \dfrac{y^2}{9} = 1$

6. $\dfrac{x^2}{8} - \dfrac{y^2}{12} = 1$

7. $x^2 + 4y^2 = 16$
8. $25x^2 + 9y^2 = 225$
9. $x^2 = 9 + y^2$
10. $y^2 = 16 + x^2$

11. $\dfrac{x}{4} = \sqrt{1 - \dfrac{y^2}{9}}$

12. $\dfrac{y}{2} = \sqrt{1 - \dfrac{x^2}{25}}$

13. $\dfrac{y}{3} = \sqrt{1 + \dfrac{x^2}{16}}$

14. $x = \sqrt{1 + \dfrac{y^2}{36}}$

15. $x^2 \le 9 - 4y^2$

16. $x^2 > 1 + 9y^2$

17. $\dfrac{x}{6} < \sqrt{1 - \dfrac{y^2}{121}},\; x \ge 0$

18. $\dfrac{y}{4} \le \sqrt{1 + \dfrac{x^2}{25}},\; y \ge 0$

Find the equations of the asymptotes of each of the following hyperbolas.

19. $\dfrac{x^2}{9} - \dfrac{y^2}{25} = 1$

20. $\dfrac{y^2}{9} - \dfrac{x^2}{16} = 1$

21. $x^2 = 16 + 4y^2$

22. $y^2 = 25 + x^2$

23. Suppose $(c, 0)$ and $(-c, 0)$ are the foci of an ellipse. Suppose the sum of the distances from any point (x, y) of the ellipse to the two foci is $2a$. (See figure.)

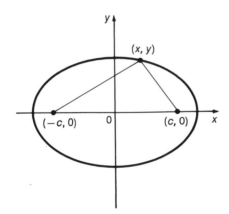

(a) Show that the equation of the resulting ellipse is

$$\dfrac{x^2}{a^2} + \dfrac{y^2}{a^2 - c^2} = 1.$$

(b) Show that a and $-a$ are the x-intercepts.

(c) Let $b^2 = a^2 - c^2$, and show that b and $-b$ are the y-intercepts.

4.7 Variation

Suppose Mert's salary is always three times Sam's salary. Then, if y represents Mert's salary, and x represents Sam's salary, we see that the two salaries satisfy a relationship of the form

$$y = kx \qquad (k \text{ a real number},\; k \ne 0).$$

(In our example, $k = 3$.) Such a relationship is called a **direct variation;** here we can say that Mert's salary **varies directly** as Sam's salary. The real number k is sometimes called the **constant of variation.**

Example 18 Suppose the area of a certain rectangle varies directly as the width. If the area is 50 when the width is 10, find the area when the width is 25.

Since we know that area varies directly as width, we have

$$A = kw,$$

where A = area, w = width, and k is a nonzero constant which we must determine. When $A = 50$ and $w = 10$, we have

$$50 = 10k,$$

or $k = 5$. Thus, in this example, the relationship between area and width can be expressed as

$$A = 5w.$$

If $w = 25$, we have

$$A = 5(25) = 125.$$

If the variables satisfy a relationship of the form

$$y = \frac{k}{x} \qquad (k \text{ a real number, } k \neq 0),$$

then x and y are said to **vary inversely.** For example, suppose that in a certain manufacturing process the cost of producing a single item varies inversely as the number of items produced. Suppose further that if 100 items are produced, each costs \$2. To find the cost per item if 400 items are produced, we can let x represent the number of items produced and y the cost per item, and write

$$y = \frac{k}{x},$$

for some nonzero constant k. We know that $y = \$2$ when $x = 100$. Hence

$$2 = \frac{k}{100}$$

or $k = 200$.

Thus, the relationship between x and y is given by

$$y = \frac{200}{x}.$$

When 400 items are produced, the cost per item is given by

$$y = \frac{200}{400} = \$.50.$$

Example 19 Suppose m varies directly as the square of p and inversely as q. Suppose also that $m = 8$ when $p = 2$ and $q = 6$. Find m if $p = 6$ and $q = 10$.

Here m depends on several variables. This is called **joint variation**. We can write

$$m = \frac{kp^2}{q}.$$

We know that $m = 8$ when $p = 2$ and $q = 6$. Thus,

$$8 = \frac{k \cdot 2^2}{6}.$$

$$k = 12$$

and so,

$$m = \frac{12p^2}{q}.$$

If $p = 6$ and $q = 10$, we have

$$m = \frac{12 \cdot 6^2}{10}$$

$$m = \frac{216}{5}.$$

4.7 Exercises

1. Suppose m varies directly as z and p. If $m = 10$ when $z = 3$ and $p = 5$, find m when $z = 5$ and $p = 7$.
2. Suppose r varies directly as the square of m, and inversely as s. If $r = 12$ when $m = 6$ and $s = 4$, find r when $m = 4$ and $s = 10$.
3. The distance a body falls from rest varies directly as the square of the time it falls (disregarding air resistance). If an object falls 1024 feet in 8 seconds, how far will it fall in 12 seconds?
4. Hooke's law for an elastic spring states that the distance a spring stretches varies directly as the force applied. If a force of 15 pounds stretches a certain spring 8 inches, how much will a force of 30 pounds stretch the spring?
5. In electric current flow it is found that the resistance (measured in units called ohms) offered by a fixed length of wire of a given material varies inversely as the square of the diameter of the wire. If a wire .01 inches in diameter has a resistance of .4 ohm, what is the resistance of a wire of the same length and material but .03 inches in diameter?
6. It is shown in engineering that the maximum load a cylindrical column of circular cross section can hold varies directly as the fourth power of the diameter and inversely as the square of the height. If a 9 foot column 3 feet in diameter will support a load of 8 tons, how much load will be supported by a column 12 feet high and 2 feet in diameter?

7. The maximum load of a horizontal beam which is supported at both ends varies directly as the width and square of the height, and inversely as the length between supports. If a beam 20 feet long, 5 inches wide, and 6 inches high can support a maximum of 1000 pounds, what is the maximum load of a beam of the same material 40 feet long, 10 inches wide, and 3 inches high?

Chapter Five

Exponential and Logarithmic Functions

Many applications of mathematics, particularly those pertaining to growth and decay of populations, involve exponential functions. Other applications, such as numerical calculations, depend on logarithmic functions. In the third section we shall see how these two types of functions are closely related.

5.1 Inverse Relations

As defined in Chapter 3, a relation is a set of ordered pairs, while a function is defined as a special type of relation in which each value from the domain corresponds to exactly one value from the range. We define the **inverse** of relation R, written R^{-1} (read "R inverse"), as

$$R^{-1} = \{(y, x) \,|\, (x, y) \in R\}.$$

For example, given $f = \{(x, y) \,|\, y = 3x + 1\}$, the ordered pair $(2, 7)$ belongs to the function (if $x = 2$, then $y = 7$), and so by the definition of inverse, we must have $(7, 2)$ as an element of f^{-1}. Also, since $(-3, -8) \in f$, $(-8, -3) \in f^{-1}$.

The definition of inverse relation provides a relatively simple method for graphing an inverse, given the original relation. If (a, b) belongs to the original relation, then (b, a) belongs to the inverse. As shown in Figure 5.1, a line from

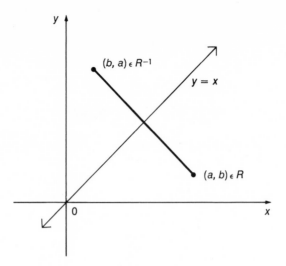

Figure 5.1

(a, b) to (b, a) is bisected by and is perpendicular to the line $y = x$. Hence, to graph R^{-1}, we reflect the graph of R about the line $y = x$. Figure 5.2 shows the graph of $f = \{(x, y) | y = 3x + 1\}$ together with the graph of its inverse f^{-1}. Note that in this case both f and f^{-1} are functions.

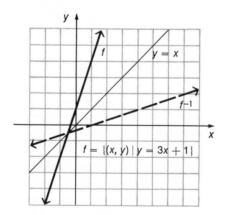

Figure 5.2

Example 1 Figure 5.3 shows the graph of three relations and their inverses. Note that in (a) both R and R^{-1} are functions. In (b), R is a function, but R^{-1} is not, while in (c) R is not a function but R^{-1} is.

Recall that we can identify a function by passing a vertical line through the graph. If any vertical line cuts the graph in more than one point, the graph

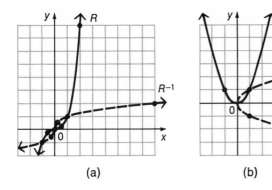

(a)	(b)	(c)

Figure 5.3

is not the graph of a function. There is a similar test to determine if a given relation has an inverse which is a function. If a horizontal line cuts the original graph in more than one point, then the inverse cannot satisfy the definition of a function. (Why?)

We have written the equation defining a function in the form $y = f(x)$, where x is a domain element. We want to express inverse functions in the same way, as $y = f^{-1}(x)$, where x is a domain element of the inverse function. To do this, we note that if $(x, y) \in f$, then $(y, x) \in f^{-1}$. Hence, to get the equation of f^{-1}, given the equation of f, we exchange x and y in the defining equation of f. For example, if $f = \{(x, y) \,|\, y = 3x + 1\}$, then we get f^{-1} by exchanging x and y in the defining equation for f. This gives

$$x = 3y + 1.$$

If we now solve for y, we get

$$y = \frac{x - 1}{3}.$$

Hence, $f^{-1}(x) = \dfrac{x - 1}{3}$, and we can now write f^{-1} as

$$f^{-1} = \left\{ (x, y) \,\Big|\, y = \frac{x - 1}{3} \right\}.$$

Example 2 Let $f = \{(x, y) \,|\, y = x^3 - 1\}$. Find f^{-1}. Graph f and f^{-1}.

We obtain the defining equation for f^{-1} by exchanging x and y in the equation for f, and solving for y, as follows.

$$x = y^3 - 1$$
$$y = \sqrt[3]{x + 1}$$

The graphs of f and f^{-1} are shown in Figure 5.4.

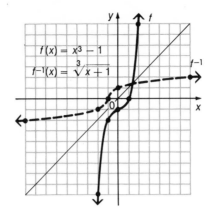

Figure 5.4

We have seen that if $f(x) = 3x + 1$, then $f^{-1}(x) = \dfrac{x-1}{3}$. Note that

$$f[f^{-1}(x)] = f\left[\frac{x-1}{3}\right] = 3\left(\frac{x-1}{3}\right) + 1 = x,$$

while $\qquad f^{-1}[f(x)] = f^{-1}[3x + 1] = \dfrac{(3x+1)-1}{3} = x.$

This example illustrates the next result, which can be proven from the definition of inverse relation.

THEOREM 5.1   If both f and f^{-1} are functions, then

$$f[f^{-1}(x)] = x \qquad \text{and} \qquad f^{-1}[f(x)] = x.$$

5.1 Exercises

For each of the following relations, (a) tell whether or not the relation is a function, (b) identify the domain and range of the relation, (c) sketch the graph of the relation, (d) sketch the graph of the inverse, (e) give the domain and range of the inverse, and (f) write an equation for the inverse.

1. $y = 3x - 4$
2. $y = 4x - 5$
3. $x + y = 0$
4. $y = 6 - x$

5. $y = x^2$
6. $y = (x - 2)^2$
7. $y = \sqrt[3]{x}$
8. $y = x^3$

Sketch the graph of the inverse of each of the following relations.

9. $x^2 + y^2 = 9$
10. $4x^2 + 9y^2 = 36$

11. $x^2 + 4y^2 = 16$
12. $x^2 - y^2 = 1$

13. $y = |x|$

14. $x = |y|$

15. $y = \sqrt{25 - x^2}$

16. $y = -\sqrt{4 + x^2}$

17. $x = \sqrt{4 - y}$

18. $y = \sqrt{2 + x}$

5.2 Exponential Functions

We know that if $a > 0$, we can define the symbol a^m for any rational value of m. In this section we want to extend the definition of a^m to include all real (and not just rational) values of the exponent m. To do this we shall need the following results.

THEOREM 5.2   If $a > 1$ and m is a positive rational number, then $a^m > 1$.

To prove this, let $m = p/q$, where p and q are positive integers. Since $a > 1$, we can use Exercises 22 and 25 of Section 1.4, to write $a^2 > 1$, and $a^3 > 1$, and in general, $a^p > 1$ for any positive integer p.

Now we must prove $a^{p/q} > 1$. Suppose it is not. Since $a^{p/q}$ is positive, we have

$$0 < a^{p/q} \le 1.$$

If we now multiply both sides of $a^{p/q} \le 1$ by the positive quantity $a^{p/q}$, we get

$$(a^{p/q})^2 \le a^{p/q}.$$

Hence,

$$(a^{p/q})^2 \le a^{p/q} \le 1.$$

In the same way,

$$(a^{p/q})^3 \le (a^{p/q})^2 \le a^{p/q} \le 1,$$

and in general,

$$(a^{p/q})^r \le 1,$$

for any positive integer r. If $r = q$, we have

$$(a^{p/q})^q \le 1,$$

or

$$a^p \le 1,$$

which contradicts the result of the previous paragraph. Hence, the assumption is false, and we must have $a^{p/q} > 1$, which completes the proof.

Suppose now $a > 1$ and m and n are rational numbers with $m < n$. If $m < n$, then $n - m > 0$, and by the result above, we can write

$$a^{n-m} > 1.$$

Multiplying both sides by a^m (which is positive) gives

$$a^n > a^m$$

or

$$a^m < a^n,$$

which proves the following result.

THEOREM 5.3   If $a > 1$ and m and n are rational numbers with $m < n$, then $a^m < a^n$.

Now we are ready to define a^x for any real value of x, $(a > 0)$. If x is a real number, then there exists some integer k such that $x \leq k$ (see Exercise 17). We know that a^k is defined and exists. If m is any rational number, with $m \leq x \leq k$, then by Theorem 5.3, $a^m \leq a^k$. Hence, a^k is an upper bound for the set

$$\{a^m \,|\, m \text{ is rational}, m \leq x\}.$$

By the completeness axiom of Section 1.4, this set has a *least* upper bound. This least upper bound we define as the number a^x, for any real number x. That is, if x is any real number, then we make the following definition.

▶ If $a > 1$, then a^x is the least upper bound of the set $\{a^m \,|\, m \text{ is rational}, m \leq x\}$.

▶ If $a = 1$, then $a^x = 1$.

▶ If $0 < a < 1$, then $\dfrac{1}{a} > 1$, and $a^x = \dfrac{1}{(1/a)^x}$.

Note that we make no attempt to define a^x for $a < 0$. Since, for example, $(-6)^x$ is not a real number if $x = \frac{1}{2}$, we do not define a^x for $a < 0$.

Using the work above, we can define a^x for any real number x and any number $a > 0$. The exponential a^x defined in this way still satisfies all properties of exponents from Chapter 2. Hence, the domain of a function such as $y = 2^x$ now includes all real numbers. We call any such function $y = a^x$ $(a > 0, a \neq 1)$ an **exponential function.** Figure 5.5 shows the graph of $f(x) = 2^x$, which is typical of the graphs of functions of the form $f(x) = a^x$, for $a > 1$,

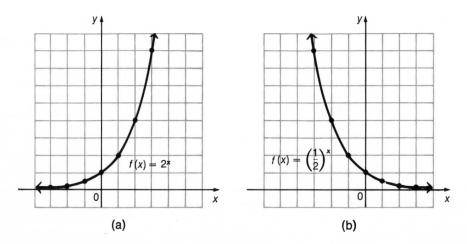

(a)   (b)

Figure 5.5

and the graph of $f(x) = (\frac{1}{2})^x$ which is typical of the graphs of functions of the form $f(x) = a^x$, for $0 < a < 1$.

Example 3   Graph $y = 2^{-x^2}$.

Since $-x^2 \leq 0$ for all real x, it follows that $0 < y \leq 1$ for real values of x. By plotting several points, we get the result shown in Figure 5.6. Functions of this type are important in probability theory.

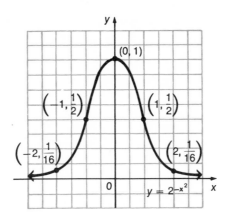

Figure 5.6

5.2 Exercises

Graph each of the following functions.

1. $y = 3^x$
2. $y = 4^x$
3. $y = (\frac{3}{2})^x$
4. $y = 3^{-x}$
5. $y = 10^{-x}$
6. $y = 10^x$
7. $y = 2^{x+1}$
8. $y = 2^{-x+1}$
9. $y = 2^{|x|}$
10. $y = 2^{-|x|}$
11. $y = 2^x + 2^{-x}$
12. $y = (\frac{1}{2})^x + (\frac{1}{2})^{-x}$

13. Using the information given in the table of values below, graph $y = 3^x$ on a large scale for $0 \leq x \leq 1$. Then estimate from the graph (a) $3^{1/3}$, (b) $3^{2/3}$.

x	0	1/4	1/2	3/4	1
3^x	1	1.3	1.7	2.3	3

14. Graph $y = a^x$ for $a = 1$.
15. Suppose the domain of $y = 2^x$ includes only rational values of x (which are the only values discussed prior to this section). Describe in words the resulting graph.

16. Prove: if $0 < a < 1$ and m and n are rational numbers such that $m < n$, then $a^m > a^n$. $\left(\text{Hint: if } 0 < a < 1, \text{ then } \dfrac{1}{a} > 1.\right)$

17. Let x be any real number. Prove that there exists an integer k such that $x \leq k$. (Hint: use an indirect proof.)

5.3 Logarithmic Functions

In the previous section we discussed exponential functions of the form $y = a^x$, for all positive a, $a \neq 1$. In this section we shall discuss the inverses of such functions. Recall that to obtain the inverse of a function such as $y = a^x$, we exchange x and y. Doing this, we obtain

$$x = a^y,$$

as the inverse of the exponential function $y = a^x$. In order to solve the equation $x = a^y$ for y, we use the following definition. For all real numbers y, all positive numbers a, $a \neq 1$,

▶ $y = \log_a x$ means the same as $x = a^y$.

Log is an abbreviation for "logarithm." Read $\log_a x$ as "the logarithm of x to the base a." A **logarithmic function** is a function of the form $y = \log_a x$ ($a > 0$, $a \neq 1$).

Example 4 In the chart of this example we show several pairs of equivalent statements. The same statement is written in both exponential and logarithmic forms.

Exponential Form	*Logarithmic Form*
$2^3 = 8$	$\log_2 8 = 3$
$(1/2)^{-4} = 16$	$\log_{1/2} 16 = -4$
$10^5 = 100,000$	$\log_{10} 100,000 = 5$
$3^{-4} = \dfrac{1}{81}$	$\log_3 \dfrac{1}{81} = -4$

Since $b^1 = b$ and $b^0 = 1$, for all positive b, we have

▶ $\log_b b = 1$ and $\log_b 1 = 0$.

Exponential and logarithmic functions are inverses of each other. Since the domain of an exponential function is the set of all real numbers, the range of a log function will also be the set of all real numbers. In the same way, the range of an exponential function and the domain of a log function are both the set of positive real numbers. Using the technique of graphing inverses discussed in Section 5.1, we can graph logarithmic functions by reflecting the graphs of exponential functions about the line $y = x$ as shown in Figure 5.7.

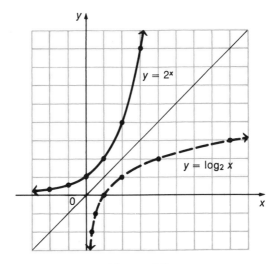

Figure 5.7

This graph suggests the following **axiom of logarithms.**

▶ If $a > 1$ and $m < n$, then $\log_a m < \log_a n$.

Example 5 Graph $y = \log_3 |x|$.

Using the equivalent expression $|x| = 3^y$, we can plot points to obtain the graph (see Figure 5.8).

x	-3	-1	$-\frac{1}{3}$	1	3
y	1	0	-1	0	1

Alternatively, we could graph $|y| = 3^x$, and then find the graph of its inverse.

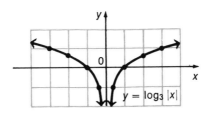

Figure 5.8

Logarithms were originally important as an aid for numerical calculations. The use of logarithms for this purpose is based on the properties discussed in the next theorem.

THEOREM 5.4   If x and y are any positive real numbers, r is any real number, and a is any positive real number, $a \neq 1$, then

(a) $\log_a xy = \log_a x + \log_a y$.

(b) $\log_a \dfrac{x}{y} = \log_a x - \log_a y$.

(c) $\log_a x^r = r \log_a x$.

To prove (a), let $m = \log_a x$, and $n = \log_a y$. Then, by the definition of logarithm,

$$a^m = x \quad \text{and} \quad a^n = y.$$

Hence, we can write

$$a^m \cdot a^n = xy.$$

By a property of exponents, we have

$$a^{m+n} = xy.$$

Now we can use the definition of logarithm to write

$$\log_a xy = m + n.$$

Since $m = \log_a x$ and $n = \log_a y$, we have

$$\log_a xy = \log_a x + \log_a y.$$

Parts (b) and (c) can be proved similarly.

Example 6   Using the results of Theorem 5.4, we have

(a) $\log_a \dfrac{mnp}{q^2} = \log_a m + \log_a n + \log_a p - 2 \log_a q$.

(b) $\log_a \sqrt[3]{m^2} = \frac{2}{3} \log_a m$.

Example 7   Assume $\log_{10} 2 = .3010$, and $\log_{10} 3 = .4771$. Find the base 10 logs of 4, 5, 6, and 12.

Using the results of Theorem 5.4, we have

(a) $\log_{10} 4 = \log_{10} 2^2 = 2 \log_{10} 2 = 2(.3010) = .6020$.

(b) $\log_{10} 5 = \log_{10} \frac{10}{2} = \log_{10} 10 - \log_{10} 2 = 1 - .3010 = .6990$.

(c) $\log_{10} 6 = \log_{10} (2 \cdot 3) = \log_{10} 2 + \log_{10} 3 = .3010 + .4771 = .7781$.

(d) $\log_{10} 12 = \log_{10} (4 \cdot 3) = \log_{10} 4 + \log_{10} 3 = .6020 + .4771 = 1.0791$.

5.3 Exercises

Change each of the following to logarithmic form.

1. $3^4 = 81$ 3. $10^4 = 10,000$ 5. $(\frac{1}{2})^{-4} = 16$

2. $2^5 = 32$ 4. $8^2 = 64$ 6. $(\frac{2}{3})^{-3} = \frac{27}{8}$

7. $10^{-4} = .0001$
8. $(\frac{1}{100})^{-2} = 10,000$

Change each of the following to exponential form.

9. $\log_6 36 = 2$ 11. $\log_{\sqrt{3}} 81 = 8$ 13. $\log_m k = n$
10. $\log_5 5 = 1$ 12. $\log_4 \frac{1}{64} = -3$ 14. $\log_2 r = y$

Write each of the following expressions as a single logarithm.

15. $\log_a x + \log_a y - \log_a m$
16. $(\log_b k - \log_b m) - \log_b a$
17. $2 \log_m a - 3 \log_m b^2$
18. $\frac{1}{2} \log_y p^3 q^4 - \frac{2}{3} \log_y p^4 q^3$
19. $-\frac{3}{4} \log_x a^6 b^8 + \frac{2}{3} \log_x a^9 b^3$
20. $\log_a (pq^2) + 2 \log_a \left(\dfrac{p}{q} \right)$

5.4   Common Logarithms

Since the number system we use is base 10, logarithms to base 10 are most convenient for numerical calculations. Base 10 logarithms are called **common logarithms.** For simplicity, we shall abbreviate $\log_{10} x$ as $\log x$. Using this notation, we have

$$\log 1000 = \log 10^3 = 3$$
$$\log 10 = \log 10^1 = 1$$
$$\log 1 = \log 10^0 = 0$$
$$\log .1 = \log 10^{-1} = -1$$
$$\log .001 = \log 10^{-3} = -3,$$

and so on.

While it can be shown that there is no rational number x such that $10^x = 6$ (and hence no rational number x such that $x = \log 6$) we can use a table of common logarithms to find a rational approximation for $\log 6$. From the excerpt of the log table included on the next page, we see that a decimal approximation for $\log 6$ is given by .7782. Hence,

$$\log 6 \approx .7782$$

(or, equivalently, $10^{.7782} \approx 6$, where $\approx$ means "approximately equal to"). Since most logs are approximations, we shall replace $\approx$ with $=$ and write $\log 6 = .7782$.

To find $\log 6240$, we first note that

$$1000 < 6240 < 10,000,$$

so that $\log 1000 < \log 6240 < \log 10,000$, by the axiom of logarithms. Since $\log 1000 = 3$ and $\log 10,000 = 4$, we have

$$3 < \log 6240 < 4. \tag{1}$$

To find the decimal part of log 6240 in the table, locate the first two significant figures, 62 in the left column of the table. Then find the third digit, 4, across the top.

	0	1	2	3	4	5	6	7	8	9
5.7	.7559	.7566	.7574	.7582	.7589	.7597	.7604	.7612	.7619	.7627
5.8	.7634	.7642	.7649	.7657	.7664	.7672	.7679	.7686	.7694	.7701
5.9	.7709	.7716	.7723	.7731	.7738	.7745	.7752	.7760	.7767	.7774
6.0	.7782	.7789	.7796	.7803	.7810	.7818	.7825	.7832	.7839	.7846
6.1	.7853	.7860	.7868	.7875	.7882	.7889	.7896	.7903	.7910	.7917
6.2	.7924	.7931	.7938	.7945	.7952	.7959	.7966	.7973	.7980	.7987
6.3	.7993	.8000	.8007	.8014	.8021	.8028	.8035	.8041	.8048	.8055
6.4	.8062	.8069	.8075	.8082	.8089	.8096	.8102	.8109	.8116	.8122
6.5	.8129	.8136	.8142	.8149	.8156	.8162	.8169	.8176	.8182	.8189
6.6	.8195	.8202	.8209	.8215	.8222	.8228	.8235	.8241	.8248	.8254

The intersection of this row and column gives .7952, and hence from (1) we have

$$\log 6240 = 3.7952.$$

The decimal part is called the **mantissa,** and the integer part the **characteristic.** To find log .00587, we note

$$.001 < .00587 < .01$$

and

$$\log .001 < \log .00587 < \log .01$$

or

$$-3 < \log .00587 < -2.$$

From the log table, the mantissa of log .00587 is .7686. Thus,

$$\log .00587 = -3 + .7686.$$

It is often desirable to write the characteristic, -3, as $7 - 10,$ so that the log becomes

$$\log .00587 = 7.7686 - 10.$$

Let us now use the table of common logarithms in the Appendix and the properties of logarithms in Theorem 5.4 to aid in a numerical calculation. For example, to evaluate

$$(\sqrt[3]{42})(76.9)(.00283),$$

we can use the properties of logarithms and the table to write

$$\begin{aligned}
\log (\sqrt[3]{42})(76.9)(.00283) &= \tfrac{1}{3} \log 42 + \log 76.9 + \log .00283 \\
&= \tfrac{1}{3}(1.6232) + 1.8859 + (7.4518 - 10) \\
&= .5411 + 1.8859 + (7.4518 - 10) \\
&= 9.8788 - 10.
\end{aligned}$$

From the log table we find that .756 is the number (to three places) whose logarithm is 9.8788 − 10, and so

$$(\sqrt[3]{42})(76.9)(.00283) \approx .756.$$

Example 8   Calculate $\sqrt[3]{.00762}$.
   We find that

$$\log \sqrt[3]{.00762} = \tfrac{1}{3}(7.8820 - 10).$$

This calculation can be simplified if we write the characteristic, −3, as 27 − 30. This gives

$$\log \sqrt[3]{.00762} = \tfrac{1}{3}(27.8820 - 30) = 9.2940 - 10.$$

Hence, from the log table,

$$\sqrt[3]{.00762} \approx .197.$$

The table of logs included in this text contains decimal approximations of common logs to four places of accuracy. More accurate tables are available; however, more accuracy can be obtained from the table included in this text by the process of **linear interpolation.** As an example, let us use linear interpolation to approximate log 75.37. Note that

$$\log 75.3 < \log 75.37 < \log 75.4.$$

Figure 5.9 shows the portion of the curve $y = \log x$ between $x = 75.3$ and $x = 75.4$. We shall use the line segment PR to approximate the log curve (this approximation is adequate for values of x relatively close to one another).

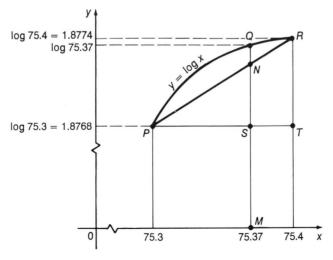

Figure 5.9

From the figure, note that log 75.37 is given by the length of segment MQ, which we cannot find. We can, however, find MN, which we shall use as our approximation to log 75.37. By properties of similar triangles, we have

$$\frac{PS}{PT} = \frac{SN}{TR}.$$

In this case, $PS = 75.37 - 75.3 = .07$, $PT = 75.4 - 75.3 = .1$, and $RT = \log 75.4 - \log 75.3 = 1.8774 - 1.8768 = .0006$. Hence,

$$\frac{.07}{.1} = \frac{SN}{.0006}$$

or

$$SN = .7(.0006) \approx .0004.$$

Since log 75.37 $= MN = MS + SN$, and since $MS = \log 75.3 = 1.8768$, we have

$$\log 75.37 = 1.8768 + .0004 = 1.8772.$$

Example 9 Find log .008726.

Here log $.00872 = 7.9405 - 10$ and log $.00873 = 7.9410 - 10$. Since .008726 is .6 of the distance between .00872 and .00873, we take .6 of the difference between log .00873 and log .00872. We have

$$.6[(7.9410 - 10) - (7.9405 - 10)] = .6(.0005) \approx .0003.$$

Hence,

$$\log .008726 = (7.9405 - 10) + .0003$$
$$= 7.9408 - 10.$$

Example 10 Find x such that log $x = .3275$.

From the log table, we see

$$\log 2.12 = .3263 < .3275 < .3284 = \log 2.13.$$

The distance between .3263 and .3275 is .0012, while the distance between .3263 and .3284 is .0021. Hence, .3275 is

$$\frac{.0012}{.0021} \approx .6$$

of the way from .3263 to .3284. From this last result, we see that x should be .6 of the way from 2.12 to 2.13, so that

$$\log 2.126 = .3275.$$

5.4 Exercises

Find the characteristic of the logarithms of each of the following.

1. 875   2. 9462   3. 2,400,000   4. 875000   5. .00023

6. .098
7. .000042
8. .000000257

Find the logarithms of the following numbers.

9.	875	13.	7.63	17.	3.876	20.	103200
10.	3750	14.	9.37	18.	2.975	21.	.0003824
11.	12800	15.	.000893	19.	68250	22.	.00008632
12.	653000	16.	.00376				

Find the numbers having each of the following logarithms.

23.	1.5366	26.	$9.2504 - 10$	29.	2.8039
24.	2.9253	27.	3.4947	30.	$8.8078 - 10$
25.	$8.8733 - 10$	28.	4.6863		

Use logarithms (and interpolation) to find approximations to three-place accuracy for each of the following.

31. $\dfrac{(79.6)(4.83)(9.3)}{43.8}$

32. $\dfrac{(123)(7.892)(.436)}{2.81}$

33. $\dfrac{(21.3)^2}{(.86)^3}$

34. $\dfrac{(.893)^2}{(.624)^3}$

35. $(8.16)^{\frac{1}{3}}$

36. $\sqrt{276}$

37. $(.463)^{.4}$

38. $8^{.614}$

39. π^π (Use $\pi = 3.14$)

40. $\dfrac{(7.06)^3}{(31.7)(\sqrt{1.09})}$

41. $\dfrac{\log 21}{\log 3}$

42. $(21.9)^2 + (4.13)^3$

43. Suppose the pressure p (pounds per cubic foot) and volume v (cubic feet) of a certain gas are related by the formula

$$pv^{1.6} = 800.$$

Find p if $v = 7.6$ cubic feet.

44. The approximate period T (in seconds) of a simple pendulum of length L (in feet) is

$$T \approx 2\pi\sqrt{\frac{L}{32}}.$$

Find the period of a pendulum of length $L = 26$ feet.

45. The number of years, n, since two independently evolving languages split off from a common ancestral language is approximated by

$$n \approx \frac{1000 \log r}{2 \log .86},$$

where r is the proportion of words from the ancestral language common to both languages.

(a) Find n if $r = .9$.

(b) Find n if $r = .3$.

(c) How many years have elapsed since the split if half of the words of the ancestral language are common to both languages?

5.5 Exponential and Logarithmic Equations

To solve equations involving logarithms and exponents, it is often helpful to state the following property of logarithms, which follows from the definition of function.

THEOREM 5.5 If $x > 0$, $y > 0$, $b > 0$, $b \neq 1$, then $x = y$ if and only if $\log_b x = \log_b y$.

We can use this result to solve the equation

$$7^x = 12.$$

If we take base 10 logs of both sides, we have

$$\log 7^x = \log 12$$
$$x \log 7 = \log 12$$
$$x = \frac{\log 12}{\log 7}.$$

To get a decimal approximation for x, we can write

$$x = \frac{\log 12}{\log 7} = \frac{1.0792}{.8451} \approx 1.3.$$

Example 11 Solve $3^{2x-1} = 4^{x+2}$.

If we take logs of both sides, we have

$$\log 3^{2x-1} = \log 4^{x+2}$$

or

$$(2x - 1) \log 3 = (x + 2) \log 4,$$

which leads to

$$2x \log 3 - \log 3 = x \log 4 + 2 \log 4$$

or

$$2x \log 3 - x \log 4 = 2 \log 4 + \log 3.$$

Since the common factor of the left member is x, we have

$$x(2 \log 3 - \log 4) = 2 \log 4 + \log 3$$

or

$$x = \frac{2 \log 4 + \log 3}{2 \log 3 - \log 4}.$$

Using the properties of logarithms, this can be expressed as

$$x = \frac{\log 16 + \log 3}{\log 9 - \log 4}$$

or finally,

$$x = \frac{\log 48}{\log \dfrac{9}{4}}.$$

This quotient could be approximated by a decimal, if desired.

Example 12 Solve $\log (x + 4) - \log (x + 2) = \log x$.
Here we have

$$\log \frac{x + 4}{x + 2} = \log x$$

$$\frac{x + 4}{x + 2} = x$$

$$x + 4 = x(x + 2)$$

$$x + 4 = x^2 + 2x$$

$$x^2 + x - 4 = 0.$$

By the quadratic formula, we have

$$x = \frac{-1 \pm \sqrt{1 + 16}}{2}$$

$$x = \frac{-1 + \sqrt{17}}{2} \qquad \text{or} \qquad x = \frac{-1 - \sqrt{17}}{2}.$$

We cannot evaluate $\log x$ for

$$x = \frac{-1 - \sqrt{17}}{2},$$

since this number is negative and hence not in the domain of $\log x$. Thus the only valid solution is the positive number

$$x = \frac{-1 + \sqrt{17}}{2}.$$

Example 13 Solve $\log (3x + 2) + \log (x - 1) = 1$.
Since $1 = \log 10$, we have

$$\log (3x + 2)(x - 1) = \log 10$$

$$(3x + 2)(x - 1) = 10$$

$$3x^2 - x - 2 = 10$$

$$3x^2 - x - 12 = 0.$$

If we now use the quadratic formula, we have

$$x = \frac{1 \pm \sqrt{1 + 144}}{6}$$

$$x = \frac{1 + \sqrt{145}}{6} \quad \text{or} \quad x = \frac{1 - \sqrt{145}}{6}.$$

If $x = \dfrac{1 - \sqrt{145}}{6}$, then $x - 1 < 0$, and so $\log (x - 1)$ does not exist. Hence, this proposed solution must be discarded and the only solution is

$$x = \frac{1 + \sqrt{145}}{6}.$$

5.5 Exercises

Solve each of the following equations.

1. $3^x = 6$
2. $4^x = 12$
3. $6^x = 8$
4. $1^x = 3$
5. $2^x = -3$
6. $(\frac{1}{4})^x = 100$
7. $3^{x-1} = 4$
8. $5^{2-x} = 12$
9. $6^{2x-1} = 8$
10. $2^{x-7} = 5$
11. $1.8^{x+4} = 9.31$
12. $3^{2x-5} = 6$
13. $\log (x - 1) = 1$
14. $\log x^2 = 1$

15. $\log (x - 3) = 1 - \log x$
16. $\log (x - 6) = 2 - \log (x + 15)$
17. $\log (x + 2) = \log 6$
18. $\log x - \log (x + 1) = \log 5$
19. $\log_2 x = 3$
20. $\log_x 10 = 3$
21. $2 + \log x = 0$
22. $\log_3 x + \log_3 (2x + 5) = 1$
23. $\log (x - 3) = 1 + \log (x + 1)$
24. $\log x + \log (3x - 13) = 1$
25. $\log x = \log (x - 4)$
26. $\left(1 + \dfrac{r}{2}\right)^5 = 9$
27. $100(1 + .02)^{3+n} = 150$

5.6 Natural Logarithms

Natural logarithms, written ln x, are logs to the base e. The number e is an irrational number, approximated by $e \approx 2.71828$, which occurs in many practical applications. Several examples of such applications are presented in this section. To convert from base 10 logs, with which we are familiar, to base e logs, we use the next theorem.

THEOREM 5.6   If x is any positive number, and if a and b are positive real numbers, $a \neq 1$, $b \neq 1$, then

$$\log_a x = \frac{\log_b x}{\log_b a}.$$

To prove this result, we can use the fact that for positive x and positive a, $a \neq 1$, we can write $y = \log_a x$, from which $x = a^y$, or $x = a^{\log_a x}$. Verify this statement from the definition of logarithm, or Theorem 5.1. If we now take base b logs of both sides of this last equation, we have

$$\log_b x = \log_b a^{\log_a x}$$

or

$$\log_b x = (\log_a x)(\log_b a),$$

from which we obtain

$$\log_a x = \frac{\log_b x}{\log_b a}.$$

We can use this result to find the natural logarithm of a number. For example, to find $\ln 47.2$, we can let $x = 47.2$, $a = e$, and $b = 10$. Using these values, we have

$$\ln 47.2 = \log_e 47.2 = \frac{\log_{10} 47.2}{\log_{10} e}$$

$$\ln 47.2 = \frac{\log 47.2}{\log e}.$$

Since $e \approx 2.71828$, we can calculate $\log e$: $\log e = .4343$. Thus,

$$\ln 47.2 = \frac{\log 47.2}{.4343}.$$

Since $1/(.4343) \approx 2.3026$, this last statement becomes

$$\ln 47.2 = (2.3026)(\log 47.2)$$
$$= (2.3026)(1.6739)$$
$$= 3.8543.$$

Using this same principle, we can prove the next result.

THEOREM 5.7   For all positive x, $\ln x = (2.3026) \log x$.

Example 14

$$\ln .427 = (2.3026)(\log .427)$$
$$= (2.3026)(-1 + .6304)$$
$$= (2.3026)(-.3696)$$
$$= -.8510$$

Natural logarithms are often used in discussions of growth and decay, as illustrated by the next examples.

Example 15   Suppose the amount, y, of a certain radioactive substance present at a time t is given by

$$y = y_0 e^{-.1t},$$

where y_0 is the amount present initially and t is measured in days. Find the half-life of the substance. (The half-life of a radioactive substance is the time it takes for exactly half the sample to decay.)

We want to know the time t that must elapse for y to be reduced to a value equal to $\frac{1}{2}y_0$. That is, we want to solve the equation

$$\tfrac{1}{2}y_0 = y_0 e^{-.1t}.$$

Here we have $\frac{1}{2} = e^{-.1t}$, and if we now take natural logs of both sides, we get

$$\ln \tfrac{1}{2} = \ln e^{-.1t}$$
$$\ln \tfrac{1}{2} = -.1t(\ln e).$$

Since $\ln e = 1$, we have

$$\ln \tfrac{1}{2} = -.1t$$
$$t = \frac{-\ln \tfrac{1}{2}}{.1}$$
$$t = -10 \ln \tfrac{1}{2}.$$

By the earlier results of this section, we have

$$\ln \tfrac{1}{2} = (2.3026) \log \tfrac{1}{2} = 2.3026(-.3010) = -.6931.$$

Hence,

$$t \approx (-10)(-.6932)$$
$$t \approx 6.9 \text{ days.}$$

Example 16   Psychologists tell us that under certain conditions the total number of a certain kind of fact remembered is approximated by

$$y = y_0\left(\frac{1 + e}{1 + e^{t+1}}\right),$$

where y is the number of facts remembered at a time t measured in days, and y_0 is the initial number remembered. Graph the function.

Here we can plot some points as an aid in graphing the function. If $t = 0$, we have

$$y = y_0\left(\frac{1 + e}{1 + e^{0+1}}\right) = y_0(1) = y_0.$$

If $t = 1$, we have

$$y = y_0\left(\frac{1 + e}{1 + e^2}\right) \approx y_0\left(\frac{3.718}{8.389}\right) \approx .44y_0.$$

By plotting several such points, we get the graph of Figure 5.10. This graph, typical of the remembering of certain things under certain conditions, is often called a **forgetting curve**.

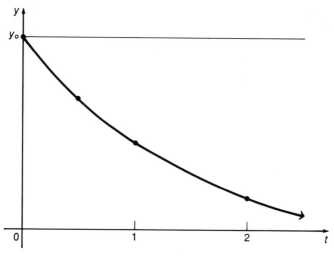

Figure 5.10

Example 17 How long will it take in Example 16 for exactly half the items to be forgotten?

We want to find the value of t that makes $y = \frac{1}{2}y_0$. That is, we want

$$\tfrac{1}{2}y_0 = y_0\left(\frac{1+e}{1+e^{t+1}}\right)$$

or

$$\tfrac{1}{2} = \frac{1+e}{1+e^{t+1}}.$$

This gives $1 + e^{t+1} = 2(1 + e) = 2 + 2e$

or

$$e^{t+1} = 1 + 2e.$$

Taking natural logs of both sides gives

$$\ln e^{t+1} = \ln(1 + 2e)$$

or

$$(t+1)\ln e = \ln(1 + 2e).$$

Since $\ln e = 1$, we have

$$t = [\ln(1 + 2e)] - 1.$$

We know $e \approx 2.718$, and so

$$t \approx [\ln(1 + 5.436)] - 1$$
$$t \approx 1.86 - 1$$
$$t \approx .86.$$

5.6 Exercises

Find each of the following to the nearest hundredth.

1. $\ln 125$	5. $\ln 98.6$	9. $\log_{15} 5$	13. $\log_{1/2} 3$
2. $\ln 63.1$	6. $\ln 355$	10. $\log_6 8$	14. $\log_{12} 62$
3. $\ln .9$	7. $\log_5 10$	11. $\log_6 5$	15. $\log_{100} 83$
4. $\ln 1.5$	8. $\log_9 12$	12. $\log_3 15$	16. $\log_{200} 375$

Suppose the number of rabbits in a colony of rabbits increases according to the relationship

$$y = y_0 e^{.4t},$$

where t represents time in months and y_0 is the initial population of rabbits.

17. Find the number of rabbits present at time $t = 4$ if $y_0 = 100$.
18. How long will it take for the number of rabbits to triple?

A certain city finds that its population is declining according to the relationship

$$P = P_0 e^{-.04t},$$

where t is time measured in years and P_0 is the population at time $t = 0$.

19. If $P_0 = 1{,}000{,}000$, find the population at time $t = 1$.
20. If $P_0 = 1{,}000{,}000$, estimate the time it will take for the population to be reduced to 750,000.
21. How long will it take for the population to be cut in half?

Carbon 14 is a radioactive isotope of carbon which has a half-life of about 5600 years. (That is, in about 5600 years half of any given sample of carbon 14 will have decayed.) The atmosphere contains much carbon, mostly in the form of carbon dioxide, with small traces of carbon 14. Most of this atmospheric carbon is in the form of the nonradioactive isotope carbon 12. The ratio of carbon 14 to carbon 12 is virtually constant in the atmosphere. However, as a plant absorbs carbon dioxide from the air in the process of photosynthesis, the carbon 12 stays in the plant while the carbon 14 is converted to nitrogen. Thus, the ratio of carbon 14 to carbon 12 is smaller in the plant than it is in the atmosphere. Even when the plant is eaten by an animal, this ratio will continue to decrease. Based on these facts, a method of dating objects called *carbon 14 dating* has been developed. It is explained in the following problems.

22. Suppose an Egyptian mummy is discovered in which the ratio of carbon 14 to carbon 12 is only about half the ratio found in the atmosphere. About how long ago did the Egyptian die?
23. Let R be the (nearly constant) ratio of carbon 14 to carbon 12 found in the atmosphere, and let r be the ratio found in an observed specimen.

It can be shown that the relationship between R and r is given by

$$\frac{R}{r} = e^{(t \ln 2)/5600},$$

where t is the age of the specimen in years. Verify the formula for $t = 0$.

24. Verify the formula in Exercise 23 for $t = 5600$ and then for $t = 11{,}200$.
25. Solve the formula of Exercise 23 for t.
26. Suppose a specimen is found in which $r = \frac{2}{3}R$. Estimate the age of the specimen.
27. If an object is fired vertically upward and is subject only to the force of gravity, g, and to air resistance, then the maximum height, H, attained by the object is

$$H = \frac{1}{K}\left(V_0 - \frac{g}{K} \ln \frac{g + V_0 K}{g}\right),$$

where V_0 is the initial velocity of the object and K is a constant. Find H if $K = 2.5$, $V_0 = 1000$ feet per second, and $g = 32$ feet per second per second.
28. The following formula* can be used to estimate the population of the United States, where t is time in years measured from 1914 (times before 1914 are negative):

$$N = \frac{197{,}273{,}000}{1 + e^{-.03134t}}.$$

Complete the following chart.

Year	Observed Population	Predicted Population
1790	3,929,000	3,929,000
1810	7,240,000	
1860	31,443,000	30,412,000
1900	75,995,000	
1970	204,000,000	

29. Turnage† has shown that the pull, P, of a tracked vehicle on dry sand under certain conditions is approximated by

$$P = W\left[.2 + .16 \ln \frac{G(bl)^{3/2}}{W}\right]$$

where G is an index of sand strength, W is the load on the vehicle, b is the width of the track, and l is the length of the track. Find P if $W = 10$, $G = 5$, $b = 30.5$ cm, and $l = 61.0$ cm.

---

* The formula is given in Alfred J. Lotka, *Elements of Mathematical Biology*, reprinted by Dover Press, New York, 1957, page 67.
† Gerald W. Turnage, *Prediction of Track Pull Performance in a Desert Sand*, unpublished MS thesis, The Florida State University, 1971.

5.7 Geometric Progressions

Recall that a sequence is a function whose domain is a subset of the set of positive integers. In Chapter 3 of this text we discussed arithmetic sequences, which are sequences where each term is obtained by adding the same constant number to the preceeding term.

A **geometric sequence** or **geometric progression** is a sequence in which each term after the first is obtained by multiplying the preceeding term by a constant nonzero real number, called the **common ratio.** Hence,

$$2, 8, 32, 128, \ldots$$

is a geometric progression whose first term is 2 and whose common ratio is 4. If the common ratio of a geometric progression is r, then by the definition of geometric progression we have

$$a_{n+1} = r \cdot a_n,$$

for every positive integer n. (Recall: a_n represents the nth term of a sequence.)

Note that in the geometric progression

$$2, 8, 32, 128, \ldots$$

we have

$$8 = 2 \cdot 4$$
$$32 = 8 \cdot 4 = (2 \cdot 4) \cdot 4 = 2 \cdot 4^2$$
$$128 = 32 \cdot 4 = (2 \cdot 4^2) \cdot 4 = 2 \cdot 4^3.$$

To generalize this, consider a geometric progression whose first term is a_1 and whose common ratio is r. The second term can be written as $a_2 = a_1 r$, the third as $a_3 = a_2 r = (a_1 r)r = a_1 r^2$, and so on. Thus, in general, the nth term of a geometric progression is given by

▶
$$a_n = a_1 r^{n-1}.$$

For example, the sixth term of the progression whose first term is -2 and whose common ratio is $-\frac{1}{2}$ is given by

$$a_6 = -2(-\tfrac{1}{2})^{6-1}$$
$$a_6 = -2(-\tfrac{1}{2})^5$$
$$a_6 = \tfrac{1}{16}.$$

Example 18 Suppose the third term of a geometric progression is 20, while the sixth term of the same progression is 160. Find the first term a_1 and the common ratio, r.

We can use the formula for the nth term of a geometric progression. Here we have

$$a_3 = a_1 r^2 = 20$$
$$a_6 = a_1 r^5 = 160.$$

Since $a_1 r^2 = 20$, then $a_1 = \dfrac{20}{r^2}$. Substituting this in the second equation we have

$$a_1 r^5 = 160$$

$$\left(\frac{20}{r^2}\right) r^5 = 160$$

$$20 r^3 = 160$$

$$r^3 = 8$$

$$r = 2.$$

Since $a_1 r^2 = 20$, and $r = 2$, we have $a_1 = 5$.

Applications of geometric progressions frequently require the sum of a certain number of terms of the progression. To find a formula for the sum, S_n, of the first n terms of a geometric progression, we can first write

$$S_n = a_1 + a_2 + a_3 + \ldots + a_n.$$

This can also be written as

$$S_n = a_1 + a_1 r + a_1 r^2 + \ldots + a_1 r^{n-1}. \tag{2}$$

If $r = 1$, we have $S_n = n a_1$, which is the correct result for this case. If $r \neq 1$, we can multiply both sides of (2) by r, obtaining

$$r S_n = a_1 r + a_1 r^2 + a_1 r^3 + \ldots + a_1 r^n. \tag{3}$$

If we subtract (3) from (2), we have

$$S_n - r S_n = a_1 - a_1 r^n$$

or
$$S_n(1 - r) = a_1(1 - r^n)$$

which yields
$$S_n = \frac{a_1(1 - r^n)}{1 - r}.$$

This proves the following result.

THEOREM 5.8   The sum of the first n terms of a geometric progression whose first term is a_1 and whose common ratio is r $(r \neq 1)$ is

$$S_n = \sum_{i=1}^{n} a_i = \frac{a_1(1 - r^n)}{1 - r}.$$

Example 19   Find the sum of the first four terms of the sequence

$$10, 2, 2/5, \ldots.$$

Here $a_1 = 10$, $r = 1/5$, and $n = 4$. Using the formula of Theorem 5.8, we have

$$S_4 = \frac{10\left[1 - \left(\frac{1}{5}\right)^4\right]}{1 - \frac{1}{5}}$$

$$S_4 = \frac{10\left[1 - \frac{1}{625}\right]}{\frac{4}{5}}$$

$$S_4 = 10\left(\frac{624}{625}\right) \cdot \frac{5}{4}$$

$$S_4 = \frac{312}{25}.$$

Example 20 Evaluate $\sum_{i=1}^{6} 4 \cdot 3^{i-1}$.

We can write the sum of this example as

$$\sum_{i=1}^{6} 4 \cdot 3^{i-1} = 4 \cdot 3^{1-1} + 4 \cdot 3^{2-1} + 4 \cdot 3^{3-1} + \cdots + 4 \cdot 3^{6-1}$$
$$= 4 \cdot 3^0 + 4 \cdot 3^1 + 4 \cdot 3^2 + \cdots + 4 \cdot 3^5.$$

These numbers represent the terms of a geometric progression with $a_1 = 4 \cdot 3^0 = 4$, $r = 3$, and $n = 6$. Using the formula from Theorem 5.8, we have

$$\sum_{i=1}^{6} 4 \cdot 3^{i-1} = \frac{4(1 - 3^6)}{1 - 3}$$
$$= \frac{4(1 - 729)}{-2}$$
$$= 1456.$$

5.7 Exercises

Find the fifth term and the sum of the first five terms for each of the following geometric progressions.

1. 3, 6, 12, 24, ...
2. 5, 20, 80, 320, ...
3. 12, −6, 3, −$\frac{3}{2}$, ...
4. 18, −3, $\frac{1}{2}$, −$\frac{1}{12}$, ...
5. 9, −6, 4, ...
6. 50, −10, 2, ...
7. $a_1 = 4$, $r = 2$
8. $a_1 = 3$, $r = 3$
9. $a_2 = \frac{1}{3}$, $r = 3$
10. $a_2 = -1$, $r = 2$

11. The second term of a geometric progression is 3 and the fifth term is $\frac{8}{9}$. Find the eighth term.
12. Find r and a_6 in a geometric progression where $a_2 = 12$ and $a_5 = \frac{3}{2}$.

The k numbers $b_1, b_2, \ldots, b_k$ are called **kth geometric means** of the numbers a and c if

$$a, b_1, b_2, \ldots b_k, c$$

form a geometric progression.

13. Find a geometric mean between 4 and 16.
14. Find a geometric mean between -2 and -8.
15. Insert three geometric means between 2 and 32.
16. Insert four geometric means between 8 and $-\frac{1}{4}$.

Evaluate each of the following.

17. $\displaystyle\sum_{i=1}^{4} 2^i$ 19. $\displaystyle\sum_{i=1}^{8} \left(\frac{1}{2}\right)^i$ 21. $\displaystyle\sum_{i=3}^{7} 3 \cdot 2^{i-1}$ 23. $\displaystyle\sum_{i=4}^{7} 5 \cdot \left(\frac{2}{3}\right)^{i-3}$

18. $\displaystyle\sum_{i=1}^{6} (-3)^i$ 20. $\displaystyle\sum_{i=1}^{4} \left(-\frac{3}{4}\right)^i$ 22. $\displaystyle\sum_{i=3}^{8} 2 \cdot \left(\frac{3}{2}\right)^{i-2}$ 24. $\displaystyle\sum_{i=3}^{8} \frac{2}{3}\left(\frac{3}{4}\right)^{i-3}$

5.8 Sums of Infinite Geometric Sequences

In the previous section we found that the sum of the first n terms of a geometric progression is given by

$$S_n = \sum_{i=1}^{n} a_i = \frac{a_1(1 - r^n)}{1 - r}$$

where a_1 is the first term and r $(r \neq 1)$ is the common ratio. Consider now the infinite sequence

$$2, 1, \frac{1}{2}, \frac{1}{4}, \frac{1}{8}, \frac{1}{16}, \ldots$$

whose first term is 2 and whose common ratio is $\frac{1}{2}$. Using the formula above, we can show

$$S_1 = 2$$
$$S_2 = 3$$
$$S_3 = \frac{7}{2}$$
$$S_4 = \frac{15}{4}$$
$$S_5 = \frac{31}{8}$$
$$S_6 = \frac{63}{16},$$

and so on. Note that this sequence of sums,

$$2, 3, \frac{7}{2}, \frac{15}{4}, \frac{31}{8}, \frac{63}{16}, \ \cdots$$

gets closer and closer to the number 4. In fact, by selecting a value of n large enough, we can make S_n as close as we wish to 4. We express this by saying

$$\lim_{n \to \infty} S_n = 4.$$

(Read: "the limit of S_n as n increases without bound is 4.") For no value of n is $S_n = 4$. However, if n is large enough, then S_n is as close to 4 as we might wish.*

Since

$$\lim_{n \to \infty} S_n = 4,$$

we say that 4 is the sum of the infinite geometric progression

$$2, 1, \frac{1}{2}, \frac{1}{4}, \ \cdots$$

and we write

$$2 + 1 + \frac{1}{2} + \frac{1}{4} + \frac{1}{8} + \cdots = 4.$$

Example 21 Find $1 + \frac{1}{3} + \frac{1}{9} + \frac{1}{27} + \cdots$.

Here we can use the formula for the first n terms of a geometric progression to write

$$S_1 = 1$$

$$S_2 = \frac{4}{3}$$

$$S_3 = \frac{13}{9}$$

$$S_4 = \frac{40}{27},$$

and in general

$$S_n = \frac{1 \left[1 - \left(\frac{1}{3} \right)^n \right]}{1 - \frac{1}{3}}.$$

Note that as n gets larger and larger, $\left(\frac{1}{3} \right)^n$ gets closer and closer to 0. That is,

$$\lim_{n \to \infty} \left(\frac{1}{3} \right)^n = 0.$$

---

* These phrases "large enough" and "as close as we wish" are not nearly precise enough for mathematicians; much of a standard calculus course is devoted to making them more precise.

Thus, it seems reasonable to write

$$\lim_{n \to \infty} S_n = \frac{1(1-0)}{1-\frac{1}{3}} = \frac{3}{2}.$$

Hence,

$$1 + \frac{1}{3} + \frac{1}{9} + \frac{1}{27} + \cdots = \frac{3}{2}.$$

In general, if a geometric progression has a first term a_1 and a common ratio r, we have

$$S_n = \frac{a_1(1-r^n)}{1-r}$$

for every positive integer n. If $|r| < 1$, we have

$$\lim_{n \to \infty} r^n = 0.$$

In this case, we can write

$$\lim_{n \to \infty} S_n = \frac{a_1(1-0)}{1-r} \qquad \text{for } |r| < 1,$$

▶ $$\lim_{n \to \infty} S_n = \frac{a_1}{1-r} \qquad \text{for } |r| < 1.$$

This quantity we define as the **sum of the infinite geometric progression** with first term a_1 and common ratio r, $|r| < 1$. We often express

$$\lim_{n \to \infty} S_n$$

by writing S_∞, or $\sum\limits_{i=1}^{\infty} a_i$, so that

$$S_\infty = \sum_{i=1}^{\infty} a_i = \lim_{n \to \infty} S_n = \frac{a_1}{1-r}, \qquad |r| < 1.$$

For example, the sum of the geometric progression with first term $a_1 = -\frac{3}{4}$ and common ratio $r = -\frac{1}{2}$,

$$-\frac{3}{4} + \frac{3}{8} - \frac{3}{16} + \frac{3}{32} - \frac{3}{64} + \cdots,$$

is given by

$$S_\infty = \frac{-\frac{3}{4}}{1 - \left(-\frac{1}{2}\right)}$$

$$S_\infty = -\frac{1}{2}.$$

We can use the formula of this section to convert repeating decimals (which represent rational numbers) to the form $\frac{p}{q}$, p and q integers, as shown in the following examples.

Example 22 Write .090909... in the form $\frac{p}{q}$, where p and q are integers.

We can write this decimal as

$$.09 + .0009 + .000009 + \ldots,$$

which is a geometric progression with $a_1 = .09$ and $r = .01$. The sum of this progression is given by

$$S_\infty = \frac{a_1}{1 - r}$$

$$= \frac{.09}{1 - .01}$$

$$= \frac{.09}{.99}$$

$$S_\infty = \frac{1}{11}.$$

Example 23 Write 2.5121212... in the form $\frac{p}{q}$.

We have

$$2.5121212\ldots = 2.5 + .012 + .00012 + .0000012 + \ldots.$$

Beginning with the second term, we have a geometric progression with $a_1 = .012$ and $r = .01$. Thus, the decimal can be written as

$$2.5 + \frac{.012}{1 - .01} = 2.5 + \frac{.012}{.990} = 2.5 + \frac{2}{165} = \frac{829}{330}.$$

Example 24 Write the repeating decimal 0.99999... in the form $\frac{p}{q}$.

We can write the decimal as an infinite geometric progression,

$$0.99999\ldots = .9 + .09 + .0009 + .00009 + \ldots,$$

where $a_1 = \frac{9}{10}$ and $r = \frac{1}{10}$. This gives

$$S_\infty = \frac{\dfrac{9}{10}}{1 - \dfrac{1}{10}}$$

$$= \frac{\dfrac{9}{10}}{\dfrac{9}{10}}$$

$$S_\infty = 1.$$

Hence, $0.99999\ldots = 1$.

5.8 Exercises

Find each of the following sums which exist by using the formula of this section, where applicable.

1. $\dfrac{3}{4} + \dfrac{3}{8} + \dfrac{3}{16} + \cdots$

2. $\dfrac{4}{5} + \dfrac{2}{5} + \dfrac{1}{5} + \cdots$

3. $3 - \dfrac{3}{2} + \dfrac{3}{4} - \cdots$

4. $9 - 3 + 1 - \cdots$

5. $\dfrac{1}{3} - \dfrac{2}{9} + \dfrac{4}{27} - \dfrac{8}{81} + \cdots$

6. $1 + \dfrac{1}{1.01} + \dfrac{1}{(1.01)^2} + \cdots$

7. $\dfrac{1}{36} + \dfrac{1}{30} + \dfrac{1}{25} + \cdots$

8. $1 + \dfrac{1}{2^2} + \dfrac{1}{2^4} + \cdots$

9. $\displaystyle\sum_{i=1}^{\infty} \left(\dfrac{1}{4}\right)^i$

10. $\displaystyle\sum_{i=1}^{\infty} \left(\dfrac{9}{10}\right)^i$

11. $\displaystyle\sum_{i=1}^{\infty} (1.2)^i$

12. $\displaystyle\sum_{i=1}^{\infty} (1.001)^{i-1}$

13. $\displaystyle\sum_{i=2}^{\infty} \left(-\tfrac{1}{4}\right)^i$

14. $\displaystyle\sum_{i=4}^{\infty} (.3)^i$

15. $\displaystyle\sum_{i=1}^{\infty} \dfrac{1}{5^i}$

16. $\displaystyle\sum_{i=1}^{\infty} \dfrac{-1}{(-2)^{i+1}}$

Express each of the following in the form $\dfrac{p}{q}$, with p and q integers.

17. $0.5555\ldots$

18. $0.3333\ldots$

19. $0.313131\ldots$

20. $0.909090\ldots$

21. $0.345555\ldots$

22. $0.568888\ldots$

23. Joann drops a ball from a height of 10 feet and notices that on each bounce the ball returns to about $\tfrac{3}{4}$ of its previous height. About how far will the ball travel before it comes to rest? (Hint: consider the sum of two progressions.)

Chapter Six

Trigonometric Functions

In previous chapters we have discussed many kinds of functions, including linear, quadratic, exponential, and logarithmic. In this chapter, we introduce the trigonometric functions, which, as we shall see, differ from those previously introduced in that they describe a periodic or repetitive relationship. There are many practical applications of the trigonometric functions, particularly to the fields of surveying and navigation. These functions also describe many natural phenomena and, hence, are important in the study of optics, heat, electronics, x-ray, acoustics, seismology, and many other areas.

6.1 The Unit Circle

A circle with center at the origin and radius one unit is called a **unit circle.** With each point (x, y) on the unit circle we can associate an arc of length s, starting at the point $(1, 0)$ and ending at the point (x, y), as shown in Figure 6.1. Positive arc lengths are measured in a counterclockwise direction from $(1, 0)$, while negative arc lengths are measured in a clockwise direction. An arc with $s > 0$ is shown in Figure 6.1.

The circumference of a circle is given by the formula $C = 2\pi r$. Using this, the circumference of the unit circle is 2π. Thus, s may take values from 0 to 2π or from -2π to 0. We can extend the domain of s to include all real numbers by allowing arcs which wrap around the circle more than once. For example, $s = 3\pi$ would correspond to the same point on the circle as $s = \pi$ or $s = -\pi$. Figure 6.2 illustrates how s can take any real number as a value by "wrapping" a real number line around the unit circle.

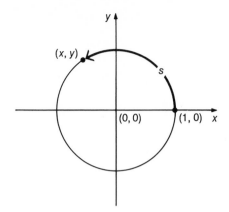

Figure 6.1

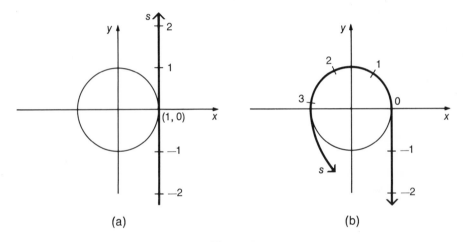

(a)   (b)

Figure 6.2

Example 1 Find points on the unit circle corresponding to arc lengths of
(a) $\pi/2$ (b) π (c) $-3\pi/2$.

The points corresponding to the given arc lengths are shown in Figure 6.3 on page 136.

Using the correspondence between an arc of length s and a unique point (x, y) on the unit circle, we can define two functions. For each value of s, we have a unique value of x and a unique value of y. The function $\{(s, y)\}$ which associates with each arc length s the y-value of the corresponding point, is called the **sine function,** and written

▶
$$y = \sin s.$$

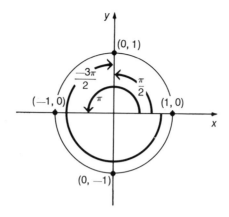

Figure 6.3

The function $\{(s, x)\}$ is called the **cosine function** and written

▶ $$x = \cos s.\,{}^{*}$$

Both the sine and cosine functions have the set of real numbers as domain. The values of x, or cos s, range from 1 to -1, as s increases from 0 to π with $x = 0$ corresponding to $s = \pi/2$. As s increases from π to 2π, x ranges from -1 to 1 with cos $s = 0$ again at $s = 3\pi/2$. Similarly, as s increases from 0 to 2π, y, or sin s, changes from 0 at $s = 0$ to 1 at $s = \pi/2$, back to 0 at $s = \pi$, to -1 at $s = 3\pi/2$, and to 0 again at $s = 2\pi$.

Note that if 2π is added to any value of s, corresponding values of sin s (or y) and cos s (or x) are unchanged. That is

▶ $$\sin s = \sin(s + 2\pi)$$

and

▶ $$\cos s = \cos(s + 2\pi).$$

A function of this type, which describes a cyclic relationship, is called a **periodic function.** Here the number 2π is called the **period** of the function because it is the smallest positive number which satisfies the two statements above. Hence, $y = \sin s$ and $x = \cos s$ are both periodic functions with a period of 2π, with all real numbers as domain and all real numbers from -1 to 1 inclusive as range.

---

* Actually, any variable names could be used to describe the sine and cosine functions, so that we might have used the more customary notation $y = \sin x$ and $y = \cos x$.

From plane geometry, for $s = \frac{\pi}{4}$, we know that the corresponding point on the unit circle has equal coordinates, so that $y = x$. Since any point (x, y) on this circle satisfies the relation $x^2 + y^2 = 1$, we have, for that point,

$$x^2 + x^2 = 1$$

$$x = \frac{1}{\sqrt{2}} = \frac{\sqrt{2}}{2}.$$

Since $x = y$, we also have

$$y = \frac{\sqrt{2}}{2}.$$

Thus, the point on the unit circle which corresponds to $s = \frac{\pi}{4}$ is $\left(\frac{\sqrt{2}}{2}, \frac{\sqrt{2}}{2}\right)$.

Hence, $\sin \frac{\pi}{4} = \frac{\sqrt{2}}{2}$ and $\cos \frac{\pi}{4} = \frac{\sqrt{2}}{2}$.

From the fact that $\left(\frac{\sqrt{2}}{2}, \frac{\sqrt{2}}{2}\right)$ is the point which corresponds to $s = \frac{\pi}{4}$, using symmetry, we can find the sine and cosine of $\frac{3\pi}{4}$ $\left(\text{or} -\frac{5\pi}{4}\right)$, $\frac{5\pi}{4}$ $\left(\text{or} -\frac{3\pi}{4}\right)$, and $\frac{7\pi}{4}$ $\left(\text{or} -\frac{\pi}{4}\right)$ as illustrated in Figure 6.4. From the figure, we see that $\cos s$ (or x) is always positive when the arc of length s ends in the first or fourth quadrants, and negative when the arc ends in the second or third quadrants. Similarly $\sin s$ (or y) is always positive in quadrants I and II, and negative in quadrants III and IV.

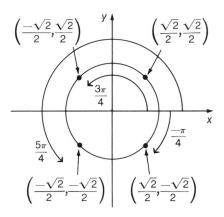

Figure 6.4

6.1 Exercises

Find sin s and cos s for the following values of s.

1. -3π
2. 4π
3. 17π
4. -2π
5. $-\pi/2$
6. $\dfrac{5}{2}\pi$
7. $\dfrac{11}{2}\pi$

8. $-\dfrac{11}{4}\pi$
9. $-\dfrac{7}{4}\pi$
10. $\dfrac{9}{4}\pi$
11. -2.25π
12. -1.5π

Find cos s, given the following values for sin s and the quadrant in which s terminates. (Hint: use the fact that for points (x, y) on the unit circle, $x^2 + y^2 = 1$.)

13. $\dfrac{\sqrt{3}}{2}, I$
14. $\dfrac{1}{\sqrt{2}}, II$
15. $-\dfrac{1}{\sqrt{2}}, IV$
16. $\dfrac{2}{3}, II$

17. $\dfrac{\sqrt{5}}{4}, I$
18. $-\dfrac{3}{5}, III$
19. $-\dfrac{5}{8}, IV$
20. $\dfrac{\sqrt{7}}{3}, II$

For each of the following, give the quadrant in which s terminates.

21. $\sin s > 0$ and $\cos s > 0$.
22. $\sin s < 0$ and $\cos s > 0$.
23. $\sin s > 0$ and $\cos s < 0$.
24. $\sin s < 0$ and $\cos s < 0$.

6.2 Angles and the Unit Circle

Associated with each arc of length s on the unit circle is an angle θ, as shown in Figure 6.5. Such an angle, with the positive x-axis as **initial side,** the ray OP as **terminal side,** and the origin as **vertex,** is said to be in **standard position.** The angle θ is positive when s is positive and negative when s is negative. (θ is the Greek letter theta.)

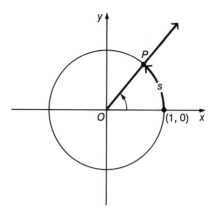

Figure 6.5 (a)

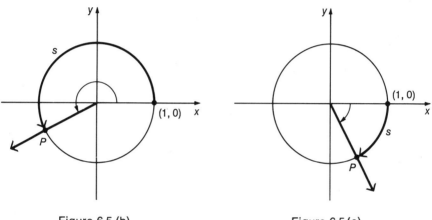

Figure 6.5 (b)   Figure 6.5 (c)

There are two common ways to measure the size of angles: degree measure and radian measure. In some mathematical work degree measure is not as convenient as radian measure of angles. We define the **radian measure** of an angle as the arc length cut off by the angle in standard position, on a unit circle. Figure 6.6, on page 140, shows an arc of length $\pi/4$; hence the associated angle θ has radian measure $\pi/4$.

A semicircle has an arc length of π, so that an angle of $180°$ has a radian measure of π, or

$$180° = \pi \text{ radians.}$$

Thus, $90° = \frac{1}{2}(180°) = \frac{1}{2}(\pi) = \pi/2$ radians, and so on. In general, if we wish to convert from degrees to radians, or from radians to degrees, we use the proportion

▶ $$\frac{\theta \text{ in radians}}{\theta \text{ in degrees}} = \frac{\pi}{180}.$$ (1)

For example, to convert $\pi/6$ radians to degrees, we write

$$\frac{\dfrac{\pi}{6}}{\theta \text{ in degrees}} = \frac{\pi}{180}$$

or
$$\theta \text{ in degrees} = 180\left(\frac{\pi}{6}\right)\left(\frac{1}{\pi}\right) = 30°.$$

Hence, $\pi/6$ radians is equivalent to 30 degrees.

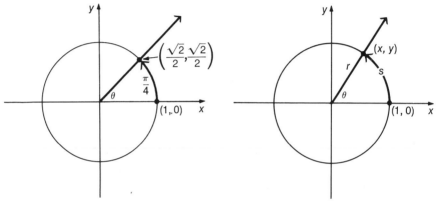

Figure 6.6   Figure 6.7

Example 2   Convert 150 degrees to radians.
Using equation (1), we have

$$\frac{\theta \text{ in radians}}{150} = \frac{\pi}{180}$$

or
$$\theta \text{ in radians} = \frac{\pi}{180}(150) = \frac{5}{6}\pi.$$

Now, in Figure 6.7, let us consider the point (x, y) on the unit circle which also lies on the terminal side of the angle θ in standard position. For simplicity, we shall often use the same symbol to refer both to an angle and its measure. Thus, in Figure 6.7 we use θ to represent both an angle and the numerical measure of the angle. From geometry, we have the relationship $\theta = rs$ where θ is the measure of any vertex angle in radians, and s is the arc intercepted by angle θ on a circle with radius r. For the unit circle, since the radius is 1, we have

$$\theta = s.$$

From this we see that we can also write the sine and cosine functions as

$$y = \sin \theta$$
and
$$x = \cos \theta,$$

where θ is measured in radians.

Example 3 Find $\sin 60°$ and $\cos 60°$.

As illustrated in Figure 6.8, we can use the relationships for a $30°–60°–90°$ triangle to find the values we need. Recall from geometry that the sides opposite the $30°$, $60°$, and $90°$ angles have lengths which are in the ratio $1 : \sqrt{3} : 2$, respectively. Thus, for a triangle with hypotenuse 1, the sides have lengths $1/2$, $\sqrt{3}/2$, and 1. Hence,

$$\cos 60° = x = \frac{1}{2}$$

and
$$\sin 60° = y = \frac{\sqrt{3}}{2}.$$

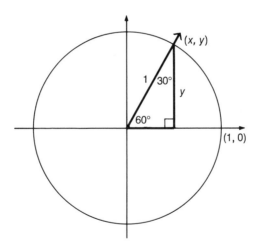

Figure 6.8

6.2 Exercises

Convert the following radian measures to degrees.

1. $\dfrac{\pi}{2}$

2. π

3. $\dfrac{\pi}{4}$

4. $\dfrac{3}{2}\pi$

5. $\dfrac{3}{4}\pi$

6. $\dfrac{5}{18}\pi$

7. $\dfrac{5}{9}\pi$

8. $\dfrac{\pi}{3}$	10. $\dfrac{\pi}{6}$	12. $\dfrac{5}{6}\pi$
9. $\dfrac{2}{3}\pi$	11. $\dfrac{5}{3}\pi$	

Convert the following degree measures to radians.

13.	240°	17.	225°	21.	450°
14.	145°	18.	10°	22.	225°
15.	540°	19.	130°	23.	210°
16.	80°	20.	50°	24.	18°

Use the methods of Example 3 to find the sine and cosine of each of the following.

25.	30°	29.	210°	32.	390°
26.	45°	30.	225°	33.	405°
27.	135°	31.	300°	34.	420°
28.	150°				

35. Find $\cos \theta$ given $\sin \theta = \dfrac{2}{3}$ and $\cos \theta < 0$.

36. Find $\sin \theta$ given $\cos \theta = \dfrac{4}{5}$ and $\sin \theta > 0$.

37. Find $\sin \theta$ given $\cos \theta = -\dfrac{5}{7}$ and θ terminates in quadrant III.

6.3 The Trigonometric Functions

In Section 6.1 we defined the sine and cosine functions as

$$\sin s = y$$
$$\cos s = x,$$

where (x, y) is the point on the unit circle which corresponds to an arc of length s. We now define four additional functions derived from these two basic functions: the **tangent, cotangent, cosecant,** and **secant** functions, abbreviated as tan, cot, csc, and sec.

▶ $$\tan s = \dfrac{y}{x} \qquad (x \neq 0)$$

▶ $$\cot s = \dfrac{x}{y} \qquad (y \neq 0)$$

▶ $$\csc s = \dfrac{1}{y} \qquad (y \neq 0)$$

▶ $$\sec s = \dfrac{1}{x} \qquad (x \neq 0)$$

These six functions are called the **trigonometric functions.** We can use the values of x and y already determined for specific values of s to find the four new functional values which correspond to those s values.

Example 4 Find $\tan \pi/3$, $\cot \pi/3$, $\csc \pi/3$, and $\sec \pi/3$.

Since $60° = \pi/3$ radians we can use the results of Example 3 to write $x = 1/2$ and $y = \sqrt{3}/2$. Then by definition

$$\tan \frac{\pi}{3} = \frac{\dfrac{\sqrt{3}}{2}}{\dfrac{1}{2}} = \sqrt{3}$$

$$\cot \frac{\pi}{3} = \frac{\dfrac{1}{2}}{\dfrac{\sqrt{3}}{2}} = \frac{1}{\sqrt{3}} = \frac{\sqrt{3}}{3}$$

$$\csc \frac{\pi}{3} = \frac{1}{\dfrac{\sqrt{3}}{2}} = \frac{2}{\sqrt{3}} = \frac{2\sqrt{3}}{3}$$

$$\sec \frac{\pi}{3} = \frac{1}{\dfrac{1}{2}} = 2.$$

From their definitions, we see that the tangent and cotangent functions are positive when x and y have the same sign, that is, in quadrants I and III; and they are negative when x and y have different signs, in quadrants II and IV. Like the sine function, the cosecant function is positive in quadrants I and II, and negative in quadrants III and IV. Like the cosine function, the secant function is positive in quadrants I and IV, and negative in quadrants II and III. The following table summarizes these facts.

Quadrant	sine	cosine	tangent	cotangent	secant	cosecant
I	+	+	+	+	+	+
II	+	−	−	−	−	+
III	−	−	+	+	−	−
IV	−	+	−	−	+	−

Because the four new functions are defined as quotients, there are restrictions on their domains. From the definition of tangent, $x \neq 0$. The condition that $x \neq 0$ implies that $s \neq \pi/2$, or $3\pi/2$. More generally, $s \neq \dfrac{\pi}{2} + k\pi$ for any integer k. In the case of the cotangent function, $y \neq 0$ implies that $s \neq k\pi$ for any integer k. Like the tangent function, the secant function requires that

$x \neq 0$, which means that $s \neq \dfrac{\pi}{2} + k\pi$, for any integer k. Similarly, the cosecant function restricts s to $s \neq k\pi$ for any integer k.

Like the sine and the cosine functions, the four new trigonometric functions are periodic. The cosecant and secant functions have the same period as sine and cosine, 2π. The tangent and cotangent functions, however, have a period of π, as indicated in the table below, which also summarizes some of the commonly used values of the trigonometric functions.

degrees	s	$\sin s$	$\cos s$	$\tan s$	$\cot s$	$\csc s$	$\sec s$
0	0	0	1	0	no value	no value	1
30	$\dfrac{\pi}{6}$	$\dfrac{1}{2}$	$\dfrac{\sqrt{3}}{2}$	$\dfrac{\sqrt{3}}{3}$	$\sqrt{3}$	2	$\dfrac{2\sqrt{3}}{3}$
45	$\dfrac{\pi}{4}$	$\dfrac{\sqrt{2}}{2}$	$\dfrac{\sqrt{2}}{2}$	1	1	$\sqrt{2}$	$\sqrt{2}$
60	$\dfrac{\pi}{3}$	$\dfrac{\sqrt{3}}{2}$	$\dfrac{1}{2}$	$\sqrt{3}$	$\dfrac{\sqrt{3}}{3}$	$\dfrac{2\sqrt{3}}{3}$	2
90	$\dfrac{\pi}{2}$	1	0	no value	0	1	no value
120	$\dfrac{2\pi}{3}$	$\dfrac{\sqrt{3}}{2}$	$\dfrac{-1}{2}$	$-\sqrt{3}$	$\dfrac{-\sqrt{3}}{3}$	$\dfrac{2\sqrt{3}}{3}$	-2
135	$\dfrac{3\pi}{4}$	$\dfrac{\sqrt{2}}{2}$	$\dfrac{-\sqrt{2}}{2}$	-1	-1	$\sqrt{2}$	$-\sqrt{2}$
150	$\dfrac{5\pi}{6}$	$\dfrac{1}{2}$	$\dfrac{-\sqrt{3}}{2}$	$\dfrac{-\sqrt{3}}{3}$	$-\sqrt{3}$	2	$\dfrac{-2\sqrt{3}}{2}$
180	π	0	-1	0	no value	no value	-1
210	$\dfrac{7\pi}{6}$	$\dfrac{-1}{2}$	$\dfrac{-\sqrt{3}}{2}$	$\dfrac{\sqrt{3}}{3}$	$\sqrt{3}$	-2	$\dfrac{-2\sqrt{3}}{3}$
225	$\dfrac{5\pi}{4}$	$\dfrac{-\sqrt{2}}{2}$	$\dfrac{-\sqrt{2}}{2}$	1	1	$-\sqrt{2}$	$-\sqrt{2}$
240	$\dfrac{4\pi}{3}$	$\dfrac{-\sqrt{3}}{2}$	$\dfrac{-1}{2}$	$\sqrt{3}$	$\dfrac{\sqrt{3}}{3}$	$\dfrac{-2\sqrt{3}}{3}$	-2
270	$\dfrac{3\pi}{2}$	-1	0	no value	0	-1	no value
300	$\dfrac{5\pi}{3}$	$\dfrac{-\sqrt{3}}{2}$	$\dfrac{1}{2}$	$-\sqrt{3}$	$\dfrac{-\sqrt{3}}{3}$	$\dfrac{-2\sqrt{3}}{3}$	2
315	$\dfrac{7\pi}{4}$	$\dfrac{-\sqrt{2}}{2}$	$\dfrac{\sqrt{2}}{2}$	-1	-1	$-\sqrt{2}$	$\sqrt{2}$
330	$\dfrac{11\pi}{6}$	$\dfrac{-1}{2}$	$\dfrac{\sqrt{3}}{2}$	$\dfrac{-\sqrt{3}}{3}$	$-\sqrt{3}$	-2	$\dfrac{2\sqrt{3}}{3}$

Given the value of any one of the trigonometric functions and the quadrant in which s terminates, we can find the values of the other five functions. For example, if $\csc s$ is $3/2$ and s terminates in quadrant II, we have the following.

$$\csc s = \frac{1}{y} = \frac{3}{2}$$

from which
$$y = \frac{2}{3}.$$

Since, for any point (x, y) on the unit circle, $x^2 + y^2 = 1$,

$$x^2 + \frac{4}{9} = 1$$

$$x^2 = \frac{5}{9}$$

$$x = \pm\frac{\sqrt{5}}{3}.$$

We know s terminates in quadrant II. Thus, we choose the negative value of x. The functions can now be written as follows.

$$\sin s = y = \frac{2}{3} \qquad\qquad \cos s = x = \frac{-\sqrt{5}}{3}$$

$$\tan s = \frac{y}{x} = \frac{-2}{\sqrt{5}} = \frac{-2\sqrt{5}}{5} \qquad \cot s = \frac{x}{y} = \frac{-\sqrt{5}}{2}$$

$$\sec s = \frac{1}{x} = \frac{-3}{\sqrt{5}} = \frac{-3\sqrt{5}}{5}$$

6.3 Exercises

Find the values of the six trigonometric functions for each of the following.

1. $s = \dfrac{\pi}{3}$

2. $s = -\dfrac{2}{3}\pi$

3. $s = \dfrac{-5}{6}\pi$

4. $s = \dfrac{13\pi}{6}$

5. $\theta = -45°$
6. $\theta = 390°$
7. $\theta = 405°$
8. $\theta = -240°$

9. $\sin s = \dfrac{3}{4}$; s terminates in quadrant I.

10. $\sin s = -\dfrac{2}{5}$; s terminates in quadrant III.

11. $\cos s = -\dfrac{4}{9}$; s terminates in quadrant II.

12. $\cos s = 0.1$; s terminates in quadrant IV.

13. $\tan \theta = \dfrac{3}{2}$; $\csc \theta = \dfrac{\sqrt{13}}{3}$.

14. $\sec \theta = -2$; $\cot \theta = \dfrac{\sqrt{3}}{3}$.

15. $\tan \theta = -\dfrac{7}{5}$; θ terminates in quadrant II. (Hint: since $x^2 + y^2 = 1$, $(\cos \theta)^2 + (\sin \theta)^2 = 1$.)

16. We defined $\sin s$ and $\cos s$ in terms of the line segments AC and OA shown in the figure. Using geometric relationships, show that the length of line segment BD equals $\tan s$. (Hint: since $\tan s = \dfrac{y}{x}$, you must show that $\dfrac{AC}{OA} = BD$.)

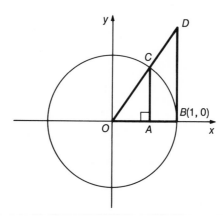

6.4 Values of the Trigonometric Functions

Up to this point, we have discussed values of the trigonometric functions only for certain special values of θ. The methods we used were primarily based on geometry. Clearly, there are many values of θ for which geometric methods are inadequate and more advanced methods, beyond the scope of this text, are required.

We have already noted, in Section 6.1, that we can find the trigonometric function values for $\pi/2 < \theta < 2\pi$ by referring to the appropriate value of θ in the interval $0 < \theta < \pi/2$ and noting the correct sign. For any value of θ in the interval $\pi/2 < \theta < 2\pi$, we shall call the first quadrant angle which has the same trigonometric function values (in absolute value) as θ the **reference angle** for θ, and denote it θ'. For example, if $\theta = 3\pi/4$, the reference angle θ' is $\pi/4$; if $\theta = 5\pi/6$, the reference angle θ' is $\pi/6$; if $\theta = 200°$, the reference angle θ' is $20°$. An easy way to determine the reference angle θ' for any angle θ is to find the measure of the *acute* angle formed by the terminal side of θ and the x-axis as illustrated in Figure 6.9. In a similar way, we can determine a **reference arc** s', $0 \le s' \le \pi/2$, for every value of s such that $s > \pi/2$.

We need values, then, only for the trigonometric functions of θ, where $0 < \theta < \pi/2$. Table 3 in the Appendix gives the values of the trigonometric functions for these values of θ. For many values of θ, the trigonometric function values are irrational numbers, so that the table actually gives four decimal place approximations for these numbers. However, for convenience, we shall

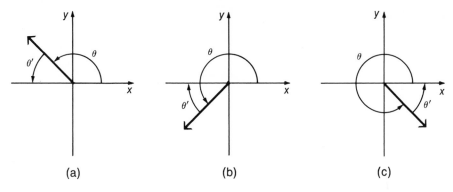

(a)　　　　　　　　(b)　　　　　　　　(c)

Figure 6.9

use the equality symbol with the understanding that the values are actually approximations. The table is designed to give the value of θ in the first two columns (in degrees, then radians) for $0 \leq \theta \leq \pi/4$, and in the last two columns for $\pi/4 \leq \theta \leq \pi/2$. Note that when locating values for $\pi/4 \leq \theta \leq \pi/2$ one must read *up* the table and refer to the names of the trigonometric functions at the *bottom* of the page.

Example 5 Find sin 10° 20′; cos 48° 10′; tan 82° 00′.

From Table 3, sin 10° 20′ = 0.1794
                            cos 48° 10′ = 0.6670
                            tan 82° 00′ = 7.115.

Example 6 Find sin .2676; cot 1.2043; sec .7447.

Again, from Table 3, sin .2676 = 0.2644
                                    cot 1.2043 = 0.3839
                                    sec .7447 = 1.360.

Example 7 Find sin 420°; cot −185°; sec −3.0369. (Use π = 3.1416 to find s′.)

For θ = 420°, θ' = 60° and sin 420° = sin 60° = .8660.
For θ = −185°, θ' = 5° and cot −185° = −cot 5° = −11.43.
We know the answer must be negative since an angle of −185° terminates in quadrant II, where cot is negative.
For θ = −3.0369, θ' = .1047 and sec −3.0369 = −sec .1047 = −1.006.

When the required value of θ is not in the table, linear interpolation may be used as illustrated in the next example.

Example 8 Find tan 40° 52′ and cos 63° 34′.

The value for tan 40° 52′ will lie .2 of the way between the values for tan 40° 50′ and tan 41° 00′.

$$\tan 40° 50' = .8642$$
$$\tan 40° 52' = ?$$
$$\tan 41° 00' = .8693$$

$$\tan 40° 52' = .2(.8693 - .8642) + .8642$$
$$= .2(.0051) + .8642$$
$$= .8652$$

We find cos 63° 34′ in a similar way.

$$\cos 63° 30' = .4462$$
$$\cos 63° 34' = ?$$
$$\cos 63° 40' = .4436$$

$$\cos 63° 34' = .4462 - (.4)(.4462 - .4436)$$
$$= .4462 - .00104$$
$$= .4452$$

Note that, in this case, we had to subtract from .4462 rather than add to it because the cosine function *decreases* as θ *increases* for $0 < \theta < \pi/2$.

6.4 Exercises

Give the reference angle or arc for each of the following. (Use $\pi = 3.14$.)

1. $\theta = 215°$
2. $\theta = 550°$
3. $\theta = -143°$
4. $\theta = -12°$
5. $\theta = 420°$
6. $s = \dfrac{10}{3} \pi$

7. $s = -\dfrac{4}{5} \pi$
8. $s = -1.9$
9. $s = 3.21$
10. $s = 2.4$

Evaluate the following trigonometric functions. (Use $\pi = 3.1416$.)

11. $\sin 215°$
12. $\cos -143°$
13. $\tan 550° 10'$
14. $\csc -12° 35'$
15. $\cot 420° 12'$

16. $\sin \dfrac{10}{3} \pi$
17. $\sec -1.9024$
18. $\cot 3.1998$
19. $\tan 2.4027$
20. $\csc -0.8\pi$

Find θ (in degrees) for each of the following.

21. $\tan \theta = 0.1944$; θ terminates in quadrant III.
22. $\sec \theta = 1.302$; θ terminates in quadrant IV.

23. $\sin \theta = 0.7771$; θ terminates in quadrant I.
24. $\cot \theta = -2.475$; θ terminates in quadrant II.
25. $\cos \theta = 0.6884$; θ terminates in quadrant IV.

Find s (in radians) for each of the following.

26. $\csc s = 1.386$; s terminates in quadrant I.
27. $\sin s = -0.3448$; s terminates in quadrant III.
28. $\cos s = 0.5878$; s terminates in quadrant IV.
29. $\tan s = -0.9110$; s terminates in quadrant II.
30. $\cot s = 2.211$; s terminates in quadrant III.

In calculus courses $\sin x$ and $\cos x$ are expressed as infinite sums, where x is measured in radians. For all real numbers x and all natural numbers n, the following hold.

$$\cos x = 1 - \frac{x^2}{2!} + \frac{x^4}{4!} - \cdots + (-1)^{n-1} \frac{x^{2n-2}}{(2n-2)!} + \cdots$$

and $\quad \sin x = x - \dfrac{x^3}{3!} + \dfrac{x^5}{5!} - \cdots + (-1)^{n-1} \dfrac{x^{2n-1}}{(2n-1)!} + \cdots$

where $\quad 2! = 2 \cdot 1; \quad 3! = 3 \cdot 2 \cdot 1; \quad 4! = 4 \cdot 3 \cdot 2 \cdot 1;$

and $\quad n! = n(n-1)(n-2)\ldots(2)(1).$

Using these formulas, $\sin x$ and $\cos x$ can be evaluated to any desired degree of accuracy. For example, let us find $\cos 0.5$ using the first three terms of the definition and compare this with the result using the first four terms. Using the first three terms we have

$$\cos 0.5 = 1 - \frac{(0.5)^2}{2!} + \frac{(0.5)^4}{4!}$$

$$= 1 - \frac{.25}{2} + \frac{.0625}{24}$$

$$= 1 - 0.125 + 0.00260$$

$$= 0.87760.$$

If we use four terms, the result is

$$\cos 0.5 = 1 - \frac{(0.5)^2}{2!} + \frac{(0.5)^4}{4!} - \frac{(0.5)^6}{6!}$$

$$= 1 - 0.125 + 0.00260 - 0.00002$$

$$= 0.87758.$$

The two results differ by only 0.00002, so that the first three terms give results which are accurate to the fourth decimal place. From the tables, using interpolation, $\cos 0.5 = .87754$.

Use the first three terms of the definitions given above for sin x and cos x to find the following. Compare your results with the values in Table 3 in the Appendix.

31. sin 1
32. sin 0.1
33. sin 0.01
34. cos 0.1
35. cos 0.01

6.5   Graphs of the Trigonometric Functions

In Section 6.1, we noted that the domain of $\sin \theta$ is the set of real numbers, while the range is $-1 \leq \sin \theta \leq 1$. Since the sine function has a period of 2π, to find its graph we can concentrate on the values $0 \leq \theta \leq 2\pi$. A good way to begin is by plotting selected points. For graphing purposes, we shall use .7 to approximate $\sin \pi/4$.

θ	0	$\pi/4$	$\pi/2$	$3\pi/4$	π	$5\pi/4$	$3\pi/2$	$7\pi/4$	2π
$\sin \theta$	0	.7	1	.7	0	$-.7$	-1	$-.7$	0

Plotting the points from the table of values above and connecting them with a smooth line, we have the solid portion of the graph of Figure 6.10. Since $\sin \theta$ is a periodic function with all real numbers as domain, the graph continues in both directions infinitely as indicated by the dashed lines. The graph of $\cos \theta$ in Figure 6.11 is determined similarly.

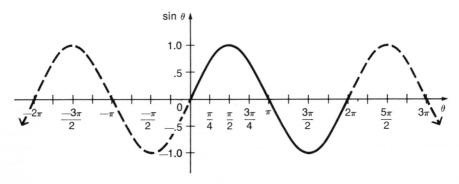

Figure 6.10

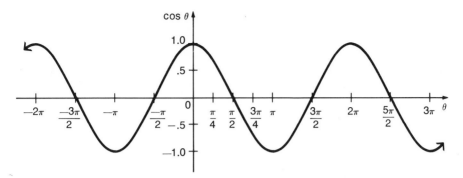

Figure 6.11

In Section 6.3, we saw that the domain of θ for the tangent function excluded values of $\theta = \pi/2 + k\pi$ (k any integer). We also noted that the period of tan θ is π, so that we need to investigate the range only within an interval of π units. A convenient interval for this purpose is $-\pi/2 < \theta < \pi/2$. Although the endpoints are not in the domain, tan θ exists for all other values in this interval. A table of values for selected points in the interval is shown below.

θ	$-\pi/3$	$-\pi/4$	$-\pi/6$	0	$\pi/6$	$\pi/4$	$\pi/3$
tan θ	-1.7	-1	$-.6$	0	.6	1	1.7

In Table 3 of the Appendix, we note that as θ increases from 0 to $\pi/2$, tan θ increases, getting large very rapidly as θ gets close to $\pi/2$. To express this we write tan $\theta \to +\infty$ as $\theta \to \pi/2$, which means that tan θ increases without bound as θ approaches $\pi/2$. In the same way, as $\theta \to -\pi/2$, tan $\theta \to -\infty$, since the tangent function is negative for $-\pi/2 < \theta < 0$. Thus, in Figure 6.12, the graph of tan θ approaches the vertical lines at $\pi/2 + k\pi$ (k any integer) but never intersects them. Such lines are called asymptotes. The graph of cot θ is similar to that for tan θ, and is left for the exercises.

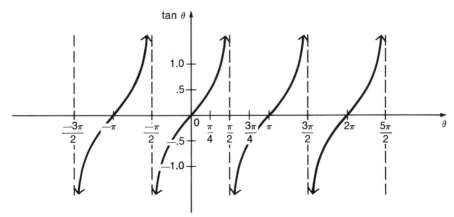

Figure 6.12

The domain of csc θ is restricted to values of $\theta \neq k\pi$ (k any integer). Thus, like tan θ, the graph will have vertical asymptotes–in this case, at $\theta = k\pi$ (k any integer). The period of the csc function is 2π, so we begin by graphing csc θ for $0 < \theta < 2\pi$. The graph is shown in Figure 6.13. The graph of sec θ, which is similar to that for csc θ, is left for the exercises.

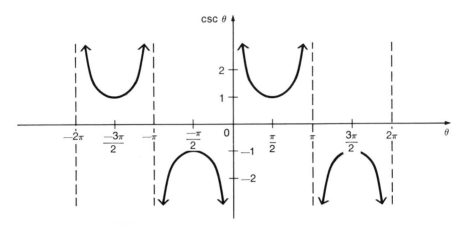

Figure 6.13

Example 9 Graph $y = 2 \sin \theta$.

For each value of θ, the value of y is twice as large as it would be for $y = \sin \theta$. Thus, the only change in the graph is in the range, which becomes $-2 \leq y \leq 2$. The graph of $y = 2 \sin \theta$ is shown in Figure 6.14.

In general, the graph of $y = a \sin \theta$ will have the same form as $y = \sin \theta$ with range $-|a| \leq y \leq |a|$. We call $|a|$ the **amplitude.**

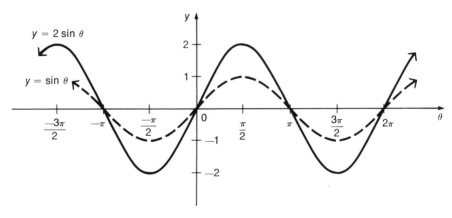

Figure 6.14

Example 10   Graph $y = \sin 2\theta$.

To obtain this graph, note the comparison between the functional values for $\sin \theta$ and $\sin 2\theta$, as shown in the table below.

θ	0	$\pi/4$	$\pi/2$	$3\pi/4$	π	$5\pi/4$	$3\pi/2$	$7\pi/4$
$\sin \theta$	0	.7	1	.7	0	$-.7$	-1	$-.7$
$\sin 2\theta$	0	1	0	-1	0	1	0	-1

We see that multiplying θ by 2 results in shortening the period to half the normal period. In general, the graph of $y = \sin b\theta$ will look like that of $\sin \theta$, but with period $|2\pi/b|$. The range is not changed. Figure 6.15 shows the graph of $y = \sin 2\theta$.

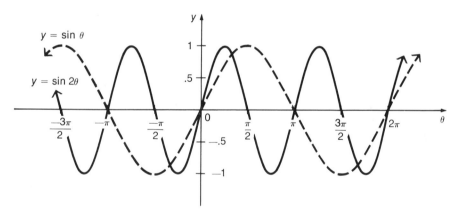

Figure 6.15

Example 11   Graph $y = \sin\left(\theta - \dfrac{\pi}{3}\right)$.

The graph of $y = \sin\left(\theta - \dfrac{\pi}{3}\right)$ in Figure 6.16, page 154, looks like the graph of $\sin \theta$ shifted to the right so that the point which was at the origin is at $\dfrac{\pi}{3}$. (When $\theta - \dfrac{\pi}{3} = 0$, $\sin\left(\theta - \dfrac{\pi}{3}\right) = 0$.) In general $y = \sin(\theta - c)$ will have the shape of a basic sine graph with the origin shifted to $(c, 0)$. We call c the **phase shift** of the graph.

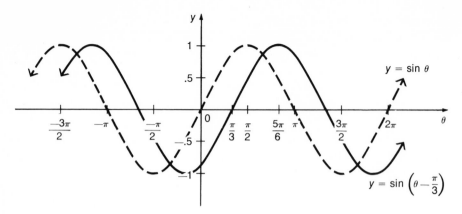

Figure 6.16

6.5 Exercises

Graph the following trigonometric functions over a two period interval. State the period for each.

1. $y = \cot \theta$
2. $y = \sec \theta$
3. $y = 2 \cos \theta$
4. $y = 3 \sin \theta$
5. $y = 5 \tan \theta$
6. $y = \cos 3\theta$
7. $y = \sin \frac{2}{3} \theta$
8. $y = \cot 2\theta$
9. $y = \csc\left(\theta + \frac{\pi}{2}\right)$

10. $y = \cos\left(\theta - \frac{\pi}{4}\right)$
11. $y = \tan\left(\theta - \frac{\pi}{4}\right)$
12. $y = \cos \theta + 4$
13. $y = \sin \frac{1}{2}\theta - 1$
14. $y = 2 \cos\left(\theta - \frac{\pi}{4}\right)$
15. $y = -2 \sin 2\left(\theta - \frac{\pi}{4}\right)$

16. Graph $y = \sin\left(\theta - \frac{\pi}{2}\right)$. Which trigonometric function has the same graph as $\sin\left(\theta - \frac{\pi}{2}\right)$?

17. Graph $y = -\sin \theta$. Compare the graph with that of $y = \sin \theta$.
18. What is true of the graphs of $y = -\cos \theta$ and $y = \cos \theta$?

6.6 Inverse Trigonometric Functions

In Chapter 5 we defined the inverse of a relation $\{(x, y)\}$ as the relation $\{(y, x)\}$. In this section we shall consider the inverse relations of the trigonometric functions. In order to conform to the notation commonly used for functions and relations, we shall write the trigonometric functions using the variables x and y. It is important not to confuse this use of x and y with our previous use of the variables to represent a point on the unit circle.

Let us begin with the function $y = \sin x$. The inverse relation is $x = \sin y$. To solve this expression for y, we shall write

$$y = \arcsin x. *$$

Recall that the domain of $y = \sin x$ is the set of real numbers while the range is $-1 \leq y \leq 1$. Thus, the domain of $y = \arcsin x$ is $-1 \leq x \leq 1$; the range is the set of real numbers. The graph of $y = \arcsin x$ is shown in Figure 6.17.

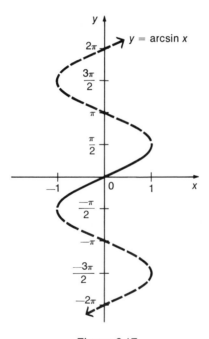

Figure 6.17

From the graph, it is clear that $y = \arcsin x$ is not a function since any x value from the domain corresponds to many values of y. However, by suitably restricting the domain of the sine function, we can define a function whose inverse, the arcsin, is also a function. This could be done in several ways. However, it is customary to restrict the domain of $y = \sin x$ to $-\pi/2 \leq x \leq \pi/2$ so that $y = \arcsin x$ will have a range of $-\pi/2 \leq y \leq \pi/2$. With this range, the arcsin relation is a function whose graph is the heavy portion of the graph in Figure 6.17.

---

* An alternate notation for arcsin x is $\sin^{-1}x$.

Example 12 Find arcsin $1/2$; arcsin $\sqrt{2}/2$.

We want to find the angle whose sin is $1/2$–that is, in the expression $\sin y = 1/2$, we wish to determine y. Since $\sin \pi/6 = 1/2$ and $\pi/6$ is in the proper interval, arcsin $1/2 = \pi/6$.

Similarly, if $\sin y = \sqrt{2}/2$ and $\sin \pi/4 = \sqrt{2}/2$, then arcsin $\sqrt{2}/2 = \pi/4$.

Each of the other five trigonometric functions also has an inverse relation which can be written as a function by a suitable restriction on the domain of the trigonometric function and, hence, on the range of the inverse, as shown below.

$$y = \arccos x \qquad (0 \leq y \leq \pi)$$

$$y = \arctan x \qquad \left(-\frac{\pi}{2} < y < \frac{\pi}{2}\right)$$

$$y = \text{arccot } x \qquad (0 < y < \pi)$$

$$y = \text{arccsc } x \qquad \left(-\frac{\pi}{2} \leq y \leq \frac{\pi}{2} \text{ and } y \neq 0\right)$$

$$y = \text{arcsec } x \qquad \left(0 \leq y \leq \pi \text{ and } y \neq \frac{\pi}{2}\right)$$

Example 13 Find arctan 1; arccos $-1/2$.

Let $y = \arctan 1$. Then $\tan y = 1$. Since $\tan \pi/4 = 1$ and $\pi/4$ is in the interval $-\pi/2 < y < \pi/2$, arctan $1 = \pi/4$.

Let $y = \arccos -1/2$. Then $\cos y = -1/2$. Both $\cos 2\pi/3$ and $\cos 4\pi/3$ equal $-1/2$. However, the range of the arccos function is limited by definition to $0 \leq y \leq \pi$, which eliminates $4\pi/3$ as a solution. Hence, arccos $-1/2 = 2\pi/3$.

Example 14 Without referring to the tables, evaluate $\tan(\text{arcsin } \sqrt{3}/2)$.

Let $u = \arcsin \sqrt{3}/2$. Then $\sin u = \sqrt{3}/2$. Since $-\pi/2 \leq u \leq \pi/2$ for the arcsin function and since $\sin u = \sqrt{3}/2$ is positive, u must lie in quadrant I. (If $-\pi/2 \leq u \leq 0$, $\sin u$ would be negative.) Thus, $u = \pi/3$ and $\tan(\text{arcsin } \sqrt{3}/2) = \tan u = \tan \pi/3$ which is $\sqrt{3}$. A sketch of the unit circle is helpful in problems of this kind. Note, in Figure 6.18, by the Pythagorean theorem,

$$x = \sqrt{1 - (\sqrt{3}/2)^2} = 1/2$$

Therefore, $\tan(\text{arcsin } \sqrt{3}/2) = \tan u$, which from the figure we can find to be

$$\tan u = \frac{\sqrt{3}/2}{x} = \frac{\sqrt{3}/2}{1/2} = \sqrt{3}.$$

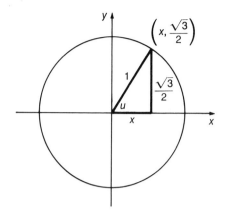

Figure 6.18

6.6 Exercises

For each of the following, give the value of y in both degrees and radians.

1. $y = \arcsin \dfrac{-\sqrt{3}}{2}$

2. $y = \arccos \dfrac{\sqrt{3}}{2}$

3. $y = \arctan 1$

4. $y = \text{arccot} \ -1$

5. $y = \arcsin -1$

6. $y = \arccos -1$

7. $y = \text{arcsec} \ 2$

8. $y = \text{arccsc} \ -\sqrt{2}$

9. $y = \arccos \dfrac{-\sqrt{2}}{2}$

10. $y = \arcsin \dfrac{-\sqrt{2}}{2}$

11. $y = \text{arccot} \ -\sqrt{3}$

12. $y = \text{arccsc} \ -2$

13. $\arcsin \ -0.1334$

14. $\arccos \ -0.1334$

15. $\arccos \ -0.3987$

16. $\arcsin \ 0.7790$

17. $\text{arccsc} \ 2.411$

18. $\text{arccot} \ 1.767$

19. $\arctan \ 1.111$

20. $\text{arcsec} \ 1.766$

Use the methods of Example 14 to evaluate the following without using the tables.

21. $\sin\left(\arctan \dfrac{\sqrt{3}}{3}\right)$

22. $\cos\left(\arctan \dfrac{\sqrt{3}}{3}\right)$

23. $\tan\left(\arcsin \ -\dfrac{1}{2}\right)$

24. $\cot\left(\arccos \dfrac{\sqrt{2}}{2}\right)$

25. $\sec \ (\arctan 1)$

26. $\sin \ (\text{arccot} \ -\sqrt{3})$

27. $\sin \ (\arcsin \ .6)$

28. $\cos \ (\arccos \ -.2)$

29. tan (arctan 3)
30. sec (arcsec −4)
31. tan (arcsin −2/3)
32. cos (arcsin −2/3)

33. cos (arctan −3)
34. sin (arctan −3)
35. cot (arctan −1)
36. csc (arcsin .6)

Graph each of the following.

37. $y = \text{arccos } x$
38. $y = \text{arctan } x$
39. $y = \text{arccot } x$

Chapter Seven

Trigonometric Identities and Conditional Equations

The equations we have discussed in this book have been primarily equations in which some replacements for the variable(s) made the equation true, while others made it false. Such equations are called **conditional equations,** since their equality is conditional upon the replacement used for the variable(s). An **identity** is a statement which is true for every value in its domain. Thus,

$$5(x + y) = 5x + 5y, \qquad x \le |x|, \qquad \text{and} \qquad x < 1 \text{ or } x > -2$$

are all identities.

In this chapter, we shall discuss both conditional equations and identities which involve trigonometric functions. The variables we shall use in the trigonometric functions may represent either angles or real numbers. Let us agree that the domain will always be all real numbers for which the given functions are defined.

7.1 Fundamental Identities

Trigonometric identities are useful in evaluating trigonometric functions; in changing an expression involving one trigonometric function to one with a different trigonometric function; and in simplifying trigonometric expressions. Their use in simplifying trigonometric expressions is of great importance in more advanced courses, particularly in calculus.

For convenience, the definitions of the six trigonometric functions are restated here. If (x, y) is a point on the unit circle corresponding to an arc of length s, then

$$\sin s = y \qquad\qquad \cos s = x$$

$$\tan s = \frac{y}{x} \qquad\qquad \cot s = \frac{x}{y}$$

$$\csc s = \frac{1}{y} \qquad\qquad \sec s = \frac{1}{x}.$$

From the definitions, we immediately have the following relationships which are true for all suitable replacements of the variable and, hence, are identities.

▶
$$\tan s = \frac{\sin s}{\cos s} \qquad\qquad \cot s = \frac{\cos s}{\sin s}$$

$$\cot s = \frac{1}{\tan s}$$

$$\csc s = \frac{1}{\sin s} \qquad\qquad \sec s = \frac{1}{\cos s}$$

Every point (x, y) on the unit circle must satisfy the equation of the unit circle $x^2 + y^2 = 1$. From the definition of sine and cosine above, $x^2 = (\cos s)^2$ and $y^2 = (\sin s)^2$. It is customary to write these as $\sin^2 s$ and $\cos^2 s$. Thus, by substitution in the equation of the unit circle, we have the identity

▶
$$\sin^2 s + \cos^2 s = 1. \qquad\qquad (1)$$

Equation (1) can also be written as

▶
$$\sin^2 s = 1 - \cos^2 s$$

or

▶
$$\cos^2 s = 1 - \sin^2 s,$$

which are both useful identities.

A little algebraic manipulation of equation (1) results in two more identities. First, we multiply both sides of equation (1) by $\frac{1}{\cos^2 s}$ (noting the necessary restrictions on s), obtaining

$$\frac{\sin^2 s}{\cos^2 s} + \frac{\cos^2 s}{\cos^2 s} = \frac{1}{\cos^2 s} \qquad (\cos s \neq 0).$$

Substituting from the identities for tan s and sec s gives

▶
$$\tan^2 s + 1 = \sec^2 s.$$

Similarly, multiplication of equation (1) by $\frac{1}{\sin^2 s}$ results in the identity

▶
$$1 + \cot^2 s = \csc^2 s.$$

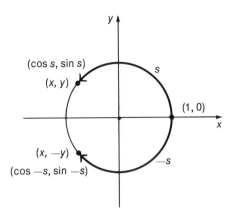

Figure 7.1

From the unit circle in Figure 7.1, we see that for any arc s terminating at the point (x, y), the arc $-s$ will terminate at the point $(x, -y)$. From the definitions of sin and cos, we have $\sin(-s) = -y = -\sin s$, so that

▶ $$\textbf{sin}(-\textbf{s}) = -\textbf{sin}\ \textbf{s}.$$

Also, $\cos(-s) = x = \cos s$, hence

▶ $$\textbf{cos}(-\textbf{s}) = \textbf{cos}\ \textbf{s}.$$

Furthermore,
$$\tan(-s) = \frac{\sin(-s)}{\cos(-s)}$$
$$= \frac{-\sin s}{\cos s}$$
$$= -\frac{\sin s}{\cos s}$$

▶ $$\textbf{tan}(-\textbf{s}) = -\textbf{tan}\ \textbf{s}.$$

One of the skills required for more advanced work in mathematics is the ability to utilize the trigonometric identities to rewrite trigonometric expressions in alternate forms. To develop this skill, we shall use the fundamental identities to verify that a trigonometric equation is itself an identity (for those values of the variable for which it is defined). The following examples illustrate the method.

Example 1 Verify that
$$\cot s + 1 = (\csc s)(\cos s + \sin s)$$
is an identity.

We must use the fundamental identities to rewrite one side of the equation so that it is identical to the other. Let us work with the right-hand side, R. Using the identities we have developed, we can write

$$R = (\csc s)(\cos s + \sin s)$$

$$= \frac{1}{\sin s}(\cos s + \sin s)$$

$$= \frac{\cos s}{\sin s} + \frac{\sin s}{\sin s}$$

$$= \cot s + 1.$$

Sometimes it is helpful to work with both sides of the equation, as illustrated in the following example, but it is important to note that this must be done separately. We must be careful not to use the techniques suitable for solving a conditional equation, such as the addition and multiplication properties of equality, because these techniques do not establish the truth of the original statement.

Example 2 Verify that the equation

$$\tan^2 s(1 + \cot^2 s) = \frac{1}{1 - \sin^2 s}$$

is an identity.

Working first with the left-hand side, L, we have

$$L = \tan^2 s(1 + \cot^2 s)$$
$$= \tan^2 s + \tan^2 s \cot^2 s$$
$$= \tan^2 s + 1$$
$$= \sec^2 s.$$

On the right-hand side,

$$R = \frac{1}{1 - \sin^2 s}$$

$$= \frac{1}{\cos^2 s}$$

$$= \sec^2 s.$$

Since $L = R$, the equation is an identity.

A useful device in "proving" identities is to change (by means of substitution) all trigonometric functions into sine and cosine as in the next example.

Example 3 Verify that the equation

$$\sin^2 s - \cos^2 s = \frac{\tan s - \cot s}{\tan s + \cot s}$$

is an identity.

We shall show that R can be made to look like L.

$$R = \frac{\tan s - \cot s}{\tan s + \cot s}$$

$$= \frac{\dfrac{\sin s}{\cos s} - \dfrac{\cos s}{\sin s}}{\dfrac{\sin s}{\cos s} + \dfrac{\cos s}{\sin s}}$$

$$= \frac{\dfrac{\sin^2 s - \cos^2 s}{\cos s \cdot \sin s}}{\dfrac{\sin^2 s + \cos^2 s}{\cos s \cdot \sin s}}$$

$$= \frac{\sin^2 s - \cos^2 s}{\sin^2 s + \cos^2 s}$$

$$= \sin^2 s - \cos^2 s$$

$$= L$$

7.1 Exercises

For each trigonometric function in column I, choose the function from column II which completes an identity.

Column I	Column II
1. $\dfrac{\cos x}{\sin x}$	a) $\tan^2 x$
2. $\tan x$	b) $\sin^2 x + \cos^2 x$
3. $-\tan x \cos x$	c) $\cot x$
4. $\cos(-x)$	d) $\sin(-x)$
5. $\tan^2 x + 1$	e) $\dfrac{\sec x}{\csc x}$
6. $\sec^2 x - 1$	f) $\dfrac{1}{\sec x}$
7. 1	g) $\dfrac{1}{\cos^2 x}$

For each of the following, find the values of the remaining four trigonometric functions.

8. $\sin s = \dfrac{2}{5}$, $\cos s = \dfrac{-\sqrt{21}}{5}$.

9. $\cos \theta = \dfrac{2}{3}$, $\sin s = \dfrac{\sqrt{5}}{3}$.

10. $\tan x = 5$, $\cos x = \dfrac{-5}{\sqrt{26}}$.

11. $\cot t = \dfrac{12}{5}$, $\sin t = \dfrac{5}{13}$.

12. $\csc \theta = \dfrac{5}{4}$, $\sec \theta = \dfrac{5}{3}$.

Verify the following trigonometric identities.

13. $1 - \sec \theta \cos \theta = \tan \theta \cot \theta - 1$
14. $\csc^4\theta = \cot^4\theta + 2 \cot^2\theta + 1$
15. $\dfrac{\sin^2\theta}{\cos^2\theta} = \sec^2\theta - 1$
16. $\cot \theta \sin \theta = \cos \theta$
17. $\sec^2\theta - \tan^2\theta = \cos^2\theta + \sin^2\theta$
18. $\sin^2x - 1 = -\cos^2x$
19. $\dfrac{\sin^2x}{\cos x} = \sec x - \cos x$
20. $(1 + \tan^2x)\cos^2x = 1$
21. $\cot s + \tan s = \sec s \csc s$
22. $\dfrac{\cos \alpha}{\sec \alpha} + \dfrac{\sin \alpha}{\csc \alpha} = \sec^2\alpha - \tan^2\alpha$

7.2 The Addition Identities

It is often helpful to express a trigonometric function of the sum of two values in terms of the trigonometric functions for those values. For example, we wish to express $\cos(s - t)$ in terms of trigonometric functions of s and t. To do this consider points $A(0, 1)$, $B(x_1, y_1)$, $C(x_2, y_2)$, and $D(x_3, y_3)$, on the unit circle of Figure 7.2. Let $\overset{\frown}{AD} = s$ and $\overset{\frown}{AB} = t$, where $0 < t < s < 2\pi$, and $\overset{\frown}{AD}$ represents the arc AD. Let $\overset{\frown}{AC} = s - t$. Since $\overset{\frown}{BD}$ also equals $s - t$, $\overset{\frown}{AC}$ has the same length as $\overset{\frown}{BD}$ and thus, chords AC and BD are of equal length. By the distance formula,

$$\sqrt{(x_2 - 1)^2 + (y_2 - 0)^2} = \sqrt{(x_3 - x_1)^2 + (y_3 - y_1)^2}.$$

Squaring both sides, we have

$$x_2{}^2 - 2x_2 + 1 + y_2{}^2 = x_3{}^2 - 2x_3x_1 + x_1{}^2 + y_3{}^2 - 2y_3y_1 + y_1{}^2. \qquad (2)$$

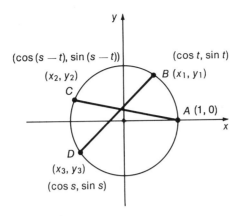

Figure 7.2

The points B, C, and D lie on the unit circle, so that

$$x_1^2 + y_1^2 = 1$$
$$x_2^2 + y_2^2 = 1$$
$$x_3^2 + y_3^2 = 1.$$

Substituting from these equations into equation (2) gives

$$2 - 2x_2 = 2 - 2x_3x_1 - 2y_3y_1$$
$$x_2 = x_3x_1 + y_3y_1.$$

Now, substituting the appropriate trigonometric functions for x_1, x_2, x_3, y_1, and y_3, the expression becomes

▶ $$\cos(s - t) = \cos s \cos t + \sin s \sin t,$$

which is the desired result. (We assumed $0 < t < s < 2\pi$. However, by considering the other cases, it can be shown that the result is valid for all real numbers s and t.)

To find a similar expression for $\cos(s + t)$, we write $s + t = s - (-t)$ and use the identity for $\cos(s - t)$.

$$\cos[s - (-t)] = \cos s \cos(-t) + \sin s \sin(-t)$$
$$= \cos s \cos t + \sin s(-\sin t)$$
$$= \cos s \cos t - \sin s \sin t$$

Hence,

▶ $$\cos(s + t) = \cos s \cos t - \sin s \sin t.$$

The formulas we have just derived are useful in determining certain values of the cosine function.

Example 4 Find (a) $\cos 15°$; (b) $\cos \dfrac{5}{12} \pi$.

(a) $\cos 15° = \cos(45° - 30°)$

$\qquad\qquad = \cos 45° \cos 30° + \sin 45° \sin 30°$

$\qquad\qquad = \dfrac{\sqrt{2}}{2} \cdot \dfrac{\sqrt{3}}{2} + \dfrac{\sqrt{2}}{2} \cdot \dfrac{1}{2}$

$\qquad\qquad = \dfrac{\sqrt{6} + \sqrt{2}}{4}$

(b) $\cos \dfrac{5}{12} \pi = \cos\left(\dfrac{\pi}{6} + \dfrac{\pi}{4}\right)$

$\qquad\qquad = \cos \dfrac{\pi}{6} \cdot \cos \dfrac{\pi}{4} - \sin \dfrac{\pi}{6} \sin \dfrac{\pi}{4}$

$\qquad\qquad = \dfrac{\sqrt{3}}{2} \cdot \dfrac{\sqrt{2}}{2} - \dfrac{1}{2} \cdot \dfrac{\sqrt{2}}{2}$

$\qquad\qquad = \dfrac{\sqrt{6} - \sqrt{2}}{4}$

Now, let us substitute $\dfrac{\pi}{2}$ for s in the formula for $\cos(s - t)$.

$$\cos\left(\dfrac{\pi}{2} - t\right) = \cos \dfrac{\pi}{2} \cos t + \sin \dfrac{\pi}{2} \sin t$$

$$= 0 \cdot \cos t + 1 \cdot \sin t$$

$$\cos\left(\dfrac{\pi}{2} - t\right) = \sin t$$

Further, if we substitute $\dfrac{\pi}{2} - t$ for t in the expression for $\cos\left(\dfrac{\pi}{2} - t\right)$, we have

$$\cos\left(\dfrac{\pi}{2} - t\right) = \sin t$$

$$\cos\left[\dfrac{\pi}{2} - \left(\dfrac{\pi}{2} - t\right)\right] = \sin\left(\dfrac{\pi}{2} - t\right)$$

$$\cos t = \sin\left(\dfrac{\pi}{2} - t\right).$$

From the two relationships just developed, we can write

$$\tan\left(\dfrac{\pi}{2} - t\right) = \dfrac{\sin\left(\dfrac{\pi}{2} - t\right)}{\cos\left(\dfrac{\pi}{2} - t\right)}$$

$$= \dfrac{\cos t}{\sin t}$$

$$= \cot t.$$

These three identities

$$\cos\left(\frac{\pi}{2} - t\right) = \sin t,$$

$$\sin\left(\frac{\pi}{2} - t\right) = \cos t,$$

$$\tan\left(\frac{\pi}{2} - t\right) = \cot t,$$

pair the sine and cosine functions and the tangent and cotangent functions. Functions forming such pairs are called **cofunctions.** A similar relationship holds for secant and cosecant, so that they form a third pair of cofunctions. Note that when t represents an angle, the cofunctions relate complementary angles.

Formulas for $\sin(s + t)$ and $\sin(s - t)$ can also be developed in a manner similar to that used above. From the cofunction relationship, we have

$$\sin(s + t) = \cos\left[\frac{\pi}{2} - (s + t)\right]$$

$$= \cos\left[\left(\frac{\pi}{2} - s\right) - t\right]$$

$$= \cos\left(\frac{\pi}{2} - s\right)\cos t + \sin\left(\frac{\pi}{2} - s\right)\sin t$$

$$\sin(s + t) = \sin s \cos t + \cos s \sin t.$$

(In the last step we substituted from the cofunction relationships.)

We now write $\sin(s - t)$ as $\sin(s + (-t))$ and use the identity for $\sin(s + t)$ to get

$$\sin(s - t) = \sin s \cos t - \cos s \sin t.$$

Using the identities for $\sin(s + t)$, $\cos(s + t)$, $\sin(s - t)$, and $\cos(s - t)$ together with the definition of the tangent, the following formulas can be derived.

$$\tan(s + t) = \frac{\tan s + \tan t}{1 - \tan s \tan t}$$

$$\tan(s - t) = \frac{\tan s - \tan t}{1 + \tan s \tan t}$$

Similar formulas can be found for the remaining trigonometric functions, but they are seldom used, so we shall not give them here.

Example 5 Find (a) sin 75°; (b) tan 105°.

(a) sin 75° = cos(90° − 75°)

   = cos 15°

   $$= \frac{\sqrt{6} + \sqrt{2}}{4},$$

as shown in Example 4. We could also find sin 75° by using the fact that sin 75° = sin(30° + 45°).

(b) tan 105° = tan(60° + 45°)

   $$= \frac{\tan 60° + \tan 45°}{1 - \tan 60° \tan 45°}$$

   $$= \frac{\sqrt{3} + 1}{1 - \sqrt{3} \cdot 1}$$

   $$= \frac{3 + 2\sqrt{3} + 1}{1 - 3}$$

   $$= -\sqrt{3} - 2.$$

7.2 Exercises

Use the addition identities to find the following without using the tables.

1. sin − 15°

2. $\cos \dfrac{7}{12} \pi$

3. tan −105°

4. sin 285°

5. $\tan \dfrac{5}{12} \pi$

6. cos 14° cos 31° − sin 14° sin 31°

7. $\dfrac{\tan 80° + \tan 55°}{1 - \tan 80° \tan 55°}$

8. sin 40° cos 50° + cos 40° sin 50°

9. cos 100° cos(−25°) − sin 100° sin(−25°)

10. $\dfrac{\tan 80° - \tan(-55°)}{1 + \tan 80° \tan(-55°)}$ (Hint: tan(−x) = − tan x.)

Write as a function of θ.

11. cos(180° + θ)
12. cot(90° − θ)
13. sin(θ − 30°)

14. cos(270° − θ)
15. tan(θ − 45°)
16. sin(180° + θ)

Verify the following identities.

17. $\sin\left(\dfrac{\pi}{2} + x\right) = \cos x.$

18. $\tan\left(\dfrac{\pi}{2} + x\right) = -\cot x$

19. $\cot\left(x - \dfrac{\pi}{4}\right) = \dfrac{1 + \tan x}{\tan x - 1}$

20. $1 + \cos 2x - \cos^2 x = \cos^2 x$

21. $\cos(x + y) + \cos(x - y) = 2 \cos x \cos y$

22. $\sin 2x = 2 \sin x \cos x$ (Hint: $2x = x + x$.)

23. $\cos 2x = \cos^2 x - \sin^2 x$

24. $\sin(x + y) + \sin(x - y) = \sin 2x$ (Hint: see Exercise 22.)

7.3 Multiple Value Identities

Some special cases of the addition identities are used often enough to be expressed as separate identities. These are the identities that result from the addition identities when $s = t$ so that $s + t = 2s$. These identities, which are called the **double-value identities,** are derived in this section.

In the formula for $\cos(s + t)$, we let $t = s$ to derive an expression for $\cos 2s$.

$$\cos 2s = \cos(s + s)$$
$$= \cos s \cos s - \sin s \sin s$$

▶ $\cos 2s = \cos^2 s - \sin^2 s$

By substitution from either $\cos^2 s = 1 - \sin^2 s$ or $\sin^2 s = 1 - \cos^2 s$, two useful alternate forms are obtained.

▶ $\cos 2s = 1 - 2 \sin^2 s$

▶ $\cos 2s = 2 \cos^2 s - 1$

From the addition identity for sine, we get the following result.

$$\sin 2s = \sin(s + s)$$
$$= \sin s \cos s + \cos s \sin s$$

▶ $\sin 2s = 2 \sin s \cos s$

In a similar manner, $\tan 2s$ results from the formula for $\tan(s + t)$.

$$\tan 2s = \tan(s + s)$$

$$= \frac{\tan s + \tan s}{1 - \tan s \tan s}$$

▶ $$\tan 2s = \frac{2 \tan s}{1 - \tan^2 s}$$

These formulas, together with the addition formulas, allow us to express the trigonometric functions for any multiple values of s in terms of functions of s.

Example 6   Express $\sin 3s$ in terms of $\sin s$.

$$\sin 3s = \sin(2s + s)$$
$$= \sin 2s \cos s + \cos 2s \sin s$$
$$= (2 \sin s \cos s) \cos s + (\cos^2 s - \sin^2 s) \sin s$$
$$= 2 \sin s \cos^2 s + \cos^2 s \sin s - \sin^3 s$$
$$= 2 \sin s(1 - \sin^2 s) + (1 - \sin^2 s) \sin s - \sin^3 s$$
$$= 2 \sin s - 2 \sin^3 s + \sin s - \sin^3 s - \sin^3 s$$
$$= 3 \sin s - 4 \sin^3 s$$

The **half-value identities,** listed below are another group of identities which are quite useful.

▶ $$\cos \frac{s}{2} = \pm\sqrt{\frac{1 + \cos s}{2}}$$

▶ $$\sin \frac{s}{2} = \pm\sqrt{\frac{1 - \cos s}{2}}$$

▶ $$\tan \frac{s}{2} = \pm\sqrt{\frac{1 - \cos s}{1 + \cos s}}$$

In these formulas the plus or minus sign is selected according to the quadrant in which $\frac{s}{2}$ terminates. For example, if s represents an angle of $324°$, $\frac{s}{2} = 162°$ and lies in the second quadrant, so that $\cos \frac{s}{2}$ and $\tan \frac{s}{2}$ would be negative, while $\sin \frac{s}{2}$ would be positive.

The above formulas can be derived from the alternate forms of the double-value identity for cosine as follows.

$$\cos 2u = 1 - 2 \sin^2 u$$

$$2 \sin^2 u = 1 - \cos 2u$$

$$\sin u = \pm\sqrt{\frac{1 - \cos 2u}{2}}$$

Now, we let $2u = s$, so that $u = \frac{s}{2}$ and substitute into the last expression.

$$\sin \frac{s}{2} = \pm\sqrt{\frac{1 - \cos s}{2}}$$

Similarly, by choosing another form of identity for $\cos 2u$ and replacing $2u$ by s,

$$\cos 2u = 2\cos^2 u - 1$$

$$2\cos^2 u = 1 + \cos 2u$$

$$\cos u = \pm\sqrt{\frac{1 + \cos 2u}{2}}$$

$$\cos \frac{s}{2} = \pm\sqrt{\frac{1 + \cos s}{2}} \ .$$

Finally, the tangent formula can be derived from the half-value formulas above.

$$\tan \frac{s}{2} = \frac{\pm\sqrt{\dfrac{1 - \cos s}{2}}}{\pm\sqrt{\dfrac{1 + \cos s}{2}}}$$

$$\tan \frac{s}{2} = \pm\sqrt{\frac{1 - \cos s}{1 + \cos s}}$$

Example 7 Find $\tan 22.5°$.

$$\tan 22.5° = \tan \frac{45°}{2}$$

$$= \sqrt{\frac{1 - \cos 45°}{1 + \cos 45°}}$$

$$= \sqrt{\frac{1 - \dfrac{1}{\sqrt{2}}}{1 + \dfrac{1}{\sqrt{2}}}}$$

$$= \sqrt{\frac{\sqrt{2} - 1}{\sqrt{2} + 1}}$$

$$= \sqrt{3 - 2\sqrt{2}}$$

Since $22.5°$ is in quadrant I, we used the $+$ value of the radical.

Example 8 Given $\cos s = \frac{2}{3}$ and s terminates in quadrant IV, find $\cos \frac{s}{2}$, $\sin \frac{s}{2}$, and $\tan \frac{s}{2}$.

Since s terminates in quadrant IV, we have $\dfrac{3\pi}{2} < s < 2\pi$, so that $\dfrac{3\pi}{4} < \dfrac{s}{2} < \pi$. Thus, $\dfrac{s}{2}$ terminates in quadrant II, and $\cos\dfrac{s}{2}$ and $\tan\dfrac{s}{2}$ are negative, while $\sin\dfrac{s}{2}$ is positive. From the half value identities,

$$\sin\frac{s}{2} = \sqrt{\frac{1 - \frac{2}{3}}{2}} = \sqrt{\frac{1}{6}} = \frac{\sqrt{6}}{6}$$

$$\cos\frac{s}{2} = -\sqrt{\frac{1 + \frac{2}{3}}{2}} = -\sqrt{\frac{5}{6}} = -\frac{\sqrt{30}}{6}$$

$$\tan\frac{s}{2} = \frac{\sqrt{\frac{1}{6}}}{-\sqrt{\frac{5}{6}}} = -\sqrt{\frac{1}{5}} = \frac{-\sqrt{5}}{5}.$$

The identities presented up to this point are summarized below.

$$\tan s = \frac{\sin s}{\cos s} \qquad \cot s = \frac{\cos s}{\sin s}$$

$$\cot s = \frac{1}{\tan s}$$

$$\csc s = \frac{1}{\sin s} \qquad \sec s = \frac{1}{\cos s}$$

$$\sin^2 s + \cos^2 s = 1$$
$$\sin^2 s = 1 - \cos^2 s$$
$$\cos^2 s = 1 - \sin^2 s$$

$$\tan^2 s + 1 = \sec^2 s \qquad 1 + \cot^2 s = \csc^2 s$$

$$\sin(-s) = -\sin s \qquad \cos(-s) = \cos s \qquad \tan(-s) = -\tan s$$

$$\cos(s - t) = \cos s \cos t + \sin s \sin t$$
$$\cos(s + t) = \cos s \cos t - \sin s \sin t$$
$$\sin(s - t) = \sin s \cos t - \cos s \sin t$$
$$\sin(s + t) = \sin s \cos t + \cos s \sin t$$

$$\tan(s - t) = \frac{\tan s - \tan t}{1 + \tan s \tan t}$$

$$\tan(s + t) = \frac{\tan s + \tan t}{1 - \tan s \tan t}$$

$$\sin\left(\frac{\pi}{2} - t\right) = \cos t$$

$$\cos\left(\frac{\pi}{2} - t\right) = \sin t$$

$$\tan\left(\frac{\pi}{2} - t\right) = \cot t$$

$$\cos 2s = \cos^2 s - \sin^2 s$$

$$\cos 2s = 1 - 2\sin^2 s \qquad \cos 2s = 2\cos^2 s - 1$$

$$\sin 2s = 2 \sin s \cos s$$

$$\tan 2s = \frac{2 \tan s}{1 - \tan^2 s}$$

$$\cos \frac{s}{2} = \pm\sqrt{\frac{1 + \cos s}{2}}$$

$$\sin \frac{s}{2} = \pm\sqrt{\frac{1 - \cos s}{2}}$$

$$\tan \frac{s}{2} = \pm\sqrt{\frac{1 - \cos s}{1 + \cos s}}$$

7.3 Exercises

Use the formulas of this section to complete the following.

1. $2 \sin \frac{\pi}{5} \cos \frac{\pi}{5} =$

4. $\sqrt{\dfrac{1 - \cos 18°}{2}} =$

7. $\dfrac{2 \tan \theta}{1 - \tan^2\theta} =$

2. $\cos^2 10° - \sin^2 10° =$

5. $4 \cos^2 x - 2 =$

8. $\sqrt{\dfrac{1 + \cos 40°}{2}} =$

3. $\sqrt{\dfrac{1 - \cos 40°}{2}} =$

6. $-\sqrt{\dfrac{1 - \cos 340°}{1 + \cos 340°}} =$

Use the formulas of this section to find values of the six trigonometric functions for each of the following.

9. $\theta = -22\frac{1}{2}°$ 10. $\theta = 15°$ 11. $\theta = 195°$ 12. $\theta = 67\frac{1}{2}°$

13. θ, given $\cos 2\theta = \dfrac{3}{5}$ and θ terminates in quadrant I.

14. x, given $\cos 2x = \dfrac{-5}{12}$ and x terminates in quadrant II.

15. t, given $\cos 2t = \dfrac{2}{3}$ and t terminates in quadrant II.

16. 2θ, given $\sin \theta = \dfrac{2}{5}$ and $\cos \theta < 0$.

17. $2x$, given $\tan x = 10$ and $\cos x > 0$.

18. 2θ, given $\cos \theta = \dfrac{-12}{13}$ and $\sin \theta > 0$.

Express in terms of functions of x, as in Example 6.

19. $\cos 3x$

21. $\sin 4x$

23. $\tan^2\dfrac{x}{2}$

20. $\tan 3x$

22. $\cos^2 2x$

Verify the following identities.

24. $(\sin \theta + \cos \theta)^2 = \sin 2\theta + 1$

27. $\tan \theta \cos^2\theta = \dfrac{2 \tan \theta \cos^2\theta - \tan \theta}{1 - \tan^2\theta}$

25. $\cos 2s = \cos^4 s - \sin^4 s$

26. $\sec 2x = \dfrac{\sec^2 x + \sec^4 x}{2 + \sec^2 x - \sec^4 x}$

28. $\cot 2\theta = \dfrac{\cot^2\theta - 1}{2 \cot \theta}$

7.4 Sum and Product Identities

The last group of identities we present, though used less frequently than those in the first three sections of this chapter, are important in further work in mathematics. One group can be used to rewrite a product of two functions as a sum; the other to restate a sum of two functions as a product. Some of these identities can also be used to change an expression involving both sine and cosine functions into one with only one of these functions. In Section 7.5, on conditional equations, the need for this kind of change will become clear.

Adding the two addition identities for $\sin(s + t)$ and $\sin(s - t)$ we have

$$\sin(s + t) = \sin s \cos t + \cos s \sin t$$
$$\sin(s - t) = \sin s \cos t - \cos s \sin t$$
$$\overline{\sin(s + t) + \sin(s - t) = 2 \sin s \cos t}$$

or

▶ $$\sin s \cos t = \frac{1}{2}[\sin(s + t) + \sin(s - t)].$$

If we now subtract $\sin(s - t)$ from $\sin(s + t)$, the result is

▶ $$\cos s \sin t = \frac{1}{2}[\sin(s + t) - \sin(s - t)].$$

Proceding in a similar manner with the identities for $\cos(s + t)$ and $\cos(s - t)$, the following identities are established.

▶ $$\cos s \cos t = \frac{1}{2}[\cos(s + t) + \cos(s - t)]$$

▶ $$\sin s \sin t = \frac{1}{2}[\cos(s - t) - \cos(s + t)]$$

Example 9 Express $\cos 2\theta \sin \theta$ as the sum of two functions.

$$\cos 2\theta \sin \theta = \frac{1}{2}(\sin 3\theta - \sin \theta)$$

$$= \frac{1}{2} \sin 3\theta - \frac{1}{2} \sin \theta$$

Example 10 Evaluate $\cos 15° \cos 45°$.

$$\cos 15° \cos 45° = \frac{1}{2}[\cos(15° + 45°) + \cos(15° - 45°)]$$

$$= \frac{1}{2}[\cos 60° + \cos(-30°)]$$

$$= \frac{1}{2}(\cos 60° + \cos 30°)$$

$$= \frac{1}{2}\left(\frac{1}{2} + \frac{\sqrt{3}}{2}\right)$$

$$= \frac{1 + \sqrt{3}}{4}$$

Now, let $s + t = u$ and $s - t = v$. Then $s = \dfrac{u + v}{2}$, while $t = \dfrac{u - v}{2}$. The identity

$$\sin s \cos t = \frac{1}{2}[\sin(s + t) + \sin(s - t)]$$

becomes

$$\sin\left(\frac{u + v}{2}\right) \cos\left(\frac{u - v}{2}\right) = \frac{1}{2}(\sin u + \sin v),$$

or

▶ $$\sin u + \sin v = 2 \sin \frac{u + v}{2} \cos \frac{u - v}{2}.$$

In the same way, the remaining three identities can be rewritten as follows.

▶ $$\sin u - \sin v = 2 \cos \frac{u + v}{2} \sin \frac{u - v}{2}$$

▶ $$\cos u + \cos v = 2 \cos \frac{u + v}{2} \cos \frac{u - v}{2}$$

▶ $$\cos v - \cos u = 2 \sin \frac{u + v}{2} \sin \frac{u - v}{2}$$

Example 11 Write $\sin 2x - \sin 4x$ as a product of two functions.

$$\sin 2x - \sin 4x = 2 \cos \frac{6x}{2} \sin \frac{-2x}{2}$$

$$= 2 \cos 3x \sin(-x)$$

$$= -2 \cos 3x \sin x.$$

Example 12 Verify the identity $\dfrac{\cos 3x - \cos x}{\cos x + \cos 3x} = \tan 2x.$

We have

$$L = \frac{\cos 3x - \cos x}{\cos x + \cos 3x}$$

$$= \frac{2 \sin \dfrac{3x + x}{2} \cos \dfrac{3x - x}{2}}{2 \cos \dfrac{x + 3x}{2} \cos \dfrac{x - 3x}{2}}$$

$$= \frac{\sin 2x \cos x}{\cos 2x \cos(-x)}$$

$$= \frac{\sin 2x}{\cos 2x}$$

$$= \tan 2x = R.$$

7.4 Exercises

Write each of the following as a sum or difference.

1. $\cos 35° \sin 25°$
2. $2 \sin 2x \sin 4x$
3. $3 \cos 5x \sin 3x$
4. $2 \sin 74° \cos 114°$
5. $\sin(-\theta) \sin(-3\theta)$
6. $4 \cos(-32°) \sin 15°$

Write each of the following as a product.

7. $\sin 60° - \sin 30°$
8. $\sin 28° + \sin(-18°)$
9. $\cos 42° + \cos 148°$
10. $\cos 2x - \cos 8x$
11. $\sin 12\theta - \sin 3\theta$
12. $\cos 5x + \cos 10x$

Verify the following identities.

13. $\tan x = \dfrac{\sin 3x - \sin x}{\cos 3x + \cos x}$

14. $\dfrac{\sin 5t + \sin 3t}{\cos 3t - \cos 5t} = \cot t$

15. $\dfrac{\cot 2\theta}{\tan 3\theta} = \dfrac{\cos 5\theta + \cos \theta}{\cos \theta - \cos 5\theta}$

16. $\dfrac{1}{\tan 2x} = \dfrac{\sin 3x - \sin x}{\cos x - \cos 3x}$

17. $\dfrac{\sin^2 5\theta - 2 \sin 5\theta \sin 3\theta + \sin^2 3\theta}{\sin^2 5\theta - \sin^2 3\theta} = \dfrac{\tan \theta}{\tan 4\theta}$

18. $\tan \theta = \dfrac{\sin 3\theta - \sin \theta}{\cos 3\theta + \cos \theta}$

19. Show that the double-value identity for the sine can be considered a special case of the first identity in this section.

7.5 Conditional Equations

Conditional equations which involve trigonometric functions can usually be solved by a combination of algebraic methods and use of the trigonometric identities to change the form of the equations. For example, the equation

$$3 \sin s = \sqrt{3} + \sin s$$

can be solved for $\sin s$ by algebraic methods as follows.

$$2 \sin s = \sqrt{3}$$

$$\sin s = \frac{\sqrt{3}}{2}$$

An infinite number of values of s satisfy the last expression. For $0 \leq s \leq 2\pi$, the solution set is $\left\{\dfrac{\pi}{3}, \dfrac{2\pi}{3}\right\}$.

Example 13 Solve $\tan^2 x + \tan x - 2 = 0$ for $0 \leq x \leq 2\pi$.

If we let $y = \tan x$, the equation becomes $y^2 + y - 2 = 0$, which is factorable as $(y - 1)(y + 2) = 0$. Substituting $\tan x$ back for y, we have

$$(\tan x - 1)(\tan x + 2) = 0$$
$$\tan x - 1 = 0 \quad \text{or} \quad \tan x + 2 = 0$$
$$\tan x = 1 \qquad\qquad \tan x = -2.$$

From the first equation, we get $x = \dfrac{\pi}{4}$ or $\dfrac{5\pi}{4}$; from the second, using Table 3, $x = 2.0333$ approximately or 5.1749. The solution set is

$$\left\{\dfrac{\pi}{4}, \dfrac{5\pi}{4}, \ 2.0333, \ 5.1749\right\}.$$

(In degrees the solutions are $45°$, $225°$, and approximately $116°30'$, and $323°30'$).

When an equation involves more than one trigonometric function, it is often helpful to use a suitable identity to rewrite the equation in terms of just one trigonometric function as in the following example.

Example 14 Solve $\sin x + \cos x = 0$ for $0 \leq x \leq 2\pi$.

We wish to substitute for either $\sin x$ or $\cos x$ an expression in terms of the other function. Let us use the identity $\sin^2 x = 1 - \cos^2 x$, which can be written as $\sin x = \pm\sqrt{1 - \cos^2 x}$, to replace $\sin x$ in the equation.

$$\sin x + \cos x = 0$$
$$\pm\sqrt{1 - \cos^2 x} + \cos x = 0$$
$$\pm\sqrt{1 - \cos^2 x} = \cos x$$
$$1 - \cos^2 x = \cos^2 x$$
$$1 = 2\cos^2 x$$
$$\cos x = \pm\dfrac{1}{\sqrt{2}} = \pm\dfrac{\sqrt{2}}{2}$$

The solutions for the last equation are

$$\dfrac{\pi}{4}, \dfrac{3\pi}{4}, \dfrac{5\pi}{4}, \text{ and } \dfrac{7\pi}{4}.$$

However, recall that whenever both sides of an equation are squared, extraneous roots may be introduced. Therefore, it is necessary to check these four candidates for the solution in the given equation $\sin x + \cos x = 0$. This can be written $\sin x = -\cos x$ from which we see that $\frac{\pi}{4}$ and $\frac{5\pi}{4}$ are extraneous solutions since the sine and cosine functions differ in sign only in the second and fourth quadrants. Thus the solution set is $\left\{\frac{3\pi}{4}, \frac{7\pi}{4}\right\}$.

Example 15   Solve $4 \sin x \cos x = \sqrt{3}$ for $0° \le x \le 360°$.
The identity $2 \sin x \cos x = \sin 2x$ is useful here.

$$4 \sin x \cos x = \sqrt{3}$$
$$2(2 \sin x \cos x) = \sqrt{3}$$
$$2 \sin 2x = \sqrt{3}$$
$$\sin 2x = \frac{\sqrt{3}}{2}$$

$$2x = 60°, 120°, 420°, \text{ or } 480°$$
$$x = 30°, 60°, 210°, \text{ or } 240°$$

In this case, the domain $0° \le x \le 360°$ implies $0° \le 2x \le 720°$ which allows four solutions for x.

7.5 Exercises

Solve the following for $0 \le x \le 2\pi$.

1. $\cos^2 x + 2 \cos x + 1 = 0$
2. $\sin 3x = 3\sqrt{2} - \sin 3x$
3. $\sin x - \sin 2x = 0$
4. $3 \tan x + 5 = 2$
5. $\sin^2 x + 4 \sin x = 5$
6. $\tan^2 x - 1 = 0$

7. $\sqrt{2} \cos^2 x - \sqrt{2} \sin^2 x = 1$
8. $\dfrac{2 \tan x}{1 - \tan^2 x} = 1$
9. $\sec^2 x + 1 = 0$
10. $\cos^2 x + \dfrac{\sqrt{3}}{2} \cos x - \dfrac{\sqrt{3}}{2} = \cos x$

Solve for $0° \le \theta \le 360°$

11. $\sec^4 2\theta = 4$
12. $\tan \theta + 6 \cot \theta = 5$
13. $\csc \theta = 2 \sin \theta + 1$
14. $\tan \theta - \cot \theta = 0$
15. $\sec^2 \theta = 2 \tan \theta + 4$
16. $\cos^2 \theta = \sin^2 \theta + 1$
17. $\cos 2\theta - \cos \theta = 0$

18. $\dfrac{\sin 2\theta \cos 2\theta}{\sqrt{3}} = \dfrac{1}{2}$
19. $\sin \dfrac{\theta}{2} = \cos \dfrac{\theta}{2}$
20. $\tan \dfrac{\theta}{2} + \cos \theta = 1$

Chapter Eight

Numerical Trigonometry

Trigonometry was developed because of the needs to survey land and to navigate over long distances. As a result of these needs, many efficient methods of **solving triangles,** or finding the values of all sides and all angles, have been developed. We shall discuss a few of these methods in this chapter.

8.1 Right Triangle Trigonometry

We have defined all trigonometric functions in terms of an angle in standard position with vertex at the center of a unit circle. Figure 8.1 shows such a unit circle, with a right triangle (triangle containing a 90° angle) located in such a way that one side lies on the positive x-axis, and one vertex is at the origin.

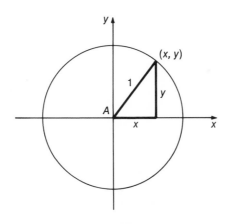

Figure 8.1

Note that the hypotenuse (longest side) of this triangle is 1, the radius of the unit circle. By the definitions given earlier, we have

$$\sin A = y = \text{length of side opposite } A$$

$$\cos A = x = \text{length of side adjacent to } A$$

$$\tan A = \frac{y}{x} = \frac{\text{length of side opposite } A}{\text{length of side adjacent to } A} \cdot$$

What about a right triangle whose hypotenuse is some length other than 1? Figure 8.2 shows a right triangle $AB'C'$, which has a hypotenuse of length c.

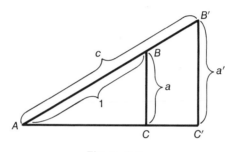

Figure 8.2

Note that right triangle ABC has hypotenuse of length 1. By the definition given above (using triangle ABC), we have

$$\sin A = a = \text{length of side opposite } A.$$

Since triangle ABC is similar to triangle $AB'C'$, we know from geometry that corresponding sides must be in proportion. Hence

$$\frac{a}{1} = \frac{a'}{c}.$$

We know $\sin A = a = a/1$. Hence, by substitution, we can write

$$\sin A = \frac{a'}{c} = \frac{\text{length of side opposite } A}{\text{length of hypotenuse}} \cdot$$

This last quotient is often abbreviated

$$\sin A = \frac{\text{side opposite}}{\text{hypotenuse}} \cdot$$

In a similar way, we can prove all the results of the next theorem.

THEOREM 8.1   If ABC is a right triangle with right angle at C, then

$$\sin A = \frac{\text{side opposite}}{\text{hypotenuse}} \qquad\qquad \cos A = \frac{\text{side adjacent}}{\text{hypotenuse}}$$

$$\tan A = \frac{\text{side opposite}}{\text{side adjacent}}.$$

Just as before, we can find the other three functions by taking reciprocals: $\cot A = 1/\tan A$, $\sec A = 1/\cos A$, and $\csc A = 1/\sin A$.

Example 1 The right triangle shown in Figure 8.3 has sides of lengths 8, 15, and 17, as indicated. Find $\sin A$, $\cos A$, and $\tan A$. Also, find the degree measures of angles A and B.

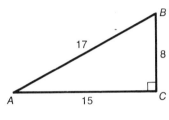

Figure 8.3

Using the results of the theorem above, we have

$$\sin A = \frac{\text{side opposite}}{\text{hypotenuse}} = \frac{8}{17}$$

$$\cos A = \frac{\text{side adjacent}}{\text{hypotenuse}} = \frac{15}{17}$$

$$\tan A = \frac{\text{side opposite}}{\text{side adjacent}} = \frac{8}{15}.$$

We can use one of these three quotients, together with the values in Table 3 to find a degree measure for A. For example, we know

$$\tan A = \frac{8}{15} = .5333.$$

From Table 3 we find $A \approx 28°$. Thus, $B = 90° - A \approx 90° - 28° = 62°$.

Usually, we shall use a to represent the length of the side opposite angle A, b to represent the length of the side opposite angle B, and so on, with C used in right triangles to represent the right angle. Although trigonometric function tables give decimal approximations, we shall use $=$ instead of $\approx$.

Example 2 Solve right triangle ABC, where $A = 34°$, and $c = 12$. See Figure 8.4.

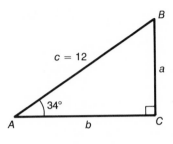

Figure 8.4

To solve the triangle, we must find the measures of the remaining sides and angles. We have $\sin A = a/c$, where $A = 34°$, $c = 12$, and a is unknown. Thus we have

$$\sin A = \frac{a}{c}$$

$$\sin 34° = \frac{a}{12}$$

or $\qquad\qquad\qquad a = 12 \sin 34.$

From Table 3, $\sin 34° = .5592$, so that

$$a = 12(.5592)$$
$$= 6.7.$$

We can use the Pythagorean theorem to find b. Since we know

$$a^2 + b^2 = c^2,$$

and $a = 6.7$, while $c = 12$, we have

$$(6.7)^2 + b^2 = 12^2$$
$$b^2 = 144 - 44.9$$
$$b^2 = 99.1.$$

We can now use logs to find b:

$$\log b = \log \sqrt{99.1} = \frac{1}{2} \log 99.1 = \frac{1}{2}(1.9961) = .9980.$$

Hence, $b = 9.95$.

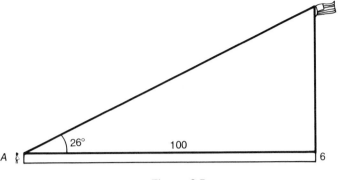

Figure 8.5

Figure 8.5 represents an observer at a point A, looking up to the top of a flagpole. The angle of view to the top, always measured from the horizontal, is called the **angle of elevation.** The angle of view of an observer looking down, again measured from the horizontal, is called the **angle of depression.** Suppose now that the observer in Figure 8.5 is 100 feet from the base of the flagpole, and that the angle of elevation is 26°. If the observer's eye level is 6 feet above the ground, let us find the height of the flagpole.

We know the length of the side adjacent to the observer, and we want to find the length of the side opposite the observer. Hence, we should use the tangent. Here we have

$$\tan A = \frac{\text{side opposite}}{\text{side adjacent}}$$

$$\tan 26 = \frac{x}{100}$$

$$x = 100 \tan 26.$$

Since $\tan 26 = .4877$, we have

$$x = 100(.4877)$$
$$\approx 48.8 \text{ feet.}$$

The eye level of the observer is 6 feet above the ground. Hence the height of the flagpole is about $48.8 + 6 = 54.8$ feet.

Example 3 Harry finds that the angle of elevation of the top of a water tower is 36°. If he moves 150 feet back from where he made the first reading, he finds that the angle of elevation is now 22°, as shown in Figure 8.6. Assume all angles are measured from ground level, and find the height of the tower.

Here we have two unknown quantities: x, the distance from the center of the base of the tower to the point where the first observation was made, and

h, the height of the tower. Since we know nothing about the length of the hypotenuse of either triangle ABC or triangle ADC, we use a function which doesn't involve hypotenuses: the tangent. We have

in triangle ABC, $\tan 36 = \dfrac{h}{x}$, or $h = x \tan 36$;

in triangle ADC, $\tan 22 = \dfrac{h}{150 + x}$, or $h = (150 + x) \tan 22$.

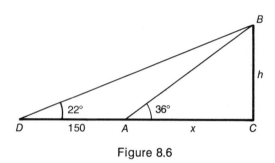

Figure 8.6

Since we have two expressions describing h, these last two quantities must be equal, or

$$x \tan 36 = (150 + x) \tan 22$$
$$x \tan 36 = 150 \tan 22 + x \tan 22$$
$$x \tan 36 - x \tan 22 = 150 \tan 22$$
$$x(\tan 36 - \tan 22) = 150 \tan 22$$
$$x = \frac{150 \tan 22}{\tan 36 - \tan 22}.$$

From Table 3 we find $\tan 36 = .7265$ and $\tan 22 = .4040$. Hence, we can write

$$x = \frac{150(.4040)}{.7265 - .4040}$$
$$= \frac{60.60}{.3225}.$$

We can use logarithms to simplify this calculation:

$$\log 60.60 = 11.7825 - 10$$
$$\log .3225 = \underline{\;\;9.5085 - 10}$$
$$\log x = \;\;2.2740.$$

From this, we see $x = 188$ feet. We know $h = x \tan 36$. Hence, we have

$$h = 188(.7265)$$
$$h = 137 \text{ feet.}$$

In calculations of this type it may be convenient to use Table 4 which gives the values of the logarithms of trigonometric functions directly. For example, from Table 4 we have

$$\log \sin 22° \ 30' = 9.5828 - 10$$
$$\log \cos 76° \ 10' = 9.3786 - 10$$
$$\log \tan 56° \ 30' = .1792.$$

8.1 Exercises

Solve each of the following right triangles.

1. $A = 28°, \ a = 15$
2. $B = 47° \ 30', \ a = 28.6$
3. $B = 58°, \ a = 18$
4. $A = 39° \ 10', \ a = 47.2$
5. $a = 12, \ b = 30$
6. $a = 9, \ b = 15$
7. $a = 18.6, \ A = 42° \ 50'$
8. $b = 36.2, \ B = 68° \ 30'$
9. $A = 19° \ 50', \ b = 473$
10. $A = 49° \ 10', \ a = 382$

11. A 30 foot ladder is leaning against a wall. Find the distance the ladder goes up the wall if it makes an angle of 42° 30' with the ground.

12. Suppose the angle of elevation of the sun is 28° 10'. Find the length of a shadow cast by a man six feet tall.

13. A swimming pool is 50 feet long, and 4 feet deep at one end. If it is 8 feet deep at the other end, find the total distance along the bottom.

14. The shadow of a vertical tower is 59 feet long when the angle of elevation of the sun is 36° 20'. Find the height of the tower.

15. Find the angle of elevation of the sun if a 50 foot flagpole casts a 73 foot long shadow.

16. From the top of a certain building, the angle of depression of a point on the ground is 34° 40'. How far is the point on the ground from the top of the building if the building is 368 feet tall? (Hint: find a hypotenuse)

17. An airplane is flying at an altitude of 10,000 feet. The angle of depression of a tree is 13° 40'. How far horizontally must the plane fly to be directly over the tree?

18. From the top of a small building, the angle of elevation to the top of a nearby taller building is 46° 30', while the angle of depression to the bottom is 14° 10'. If the smaller building is 80 feet high, find the height of the taller building.

19. From a certain point, the angle of elevation to the top of a pyramid is 35° 30'. From a point 135 feet further back, the angle of elevation of the top of the pyramid is 21° 10'. Find the height of the pyramid.

20. A lighthouse keeper is watching a boat approach directly to the lighthouse. When he first begins watching the boat, the angle of depression of the boat is 15° 50'. Just as the boat turns away from the lighthouse, the angle of depression is 35° 40'. If the height of the lighthouse is 68 feet, find the distance traveled by the boat as it approaches the lighthouse.

In navigation, **bearing** *is often defined as the degree measure of the angle between due north and the line representing the direction of travel. This angle is measured in a clockwise direction. The figure shows several such bearings.*

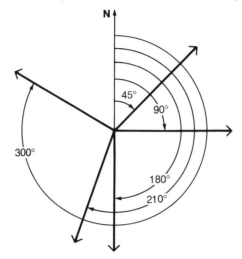

21. A ship leaves port and sails on a bearing of 28°. Another ship leaves the same port at the same time and sails on a bearing of 118°. If the first ship sails at 20 miles per hour, and the second sails at 24 miles per hour, find the distance between the two ships after five hours.

22. Radio direction finders are set up at points A and B, which are 2 miles apart. From A it is found that the bearing of the signal from a certain radio transmitter is 36° 20′, while from B the bearing of the same signal is 306° 20′. Find the distance of the transmitter from B.

8.2 The Law of Sines

The work of the previous section applied only to right triangles. In the next few sections we shall generalize this work to include all triangles. Perhaps the simplest formula valid for any general triangle is given by the law of sines. To obtain this result, we consider any triangle. An example is shown in Figure 8.7.

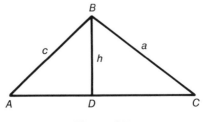

Figure 8.7

First, we construct the perpendicular from B to side AC. Let h represent the length of this perpendicular. As shown in the figure, c is the hypotenuse of right triangle ADB, while a is the hypotenuse of right triangle BDC. By the results of the last section, we have

$$\sin A = \frac{h}{c} \qquad\qquad \sin C = \frac{h}{a}$$

(from triangle ADB) $\qquad$ (from triangle BDC).

These last equalities can be written as

$$h = c \sin A \qquad\qquad h = a \sin C.$$

Since $h = c \sin A$ and $h = a \sin C$, we have

$$a \sin C = c \sin A,$$

or, upon dividing both sides by $\sin A \sin C$,

$$\frac{a}{\sin A} = \frac{c}{\sin C}.$$

In the same way, we can show

$$\frac{a}{\sin A} = \frac{b}{\sin B}.$$

Thus we have proved the next theorem.

THEOREM 8.2 THE LAW OF SINES

In any triangle ABC, with sides a, b, and c, we have

$$\frac{a}{\sin A} = \frac{b}{\sin B} = \frac{c}{\sin C}.$$

Example 4 Solve triangle ABC, with $A = 32°$, $B = 81°$, and $a = 12$.

Since we know A, B, and a, we use the part of the law of sines that involves these variables.

$$\frac{a}{\sin A} = \frac{b}{\sin B}$$

Substituting the known values gives

$$\frac{12}{\sin 32} = \frac{b}{\sin 81}$$

$$b = \frac{12 \sin 81}{\sin 32}.$$

This calculation will be simpler if we use logarithms. Using both Tables 2 and 4, we have

$$\log 12 = 1.0792$$
$$\log \sin 81 = \underline{9.9946 - 10}$$
$$11.0738 - 10$$
$$\log \sin 32 = \underline{9.7242 - 10}$$
$$\log b = 1.3496.$$

Note that we added log 12 and log sin 81, and then subtracted log sin 32. From the result of this calculation, we find that

$$b = 22.$$

Using the fact that the sum of the angles of any triangle is 180°, we find C:

$$C = 180 - (A + B)$$
$$= 180 - (32 + 81)$$
$$= 180 - 113$$
$$= 67°.$$

We can now use the law of sines to find c (why can't we use the Pythagorean theorem?). We have

$$\frac{a}{\sin A} = \frac{c}{\sin C}$$

$$\frac{12}{\sin 32} = \frac{c}{\sin 67}$$

$$c = \frac{12 \sin 67}{\sin 32}.$$

By using logarithms as above, verify that $c = 21$.

Example 5 Mabel wishes to measure the distance across the Big Muddy river (see Figure 8.8). She finds that $C = 112°$, $A = 31°$, and $b = 350$ feet. She wants to find a, the required distance.

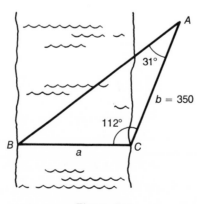

Figure 8.8

Before we can use the law of sines to find a, we must first find angle B. We have

$$B = 180 - (A + C)$$
$$= 37°.$$

Hence,

$$\frac{a}{\sin A} = \frac{b}{\sin B}$$

$$\frac{a}{\sin 31} = \frac{350}{\sin 37}$$

$$a = \frac{350 \sin 31}{\sin 37}$$

$$a = 300 \text{ feet.}$$

Example 6 Solve triangle ABC, if $A = 55°$, $a = 8$, and $b = 25$.

Let us first look for B. We have

$$\frac{a}{\sin A} = \frac{b}{\sin B}$$

$$\frac{8}{\sin 55} = \frac{25}{\sin B}$$

$$\sin B = \frac{25 \sin 55}{8}.$$

By logarithms, we can write

$$\begin{aligned}
\log 25 &= 1.3979 \\
\log \sin 55 &= \underline{9.9134 - 10} \\
& 11.3113 - 10 \\
\log 8 &= \underline{.9031} \\
\log \sin B &= 10.4082 - 10.
\end{aligned}$$

The fact that $\log \sin B = 10.4082 - 10$, or simply $.4082$, implies that $\sin B$ is greater than 1 ($\log 1 = 0$), which we know is impossible. Hence, triangle ABC cannot exist. This is perhaps made plausible in Figure 8.9.

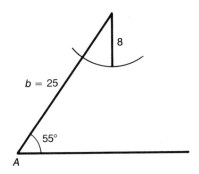

Figure 8.9

When any two angles and the length of a side of a triangle are given, the law of sines can be applied directly. However, if one angle and two sides are given, the triangle may not even exist, as in Example 6, or there may be more than one such triangle. This last situation is discussed in the next section. Furthermore, if two sides and the angle between the two sides are given, the law of sines is of no help. For example, if $a = 10$, $b = 15$, and $C = 82°$, we have

$$\frac{10}{\sin A} = \frac{15}{\sin B} = \frac{c}{\sin 82}$$

which contains too many variables for us to solve. Such problems require the law of cosines, which is discussed in a later section.

8.2 Exercises

Solve each of the following triangles.

1. $A = 35°$, $B = 46°$, $c = 14$
2. $B = 57°$, $C = 29°$, $a = 32$
3. $A = 46° \ 10'$, $B = 32° \ 40'$, $b = 87.3$
4. $A = 59° \ 30'$, $B = 48° \ 20'$, $b = 32.9$
5. $C = 110° \ 30'$, $A = 25° \ 40'$, $c = 36.9$
6. $B = 124° \ 10'$, $C = 18° \ 40'$, $c = 94.6$
7. To find the distance AB across a river, a distance $BC = 350$ feet is laid off on one side of the river. It is found that $B = 112° \ 10'$ and $C = 15° \ 30'$. Find AB.
8. To determine the distance RS across a deep canyon, Joanna lays off a distance $TR = 582$ yards. She then finds that $T = 32° \ 40'$, and $R = 102° \ 20'$. Find RS.
9. Radio direction finders are placed at points A and B, which are 3.46 miles apart on an east-west line. From A the bearing (see Exercise 21, Section 8.1) of a certain radio transmitter is $47° \ 30'$, while from B the bearing is $302° \ 30'$. Find the distance of the transmitter from point A.
10. Captain Crunch sails 300 miles on a bearing of $32° \ 40'$. He then changes course and sails 200 miles on a bearing of $110° \ 10'$. How far is he then from his starting point?
11. Betsy flies on a bearing of $205° \ 20'$ for 400 miles, and then changes course to a bearing of $172° \ 30'$, and flies for an additional 350 miles. How far is she then from her starting point?
12. A ship is sailing due north. The captain notices that the bearing of a lighthouse 12 miles distant is $38° \ 40'$. Later on, the captain notices that the bearing of the lighthouse has become $135° \ 40'$. How far did the ship travel between the two observations of the lighthouse?

13. Find the distance between the ship and lighthouse on the second observation in Exercise 12.

14. A hill slopes at an angle of 12° 20' with the horizontal. From the base of the hill, the angle of inclination of a 450 foot tower at the top of the hill is 35° 50'. Find the distance from the base of the hill to the base of the tower.

15. Harry notices that the bearing of a tree on the opposite side of a river is 115° 30'. Joan, who is on the same side of the river as Harry, but 450 yards away, notices that the bearing of the same tree is 45° 30'. If the river is flowing north and south between parallel banks, and if both Harry and Joan are on the bank, find the distance across the river.

8.3   The Ambiguous Case of the Law of Sines

As mentioned in the previous section, we must be very careful when working with a triangle in which two sides and an angle not included between the two sides are given. Suppose, for example, we know an acute angle A, side a, and side b. To draw a figure illustrating this situation, we might draw angle A, with side b along the terminal side of the angle. From the end of the side of length b, we can try to draw a side of length a. The possible outcomes are shown in Figure 8.10, page 192. Because of these various possibilities, we call this situation (two sides and a nonincluded angle) the **ambiguous case** of the law of sines. If angle A is obtuse, there are two possible cases, as shown in Figure 8.11, page 193.

It is possible to obtain formulas that show which of the various cases exist in a particular problem, but such work is unnecessary since all needed information can be obtained from the law of sines. As we saw in Example 6 of the previous section, if we get log sin $B > 0$ (equivalently, sin $B > 1$), we have no triangle at all. The case where we get two different triangles is illustrated in the next example.

Example 7   Solve triangle ABC if $A = 55°$, $a = 22$, and $b = 25$.

Using the law of sines we have

$$\frac{a}{\sin A} = \frac{b}{\sin B}$$

$$\frac{22}{\sin 55} = \frac{25}{\sin B}$$

$$\sin B = \frac{25 \sin 55}{22}.$$

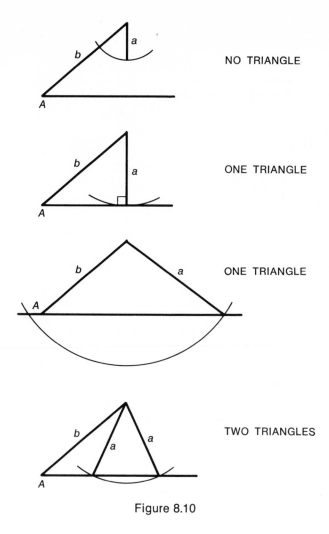

NO TRIANGLE

ONE TRIANGLE

ONE TRIANGLE

TWO TRIANGLES

Figure 8.10

Using Tables 2 and 4 we have

$$
\begin{aligned}
\log 25 &= 1.3979 \\
\log \sin 55 &= \underline{9.9134 - 10} \\
&\ \ 11.3113 - 10 \\
\log 22 &= \underline{1.3424} \\
\log \sin B &= 9.9689 - 10.
\end{aligned}
$$

There are two values of B for which $\log \sin B = 9.9689 - 10$. From Table 4 we see that one value is $B = 68°\ 30'$, and since $\sin(180 - B) = \sin B$, the other possible value of $B = 180 - 68°\ 30' = 111°\ 30'$. To keep track of these two values, let $B_1 = 111°\ 30'$, and $B_2 = 68°\ 30'$. Our two different triangles are shown in Figure 8.12.

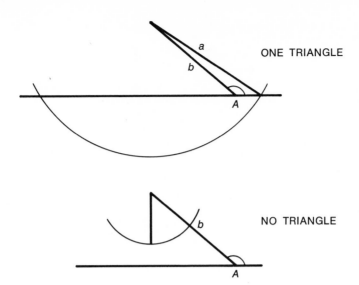

ONE TRIANGLE

NO TRIANGLE

Figure 8.11

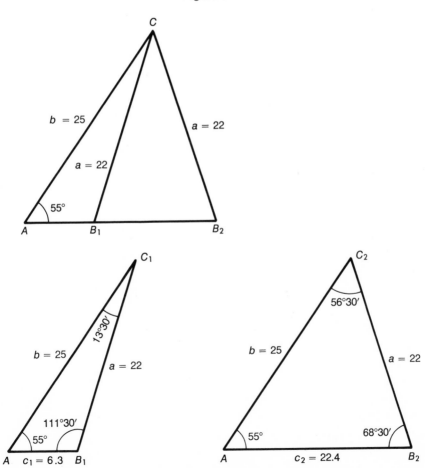

Figure 8.12

Let us first solve triangle AB_1C_1. We have

$$C_1 = 180 - (A + B_1)$$
$$= 13° \; 30'.$$

(If this answer had been negative, we would have the situation shown in Figure 8.10 above and there would be only one triangle.) Since

$$\frac{a}{\sin A} = \frac{c_1}{\sin C_1},$$

we have

$$\frac{22}{\sin 55} = \frac{c_1}{\sin 13° \; 30'}$$

$$c_1 = \frac{22 \sin 13° \; 30'}{\sin 55}$$

$$c_1 = 6.3.$$

To solve triangle AB_2C_2, we first find C_2.

$$C_2 = 180 - (A + B_2)$$
$$= 56° \; 30'.$$

Using the law of sines, we have

$$\frac{22}{\sin 55} = \frac{c_2}{\sin 56° \; 30'}$$

$$c_2 = \frac{22 \sin 56° \; 30'}{\sin 55}$$

$$c_2 = 22.4.$$

8.3 Exercises

Solve each of the following triangles.

1. $A = 42° \; 30'$, $a = 15.6$, $b = 8.1$
2. $C = 52° \; 20'$, $a = 32.5$, $c = 58.9$
3. $B = 68° \; 30'$, $b = 78.3$, $c = 145$
4. $C = 72° \; 10'$, $c = 258$, $b = 386$
5. $A = 112° \; 20'$, $b = 3.5$, $a = 5.8$
6. $C = 125° \; 30'$, $b = 56$, $c = 112$
7. $A = 38° \; 40'$, $a = 9.7$, $b = 11.8$
8. $C = 29° \; 50'$, $a = 8.6$, $c = 5.3$
9. $B = 32° \; 40'$, $a = 75.3$, $b = 51.8$
10. $C = 22° \; 50'$, $b = 158$, $c = 132$

11. A surveyor reports the following data about a piece of property: "The property is triangular in shape, with dimensions as shown in the figure." Use the law of sines to see if such a piece of property could exist.

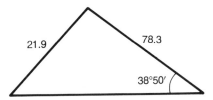

12. The surveyor tries again: "A second triangular piece of property has dimensions as shown." This time it turns out the surveyor did not consider every possible case. Use the law of sines to show why.

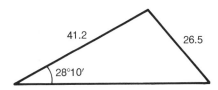

8.4 The Law of Cosines

The law of sines involves only products and quotients, and thus logarithms can be easily used to find unknown values. However, the law of sines cannot always be used. For example, if two sides and the angle between the two sides are given, the law of sines cannot be used. Also, if all three of the sides of a triangle are given, the law of sines again cannot be used easily to find the unknown angles. For both these situations we need another result, the law of cosines.

To find this law, let ABC be any arbitrary triangle. Choose a coordinate system so that the origin is at vertex B, and side BC is on the positive x-axis, as shown in Figure 8.13, page 196. Note that vertex C has coordinates $(a, 0)$. Select a point A' along AB so that $A'B = 1$. Since point A' is on the unit circle, it will have coordinates $(\cos B, \sin B)$. By similar triangles, point A has coordinates $(c \cos B, c \sin B)$.

We know that AC has length b. Using the distance formula we have

$$b = \sqrt{(c \cos B - a)^2 + (c \sin B)^2}.$$

Squaring both sides and simplifying gives

$$b^2 = (c \cos B - a)^2 + (c \sin B)^2$$
$$b^2 = c^2 \cos^2 B - 2ac \cos B + a^2 + c^2 \sin^2 B.$$

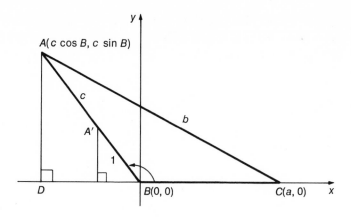

Figure 8.13

Using the distributive property and a trigonometric identity,

$$c^2 \cos^2 B + c^2 \sin^2 B = c^2(\cos^2 B + \sin^2 B) = c^2(1) = c^2.$$

Thus, we have

$$b^2 = a^2 + c^2 - 2ac \cos B.$$

This result is one form of the law of cosines. In the derivation above we could just as easily have placed A or C at the origin. This would have given the same result, but one which is different in appearance. These various forms of the law of cosines are summarized in the following theorem.

THEOREM 8.3 THE LAW OF COSINES

In any triangle, the square of any side equals the sum of the squares of the other two sides, minus twice the product of the other two sides times the cosine of the angle included between the two sides. In the triangle ABC, we have

$$a^2 = b^2 + c^2 - 2bc \cos A$$
$$b^2 = a^2 + c^2 - 2ac \cos B$$
$$c^2 = a^2 + b^2 - 2ab \cos C.$$

Example 8 Solve triangle ABC if $A = 42°$, $b = 12$ and $c = 15$.
We can find a by using the law of cosines.

$$a^2 = b^2 + c^2 - 2bc \cos A$$
$$a^2 = 12^2 + 15^2 - 2(12)(15) \cos 42$$

Logarithms are of little help with this calculation (which explains why it is wise to see if the law of sines can be applied first). Performing these calculations we have

$$a^2 = 144 + 225 - 360(.7431)$$
$$= 369 - 267.5$$
$$= 101.5.$$

We can, if we want, use logs to find a. Verify that

$$a = 10.1.$$

We now know a, b, c, and A. We can use the law of sines to find B. We have

$$\frac{10.1}{\sin 42} = \frac{12}{\sin B},$$

from which

$$\sin B = \frac{12 \sin 42}{10.1}$$

and

$$B = 52° \ 40'.$$

We can now find C.

$$C = 180 - (A + B)$$
$$= 85° \ 20'$$

Example 9 Solve triangle ABC if $a = 9.3$, $b = 14.2$, and $c = 21.1$.

Again we must use the law of cosines. It doesn't matter which angle we find first, so let's look for C. We have

$$c^2 = a^2 + b^2 - 2ab \cos C,$$

or

$$\cos C = \frac{a^2 + b^2 - c^2}{2ab}.$$

Substituting the given values yields

$$\cos C = \frac{9.3^2 + 14.2^2 - 21.1^2}{2(9.3)(14.2)}.$$

After much calculation, we obtain

$$\cos C = -.5947.$$

From this last result, we have

$$C = 126° \ 30'.$$

We can now use the law of sines to find B. Verify that

$$\sin B = \frac{14.2 \sin 126° \ 30'}{21.1},$$

and finally that

$$B = 32° \ 40'.$$

Since $A = 180 - (B + C)$, we have $A = 20° \ 50'$.

8.4 Exercises

Solve each of the following triangles.

1. $A = 28° \ 50'$, $b = 6$, $c = 5$
2. $B = 35° \ 10'$, $a = 3$, $c = 5$
3. $C = 45° \ 40'$, $b = 8$, $a = 7$
4. $A = 67° \ 20'$, $b = 30$, $c = 40$
5. $A = 80° \ 40'$. $b = 100$, $c = 80$
6. $C = 72° \ 40'$, $a = 300$, $b = 250$
7. $B = 125° \ 30'$, $a = 8$, $c = 6$
8. $C = 142° \ 50'$, $a = 3$, $b = 4$
9. $A = 112° \ 50'$, $b = 6$, $c = 12$
10. $B = 168° \ 10'$, $a = 3$, $c = 3$
11. $a = 2$, $b = 3$, $c = 4$
12. $a = 3$, $b = 4$, $c = 6$
13. $a = 9$, $b = 11$, $c = 15$
14. $a = 30$, $b = 45$, $c = 60$
15. Points A and B are on opposite sides of a lake. From a third point C the angles between the lines of sight to A and B is $46° \ 20'$. If AC is 350 yards and BC is 286 yards, find AB.
16. How much rope will be required to reach from the top of the tower in Exercise 14 of Section 8.2 to the bottom of the hill?
17. The sides of a parallelogram are 4 and 6. One angle of the figure is $58°$ while another is $122°$. Find the lengths of the diagonals of the figure.
18. Airports A and B are 450 miles apart, on an east-west line. Tom flies from A on a bearing of $46° \ 20'$ to airport C. From C he flies on a bearing of $128° \ 40'$ to B. How far is C from A?
19. Two ships leave a harbor together, traveling on courses that have an angle of $135° \ 40'$ between them. If they both travel 400 miles, how far apart are they?
20. Pearl flew from A to B, a distance of 350 miles. Then she flew from B to C, a distance of 400 miles. Finally she flew back to A, a distance of 300 miles. If A and B are on an east-west line, find the bearing from A to C. (Assume C is north of the line through A and B.)

21. The aircraft carrier Tallahassee is traveling at sea on a steady course with a bearing of 30° at 32 miles per hour. Patrol planes on the carrier carry enough fuel for a flight of 2.6 hours, and travel at a cruising speed of 520 miles per hour. One of the pilots takes off, bearing 338°, and then turns and heads in a straight line, so as to be able to catch the carrier, landing on the deck at the exact second his fuel runs out. If the pilot left at 2 pm, at what time did he turn to head for the carrier, and on what bearing did he fly in order to reach the carrier?*

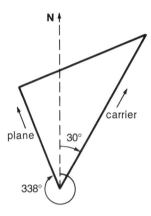

22. We know, by the law of cosines, that in any right triangle with sides a, b, and c, and angles A, B, and C, that

$$c^2 = b^2 + a^2 - 2ab \cos C.$$

Suppose ABC is a right triangle, with $C = 90°$. Then we have

$$c^2 = a^2 + b^2 - 2ab \cos 90$$
$$= a^2 + b^2 - 2ab(0)$$
$$c^2 = a^2 + b^2,$$

which is the statement of the Pythagorean theorem. Is this a legitimate proof of the Pythagorean theorem? (Hint: how did we prove the law of cosines in the first place?)

8.5 Area of a Triangle

The standard formula for the area of a triangle is

$$\text{Area} = \tfrac{1}{2} \times \text{base} \times \text{height}.$$

---

* This problem was supplied by Professor Hal Oxsen, Diablo Valley College, Pleasant Hill, California.

This formula is not always useful, since the height of a triangle is frequently unknown in practice. In this section we shall develop two other formulas that may be used to find the area of a triangle. Figure 8.14 shows a triangle ABC, with base b and height h. Since ABD is a right triangle, we can write

$$\sin A = \frac{h}{c},$$

or
$$h = c \sin A.$$

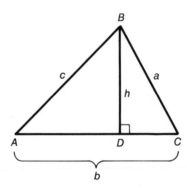

Figure 8.14

Substituting this into the standard formula for the area of a triangle gives

$$\text{Area} = \tfrac{1}{2}bc \sin A.$$

By generalizing this result, we have the next theorem.

THEOREM 8.4   The area of any triangle is given by half the product of the lengths of two sides times the sine of the included angle.

Example 10   Find the area of triangle ABC if $A = 24°$, $b = 27.3$, and $C = 52°$.

Before we can use the result of Theorem 8.4 we must first find the length of either side c or side a. Let us find a, using the law of sines. We have

$$\frac{a}{\sin 24} = \frac{27.3}{\sin 104}.$$

(Where did we get 104°?) This leads to

$$a = \frac{27.3 \sin 24}{\sin 104}$$

$$a = 11.4.$$

Now we can find the area. We have

$$\text{Area} = \tfrac{1}{2}ab \sin C$$
$$= \tfrac{1}{2}(11.4)(27.3)\sin 52$$
$$= 122.6.$$

Given the lengths of all three sides of a triangle we could find the area by first finding one angle, using the law of cosines. Then we could use Theorem 8.4. However, there is a simpler method, called Heron's Area Formula, that we can use in such cases. While the derivation of the formula is left for the exercises, we state the result as our next theorem.

THEOREM 8.5   HERON'S AREA FORMULA
If ABC is a triangle having sides a, b, and c, and if $s = \tfrac{1}{2}(a + b + c)$, then the area of the triangle is given by

$$\textbf{Area} = \sqrt{s(s - a)(s - b)(s - c)}.$$

Example 11   Find the area of the triangle having sides $a = 12.9$, $b = 18.6$, and $c = 25.3$.

To use Heron's Formula, we must first find s. We have

$$s = \tfrac{1}{2}(a + b + c)$$
$$= \tfrac{1}{2}(12.9 + 18.6 + 25.3)$$
$$= \tfrac{1}{2}(56.8.)$$
$$= 28.4.$$

Hence, the area is given by

$$\text{Area} = \sqrt{s(s - a)(s - b)(s - c)}$$
$$= \sqrt{(28.4)(28.4 - 12.9)(28.4 - 18.6)(28.4 - 25.3)}$$
$$= \sqrt{(28.4)(15.5)(9.8)(3.1)}.$$

We can now use logarithms.

$$\log 28.4 = 1.4533$$
$$\log 15.5 = 1.1903$$
$$\log 9.8 = 0.9912$$
$$\log 3.1 = \underline{0.4914}$$
$$4.1262$$
$$\log \text{area} = 2.0631.$$

From this we find the area to be about 116.

8.5 Exercises

Find the area of each of the following triangles.

1. $A = 46° 30'$, $b = 12$, $c = 8$
2. $C = 67° 40'$, $b = 46$, $a = 38$
3. $C = 124° 30'$, $a = 30$, $b = 28$
4. $B = 142° 40'$, $a = 20$, $c = 24$
5. $A = 56° 40'$, $b = 32.6$, $c = 52.8$
6. $A = 34° 50'$, $b = 35.2$, $c = 28.6$
7. $a = 5$, $b = 7$, $c = 10$
8. $a = 2$, $b = 3$, $c = 4$
9. $a = 45$, $b = 56$, $c = 38$
10. $a = 90$, $b = 100$, $c = 110$
11. $A = 24° 30'$, $B = 56° 20'$, $c = 78.4$
12. $B = 48° 30'$, $C = 56° 30'$, $a = 462$
13. Sam wants to paint a triangular region 75 by 68 by 85 feet. If one can of paint covers 125 square feet, how many cans (to the next higher can) will Sam need?
14. Give a reason for each step in the following proof of Heron's Area Formula. Let M represent the area of any triangle ABC.

$$M = \tfrac{1}{2}bc \sin A \qquad \underline{\hspace{3cm}}$$
$$= \sqrt{\tfrac{1}{4}b^2c^2 \sin^2 A} \qquad \underline{\hspace{3cm}}$$
$$= \sqrt{\tfrac{1}{4}b^2c^2(1 - \cos^2 A)} \qquad \underline{\hspace{3cm}}$$
$$= \sqrt{\tfrac{1}{2}bc(1 + \cos A) \cdot \tfrac{1}{2}bc(1 - \cos A)} \qquad \underline{\hspace{3cm}}$$

$$\tfrac{1}{2}bc(1 + \cos A) = \tfrac{1}{2}bc\left(1 + \frac{b^2 + c^2 - a^2}{2bc}\right) \qquad \underline{\hspace{3cm}}$$
$$= \frac{2bc + b^2 + c^2 - a^2}{4} \qquad \underline{\hspace{3cm}}$$
$$= \frac{(b + c)^2 - a^2}{4} \qquad \underline{\hspace{3cm}}$$
$$= \left(\frac{(b + c) + a}{2}\right)\left(\frac{(b + c) - a}{2}\right) \qquad \underline{\hspace{3cm}}$$

$$\tfrac{1}{2}bc(1 - \cos A) = \tfrac{1}{2}bc\left(1 - \frac{b^2 + c^2 - a^2}{2bc}\right) \qquad \underline{\hspace{3cm}}$$
$$= \frac{2bc - b^2 - c^2 + a^2}{4} \qquad \underline{\hspace{3cm}}$$
$$= \left(\frac{a - b + c}{2}\right)\left(\frac{a + b - c}{2}\right) \qquad \underline{\hspace{3cm}}$$

$$M = \sqrt{\left(\frac{b + c + a}{2}\right)\left(\frac{b + c - a}{2}\right)\left(\frac{a - b + c}{2}\right)\left(\frac{a + b - c}{2}\right)} \qquad \underline{\hspace{3cm}}$$

Let $s = \frac{1}{2}(a + b + c)$. Then $s - a = \dfrac{b + c - a}{2}$.

_____

$s - b = \dfrac{a - b + c}{2}$.

_____

$s - c = \dfrac{a + b - c}{2}$.

_____

$M = \sqrt{s(s - a)(s - b)(s - c)}$.

_____

Chapter Nine

Matrix Theory

A **matrix** (plural: matrices) is a rectangular ordered array of numbers which are called the **elements** of the matrix. The study of matrices as a mathematical system has been of interest to mathematicians for some time. Recently however, the use of matrices, along with other mathematical methods, has assumed greater importance in the fields of business administration and the social sciences. This, together with the advent of the computer, has led to a new emphasis on the use of matrices as a mathematical tool, particularly for working with systems of equations.

9.1 Basic Properties of Matrices

We shall write a matrix by enclosing the array of numbers in parentheses. Matrices are classified by their **dimension** or **order,** that is, the number of rows and columns they contain. Thus, an $m \times n$ (read "m by n") matrix has m rows and n columns, and is said to be of order $m \times n$. For example,

$$\begin{pmatrix} 2 & 3 & 5 \\ 7 & 1 & 2 \end{pmatrix}, \quad \begin{pmatrix} -1 \\ 1 \\ 2 \\ 0 \end{pmatrix}, \quad \text{and} \quad \begin{pmatrix} 8 & -1 & 0 \\ 2 & 1 & 6 \\ 0 & 5 & -3 \end{pmatrix}$$

are, respectively, a 2×3 matrix, a 4×1 matrix, and a 3×3 matrix. (Note that the number of rows is named first, and the number of columns second.) A **square matrix** has the same number of rows as columns.

A matrix with only one row or column, such as

$$
\begin{pmatrix} 2 \\ -1 \\ 4 \end{pmatrix}, \quad (4 \quad 5), \quad \text{or} \quad (2 \quad 8 \quad -7 \quad 6 \quad -3)
$$

is called a **row** or **column matrix.**

It is customary to use capital letters to denote matrices and subscript notation to represent the elements of a matrix as follows.

$$
A = \begin{pmatrix}
a_{11} & a_{12} & a_{13} & \cdots & a_{1n} \\
a_{21} & a_{22} & a_{23} & \cdots & a_{2n} \\
a_{31} & a_{32} & a_{33} & \cdots & a_{3n} \\
\cdot & \cdot & \cdot & & \cdot \\
\cdot & \cdot & \cdot & & \cdot \\
\cdot & \cdot & \cdot & & \cdot \\
a_{m1} & a_{m2} & a_{m3} & & a_{mn}
\end{pmatrix}
$$

Using this notation, the first row, first column element is denoted a_{11}, the second row, third column element is denoted a_{23}, and in general, the ith row, jth column element is denoted a_{ij}.

Two matrices are **equal** if they are of the same order and if their corresponding elements are equal. By this definition, the matrices

$$
\begin{pmatrix} 2 & 1 \\ 3 & -5 \end{pmatrix} \quad \text{and} \quad \begin{pmatrix} 1 & 2 \\ -5 & 3 \end{pmatrix}
$$

are not equal (even though they contain the same elements and are of the same order), since the corresponding elements differ. From the definition of equality, we see that if

$$
\begin{pmatrix} 2 & 1 \\ m & n \end{pmatrix} = \begin{pmatrix} x & y \\ -1 & 0 \end{pmatrix},
$$

then $x = 2$, $y = 1$, $m = -1$, and $n = 0$.

To add two matrices, we use the following definition. The **sum** of two $m \times n$ matrices X and Y is the $m \times n$ matrix $X + Y$, in which each element of $X + Y$ is the sum of the corresponding elements of X and Y. For example, if

$$
A = \begin{pmatrix} 1 & 2 & 3 \\ 0 & -1 & 5 \end{pmatrix} \quad \text{and} \quad B = \begin{pmatrix} -2 & 3 & 0 \\ 1 & -7 & 2 \end{pmatrix},
$$

then

$$
A + B = \begin{pmatrix} -1 & 5 & 3 \\ 1 & -8 & 7 \end{pmatrix}.
$$

Note that by this definition only matrices of the same order can be added.

Each real number has an additive inverse. The additive inverse of a matrix can be defined using the definition of the additive inverse of a real number. The **additive inverse,** or **negative,** of a matrix X is the matrix $-X$, in which each element of $-X$ is the additive inverse of the corresponding element of X. By this definition, the additive inverses of matrices A and B above are

$$-A = \begin{pmatrix} -1 & -2 & -3 \\ 0 & 1 & -5 \end{pmatrix} \quad \text{and} \quad -B = \begin{pmatrix} 2 & -3 & 0 \\ -1 & 7 & -2 \end{pmatrix}.$$

Note that

$$A + (-A) = \begin{pmatrix} 0 & 0 & 0 \\ 0 & 0 & 0 \end{pmatrix}.$$

We can now define subtraction for matrices in a manner analogous to the way we defined subtraction for real numbers. That is, for matrices X and Y, we shall define

▶
$$X - Y = X + (-Y).$$

Using A and B as defined above we have

$$A - B = A + (-B) = \begin{pmatrix} 1 & 2 & 3 \\ 0 & -1 & 5 \end{pmatrix} + \begin{pmatrix} 2 & -3 & 0 \\ -1 & 7 & -2 \end{pmatrix}$$

$$= \begin{pmatrix} 3 & -1 & 3 \\ -1 & 6 & 3 \end{pmatrix}.$$

9.1 Exercises

Mark each of the following true or false. If false, tell why.

1. $\begin{pmatrix} 1 & 3 \\ 5 & 7 \end{pmatrix} = \begin{pmatrix} 1 & 5 \\ 3 & 7 \end{pmatrix}$

2. $\begin{pmatrix} 1 \\ 2 \\ 3 \end{pmatrix} = (1 \quad 2 \quad 3)$

3. $\begin{pmatrix} x \\ y \end{pmatrix} = \begin{pmatrix} 3 \\ 5 \end{pmatrix}$ if $x = 3$ and $y = 5$.

4. $\begin{pmatrix} 3 & 5 & 2 & 8 \\ 1 & -1 & 4 & 0 \end{pmatrix}$ is a 4×2 matrix.

5. $\begin{pmatrix} 1 & 9 & -4 \\ 3 & 7 & 2 \\ -1 & 1 & 0 \end{pmatrix}$ is a square matrix.

6. $\begin{pmatrix} 2 & 4 & -1 \\ 3 & 7 & 2 \\ 0 & 0 & 0 \end{pmatrix} = \begin{pmatrix} 2 & 4 & -1 \\ 3 & 7 & 2 \end{pmatrix}$

Perform the indicated operations.

7. $\begin{pmatrix} 1 & 2 & 5 & -1 \\ 3 & 0 & 2 & -4 \end{pmatrix} + \begin{pmatrix} 8 & 10 & -5 & 3 \\ -2 & -1 & 0 & 0 \end{pmatrix}$

8. $\begin{pmatrix} 1 & 5 \\ 2 & -3 \\ 3 & 7 \end{pmatrix} + \begin{pmatrix} 2 & 3 \\ 8 & 5 \\ -1 & 9 \end{pmatrix}$

9. $\begin{pmatrix} 1 & 5 & 7 \\ 2 & 2 & 3 \end{pmatrix} + \begin{pmatrix} 4 & 8 \\ 1 & -1 \end{pmatrix}$

10. $\begin{pmatrix} 2 & 4 \\ -8 & 1 \end{pmatrix} + \begin{pmatrix} 9 & -3 \\ 8 & 5 \end{pmatrix}$

11. $\begin{pmatrix} 1 & 3 & -2 \\ 4 & 7 & 1 \end{pmatrix} - \begin{pmatrix} 3 & 6 & -5 \\ 0 & 4 & 2 \end{pmatrix}$

12. $\begin{pmatrix} 2 & 8 & 12 & 0 \\ 7 & 4 & -1 & 5 \\ 1 & 2 & 0 & 10 \end{pmatrix} - \begin{pmatrix} 1 & 3 & 6 & 9 \\ 2 & -3 & -3 & 4 \\ 8 & 0 & -2 & 17 \end{pmatrix}$

Show that each of the following statements is true, where

$$P = \begin{pmatrix} m & n \\ p & q \end{pmatrix}, \ T = \begin{pmatrix} r & s \\ t & u \end{pmatrix}, \ X = \begin{pmatrix} x & y \\ z & w \end{pmatrix}, \ \text{and } 0 = \begin{pmatrix} 0 & 0 \\ 0 & 0 \end{pmatrix}.$$

13. $X + T = T + X$
14. $X + (T + P) = (X + T) + P$
15. $X + (-X) = 0$
16. $P + 0 = P$
17. Would the statements above still be true if P, T, X and 0 were not square matrices? Why?

9.2 Multiplication of Matrices

Finding the product of two matrices is more complicated than finding the sum. However, this complicated kind of multiplication is justified by applications which are helpful in solving practical problems. The **product** XY of an $m \times n$ matrix X and an $n \times k$ matrix Y is defined as follows. Multiply each element of the first row of X by the corresponding element in the first column of Y. The sum of these n products is the first row, first column element of XY. Similarly, the sum of the products found by multiplying the elements of the first row of X times the corresponding elements of the second column of Y

gives the first row, second column element of XY, and so on. In general, to find the ith row, jth column element of XY, multiply each element in the ith row of X by the corresponding element in the jth column of Y. The sum of these products will give the desired ijth element of XY.

To illustrate this definition, let us find the product AB of the matrices

$$A = \begin{pmatrix} 1 & -3 \\ 7 & 2 \end{pmatrix} \quad \text{and} \quad B = \begin{pmatrix} 1 & 0 & -1 & 2 \\ 3 & 1 & 4 & -1 \end{pmatrix}.$$

We first note that A is a 2×2 matrix and B is a 2×4 matrix so that the number of elements in each row of A equals the number of elements in each column of B as required. Also note that the product matrix AB will be a 2×4 matrix, as illustrated in Figure 9.1.

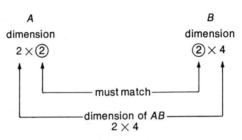

Figure 9.1

Multiplying the first row elements of A times the corresponding first column elements of B and summing up the products, we get the first row, first column element of AB.

$$\begin{pmatrix} [1(1) + (-3)3] & & & \\ & & & \end{pmatrix}$$

Now we multiply the first row elements of A times the corresponding second column elements of B, again summing up the products, to get the first row, second column element of AB.

$$\begin{pmatrix} [1(1) + (-3)3] & [1(0) + (-3)1] & & \\ & & & \end{pmatrix}$$

Continuing in this manner, we complete the product matrix as shown below.

$$\begin{pmatrix} [1(1) + (-3)3] & [1(0) + (-3)1] & [1(-1) + (-3)4] & [1(2) + (-3)(-1)] \\ [7(1) + 2(3)] & [7(0) + 2(1)] & [7(-1) + 2(4)] & [7(2) + 2(-1)] \end{pmatrix}$$

$$= \begin{pmatrix} -8 & -3 & -13 & 5 \\ 13 & 2 & 1 & 12 \end{pmatrix}$$

Can we find the product BA? Since B is 2×4 while A is 2×2, the answer is no.

Example 1 If

$$M = \begin{pmatrix} 1 & 2 & 3 \\ -1 & 0 & -2 \end{pmatrix} \quad \text{and} \quad N = \begin{pmatrix} -3 & -1 \\ 2 & 1 \\ 1 & 2 \end{pmatrix},$$

find MN.

Using the definition from above, we have

$$MN = \begin{pmatrix} [1(-3) + 2(2) + 3(1)] & [1(-1) + 2(1) + 3(2)] \\ [(-1)(-3) + 0(2) + (-2)1] & [(-1)(-1) + 0(1) + (-2)2] \end{pmatrix}$$

$$= \begin{pmatrix} 4 & 7 \\ 1 & -3 \end{pmatrix}.$$

Example 2 A contractor builds three kinds of houses, models A, B, and C, with a choice of two exteriors, colonial or contemporary. Matrix P below shows the number of each kind of house the contractor is planning to build for a new 100 home subdivision. The amounts for each of the main materials he uses depend primarily on the exterior of the house. These amounts are shown in matrix Q below, while matrix R gives the cost for each kind of material. Concrete is measured here in cubic yards, lumber in 1000 board feet, brick in 1000's, and shingles in 100 square feet.

$$\begin{matrix} & \text{colonial} & \text{contemporary} \\ \text{model A} & 0 & 30 \\ \text{model B} & 10 & 20 \\ \text{model C} & 20 & 20 \end{matrix} = P$$

$$\begin{matrix} & \text{concrete} & \text{lumber} & \text{brick} & \text{shingles} \\ \text{colonial} & 10 & 2 & 0 & 2 \\ \text{contemporary} & 50 & 1 & 20 & 2 \end{matrix} = Q$$

$$\begin{matrix} & \text{cost per unit} \\ \text{concrete} & 20 \\ \text{lumber} & 180 \\ \text{brick} & 60 \\ \text{shingles} & 25 \end{matrix} = R$$

(a) What is the total cost for each model house?
(b) How much of each of the four kinds of material must be ordered?
(c) What is the total cost of the material?

(a) To find the cost for each model, we must first find matrix PQ, which will show the total amount of each material needed for each model house.

$$PQ = \begin{pmatrix} 0 & 30 \\ 10 & 20 \\ 20 & 20 \end{pmatrix} \begin{pmatrix} 10 & 2 & 0 & 2 \\ 50 & 1 & 20 & 2 \end{pmatrix} = \begin{pmatrix} 1500 & 30 & 600 & 60 \\ 1100 & 40 & 400 & 60 \\ 1200 & 60 & 400 & 80 \end{pmatrix} \begin{matrix} \text{model } A \\ \text{model } B \\ \text{model } C \end{matrix}$$

If we now multiply PQ times the cost matrix, R, we will get the total cost for each model.

$$\begin{pmatrix} 1500 & 30 & 600 & 60 \\ 1100 & 40 & 400 & 60 \\ 1200 & 60 & 400 & 80 \end{pmatrix} \begin{pmatrix} 20 \\ 180 \\ 60 \\ 25 \end{pmatrix} = \begin{pmatrix} 72{,}900 \\ 54{,}700 \\ 60{,}800 \end{pmatrix} \begin{matrix} \text{model } A \\ \text{model } B \\ \text{model } C \end{matrix}$$

(b) The totals of the columns of matrix PQ will give a matrix whose elements represent the total amounts of each material needed for the subdivision. Let us call this matrix T and write it as a row matrix.

$$T = (3800 \quad 130 \quad 1400 \quad 200).$$

(c) To find the total cost of all the materials, we need the product of the cost matrix, R, and matrix T, the total amounts matrix. To multiply these and get a 1×1 matrix, representing the total cost, we must multiply a 1×4 matrix times a 4×1 matrix. This is why in (b) we chose to write a row matrix. Since the total materials cost is given by TR, we can write

$$TR = (3800 \quad 130 \quad 1400 \quad 200) \begin{pmatrix} 20 \\ 180 \\ 60 \\ 25 \end{pmatrix} = (188{,}400).$$

In work with matrices, a real number is called a **scalar** to distinguish it from a matrix. The product of a scalar k and a matrix X is the matrix kX each of whose elements is k times the corresponding element of X. Thus,

$$-3 \begin{pmatrix} 2 & -5 \\ 1 & 7 \end{pmatrix} = \begin{pmatrix} -6 & 15 \\ -3 & -21 \end{pmatrix}.$$

9.2 Exercises

Answer true or false for each of the following. If false, tell why.

1. The product of a 1×2 matrix and a 2×5 matrix is a 1×5 matrix.
2. The product of a 3×2 matrix and a 4×3 matrix is a 2×4 matrix.
3. If both AB and BA exist, then A and B must both be square matrices.
4. For any matrix A with ith row all zeros and any matrix B for which AB exists, the ith row of AB will contain all zeros.

5. If A and B are $n \times n$ matrices and 0 is the $n \times n$ matrix whose entries are all zeros, and if $AB = 0$, then $A = 0$ or $B = 0$.

Compute the following products.

6. $\begin{pmatrix} 1 & 2 \\ 3 & 4 \end{pmatrix}\begin{pmatrix} -1 & 5 \\ 7 & 0 \end{pmatrix}$

7. $\begin{pmatrix} -1 & 5 \\ 7 & 0 \end{pmatrix}\begin{pmatrix} 1 & 2 \\ 3 & 4 \end{pmatrix}$

8. $\begin{pmatrix} -2 & -3 & 7 \\ 1 & 5 & 6 \end{pmatrix}\begin{pmatrix} 1 \\ 2 \\ 3 \end{pmatrix}$

9. $\begin{pmatrix} 6 \\ 5 \\ 4 \end{pmatrix}(-1 \quad 1 \quad 1)$

10. $\begin{pmatrix} 4 & 3 \\ 1 & 2 \\ 0 & -5 \end{pmatrix}\begin{pmatrix} 2 & -2 \\ 1 & -1 \end{pmatrix}$

11. $(-1 \quad 1 \quad 1)\begin{pmatrix} 6 \\ 5 \\ 4 \end{pmatrix}$

12. $\begin{pmatrix} 2 & -2 \\ 1 & -1 \end{pmatrix}\begin{pmatrix} 4 & 3 \\ 1 & 2 \\ 0 & -5 \end{pmatrix}$

13. $\begin{pmatrix} 1 & 0 & 1 & 5 & 2 \\ 0 & 1 & -3 & 4 & -1 \end{pmatrix}\begin{pmatrix} 1 & 0 & 0 \\ 0 & 1 & 0 \\ 0 & 0 & 1 \end{pmatrix}$

Using matrices $P = \begin{pmatrix} m & n \\ p & q \end{pmatrix}$, $X = \begin{pmatrix} x & y \\ z & w \end{pmatrix}$, *and* $T = \begin{pmatrix} r & s \\ t & u \end{pmatrix}$, *show that the following statements are true.*

14. $(PX)T = P(XT)$
15. $P(X + T) = PX + PT$
16. $P(X - T) = PX - PT$
17. $k(X + T) = kX + kT$ for any real number k
18. $(k + h)P = kP + hP$ for any real numbers k and h

19. The Bread Box, a small neighborhood bakery, sells four main items: sweet rolls, bread, cakes, and pies. The amount of each ingredient required for these items is given by matrix A.

$$
\begin{array}{c}
\\
\text{rolls}\\ \text{(doz)}\\[4pt]
\text{bread}\\ \text{(loaves)}\\[4pt]
\text{cakes}\\[4pt]
\text{pies}\\ \text{(crust)}
\end{array}
\begin{array}{ccccc}
\text{eggs} & \text{flour}^* & \text{sugar}^* & \text{shortening}^* & \text{milk}^* \\
1 & 4 & \frac{1}{4} & \frac{1}{4} & 1 \\[6pt]
0 & 3 & 0 & \frac{1}{4} & 0 \\[6pt]
4 & 3 & 2 & 1 & 1 \\[6pt]
0 & 1 & 0 & \frac{1}{3} & 0
\end{array} = A
$$

The cost (in cents) for each ingredient when purchased in large lots or small lots is given by matrix B.

$$
\begin{array}{c}
\\
\\
\text{eggs}\\
\text{flour}\\
\text{sugar}\\
\text{shortening}\\
\text{milk}
\end{array}
\begin{array}{cc}
\multicolumn{2}{c}{\text{cost}} \\
\text{large lot} & \text{small lot} \\
1 & 1 \\
2 & 3 \\
3 & 4 \\
12 & 15 \\
5 & 6
\end{array} = B
$$

(a) Use matrix multiplication to find a matrix representing the comparative cost per item for the two purchase options.

Suppose a day's orders consist of 20 dozen sweet rolls, 200 loaves of bread, 50 cakes and 60 pies.

(b) Writing the orders as a 1×4 matrix and using matrix multiplication write as a matrix the amount of each ingredient required to fill the day's orders.

(c) Use matrix multiplication to find a matrix representing the costs under the two purchase options to fill the day's orders.

9.3 The Algebra of Matrices

Now that we have defined matrix addition and multiplication, let us see whether or not a mathematical system whose elements are matrices of the same order satisfies the definition of a field, as given in Section 1.4. To begin, let us consider the operation of addition of matrices. The sum of any two matrices of the same order is a matrix of that order by the definition of addition, so that the closure property for addition is satisfied. It is easy to confirm that the commutative and associative properties for addition are also satisfied, since addition of matrices depends only on addition of real numbers (see Exercises 13 and 14 of Section 9.1). We have defined an additive inverse, $-A$, for

* in cups

each matrix A, and to complete the field axioms for addition, we need only to define an identity matrix for addition. For any matrix A, the sum $A + (-A)$ gives a matrix whose elements are all zeros. We shall define this matrix to be a **zero matrix** and denote it 0. Note that there is a zero matrix for each order matrix and that it serves as the identity element for that order, since if 0 and A are of the same order, $0 + A = A + 0 = A$ for any matrix A.

Now let us consider the operation of multiplication for matrices. From the definition of multiplication, it follows that only square matrices can be considered, since the closure axiom does not hold for other kinds of matrices for which multiplication is defined. (Recall that an $m \times n$ matrix must multiply an $n \times k$ matrix with the resulting product an $m \times k$ matrix.) For any square matrices of the same order, the closure axiom will hold, since an $n \times n$ matrix multiplied by an $n \times n$ matrix results in an $n \times n$ matrix.

Does multiplication of matrices satisfy the commutative axiom? Let us consider an example. If

$$A = \begin{pmatrix} 1 & 2 \\ 3 & 4 \end{pmatrix} \quad \text{and} \quad B = \begin{pmatrix} 0 & -1 \\ -2 & 1 \end{pmatrix},$$

we can verify that

$$AB = \begin{pmatrix} -4 & 1 \\ -8 & 1 \end{pmatrix} \quad \text{while} \quad BA = \begin{pmatrix} -3 & -4 \\ 1 & 0 \end{pmatrix}.$$

This one counterexample is enough to show that multiplication of $n \times n$ matrices is noncommutative. The associative property however, does hold true for multiplication of $n \times n$ matrices (see Exercise 14 of Section 9.2).

Is there an identity element for multiplication of $n \times n$ matrices? If so, the product of such an $n \times n$ matrix with any $n \times n$ matrix A should be A itself. That is, if

$$\begin{pmatrix} x_{11} & x_{12} \\ x_{21} & x_{22} \end{pmatrix}$$

is to be the identity matrix, then we must have

$$\begin{pmatrix} x_{11} & x_{12} \\ x_{21} & x_{22} \end{pmatrix}\begin{pmatrix} a_{11} & a_{12} \\ a_{21} & a_{22} \end{pmatrix} = \begin{pmatrix} a_{11} & a_{12} \\ a_{21} & a_{22} \end{pmatrix} \tag{1}$$

for any matrix

$$A = \begin{pmatrix} a_{11} & a_{12} \\ a_{21} & a_{22} \end{pmatrix}.$$

Multiplying the two matrices on the left side of equation (1) and setting the elements of this product matrix equal to the corresponding elements of A, we have the following four restrictions on x_{11}, x_{12}, x_{21}, and x_{22}.

$$x_{11}a_{11} + x_{12}a_{21} = a_{11}$$
$$x_{21}a_{11} + x_{22}a_{21} = a_{21}$$
$$x_{11}a_{12} + x_{12}a_{22} = a_{12}$$
$$x_{21}a_{12} + x_{22}a_{22} = a_{22}$$

These equations are solved when $x_{11} = 1$, $x_{12} = x_{21} = 0$, and $x_{22} = 1$. (It can be shown that these are the only solutions.) Hence, the 2×2 identity matrix is

$$\begin{pmatrix} 1 & 0 \\ 0 & 1 \end{pmatrix}.$$

Generalizing for a system of $n \times n$ matrices, the **identity matrix,** which we shall denote I, is

$$I = \begin{pmatrix} 1 & 0 & \cdots & 0 \\ 0 & 1 & \cdots & 0 \\ \cdot & \cdot & & \cdot \\ \cdot & \cdot & a_{ij} & \cdot \\ \cdot & \cdot & & \cdot \\ 0 & 0 & \cdots & 1 \end{pmatrix}$$

where $a_{ij} = 1$ when $i = j$, and $a_{ij} = 0$ otherwise.

Now that we have defined an identity matrix, we can investigate to see whether or not every $n \times n$ matrix A has a multiplicative inverse, which we shall denote A^{-1}. For 2×2 matrices, if the inverse matrix for

$$A = \begin{pmatrix} a_{11} & a_{12} \\ a_{21} & a_{22} \end{pmatrix}$$

exists, it should satisfy the equation $AA^{-1} = I$, or

$$\begin{pmatrix} a_{11} & a_{12} \\ a_{21} & a_{22} \end{pmatrix} \begin{pmatrix} x_{11} & x_{12} \\ x_{21} & x_{22} \end{pmatrix} = \begin{pmatrix} 1 & 0 \\ 0 & 1 \end{pmatrix},$$

where $\begin{pmatrix} x_{11} & x_{12} \\ x_{21} & x_{22} \end{pmatrix} = A^{-1}.$

Multiplying the two matrices on the left side and setting the elements of the product matrix equal to the corresponding elements of I, we have

$$a_{11}x_{11} + a_{12}x_{21} = 1$$
$$a_{11}x_{12} + a_{12}x_{22} = 0$$
$$a_{21}x_{11} + a_{22}x_{21} = 0$$
$$a_{21}x_{12} + a_{22}x_{22} = 1.$$

Solving these four equations for x_{11}, x_{12}, x_{21}, and x_{22}, we can identify the **multiplicative inverse** A^{-1} as the matrix

$$A^{-1} = \begin{pmatrix} \dfrac{a_{22}}{a_{11}a_{22} - a_{21}a_{12}} & \dfrac{-a_{12}}{a_{11}a_{22} - a_{21}a_{12}} \\ \dfrac{-a_{21}}{a_{11}a_{22} - a_{21}a_{12}} & \dfrac{a_{11}}{a_{11}a_{22} - a_{21}a_{12}} \end{pmatrix}$$

which can be written as follows,

$$\frac{1}{a_{11}a_{22} - a_{21}a_{12}} \begin{pmatrix} a_{22} & -a_{12} \\ -a_{21} & a_{11} \end{pmatrix}.$$

We shall denote $a_{11}a_{22} - a_{21}a_{12}$ as δ (the Greek letter "delta") and write

$$A^{-1} = \frac{1}{\delta} \begin{pmatrix} a_{22} & -a_{12} \\ -a_{21} & a_{11} \end{pmatrix}.$$

Example 3   Let

$$M = \begin{pmatrix} 2 & -1 \\ 4 & -1 \end{pmatrix}.$$

Find M^{-1} and show $MM^{-1} = I$.

Here $a_{11} = 2$, $a_{12} = -1$, $a_{21} = 4$, and $a_{22} = -1$. Thus, by definition, $\delta = (2)(-1) - 4(-1) = -2 + 4 = 2$, so that

$$M^{-1} = \frac{1}{2} \begin{pmatrix} -1 & 1 \\ -4 & 2 \end{pmatrix} = \begin{pmatrix} -\frac{1}{2} & \frac{1}{2} \\ -2 & 1 \end{pmatrix}.$$

Using the definition of matrix multiplication, we have

$$MM^{-1} = \begin{pmatrix} 2 & -1 \\ 4 & -1 \end{pmatrix} \begin{pmatrix} -\frac{1}{2} & \frac{1}{2} \\ -2 & 1 \end{pmatrix} = \begin{pmatrix} 1 & 0 \\ 0 & 1 \end{pmatrix} = I.$$

Note that whenever $\delta = 0$, A^{-1} does not exist, so that not every 2×2 matrix has an inverse. Although the method shown here for finding the inverse can be extended to any $n \times n$ matrix, it rapidly becomes complex. For example, to find the inverse of a 3×3 matrix we would need to solve a system of nine equations.

We know that for all real numbers a, b, and c, $a(b + c) = ab + ac$. Is there a similar distributive property for matrices? That is, does

$$A(B + C) = AB + AC$$

for all square matrices A, B, and C of the same order? Let us consider an example. If

$$A = \begin{pmatrix} 1 & -1 \\ 2 & 1 \end{pmatrix}, \qquad B = \begin{pmatrix} 2 & -1 \\ 0 & 3 \end{pmatrix}, \qquad \text{and} \qquad C = \begin{pmatrix} -1 & 3 \\ 2 & 1 \end{pmatrix},$$

then we have

$$A(B + C) = \begin{pmatrix} 1 & -1 \\ 2 & 1 \end{pmatrix}\left[\begin{pmatrix} 2 & -1 \\ 0 & 3 \end{pmatrix} + \begin{pmatrix} -1 & 3 \\ 2 & 1 \end{pmatrix}\right]$$

$$= \begin{pmatrix} 1 & -1 \\ 2 & 1 \end{pmatrix}\begin{pmatrix} 1 & 2 \\ 2 & 4 \end{pmatrix}$$

$$= \begin{pmatrix} -1 & -2 \\ 4 & 8 \end{pmatrix}.$$

On the other hand,

$$AB + AC = \begin{pmatrix} 1 & -1 \\ 2 & 1 \end{pmatrix}\begin{pmatrix} 2 & -1 \\ 0 & 3 \end{pmatrix} + \begin{pmatrix} 1 & -1 \\ 2 & 1 \end{pmatrix}\begin{pmatrix} -1 & 3 \\ 2 & 1 \end{pmatrix}$$

$$= \begin{pmatrix} 2 & -4 \\ 4 & 1 \end{pmatrix} + \begin{pmatrix} -3 & 2 \\ 0 & 7 \end{pmatrix}$$

$$= \begin{pmatrix} -1 & -2 \\ 4 & 8 \end{pmatrix}.$$

While such an example does not, of course, prove anything, it can be shown that

$$A(B + C) = AB + AC$$

for all square matrices A, B, and C of the same order. This property of matrices is called the left distributive property of matrix multiplication over matrix addition, while

$$(B + C)A = BA + CA$$

is called the right distributive property.

In summary, for any system of square matrices of a given order, the following axioms hold.

Addition	Multiplication
closure	closure
commutative	
associative	associative
identity	identity
inverse	inverse ($\delta \neq 0$)*
distributive	

Since multiplication of $n \times n$ matrices is noncommutative, we see that the system is not a field.

---

* We have defined δ only for 2×2 matrices; an analogous definition can be given for larger square matrices. (See Section 9.4.)

The properties which do hold true for $n \times n$ matrices, however, enable us to solve matrix equations in much the same way that we solve algebraic equations. To solve matrix equations, in addition to the properties above, we need to assume a multiplication axiom of equality for matrices.

Example 4 Given $A = \begin{pmatrix} 2 & 2 \\ -1 & -2 \end{pmatrix}$ and $B = \begin{pmatrix} 2 \\ 3 \end{pmatrix}$, find a matrix X so that $AX = B$.

To solve the matrix equation $AX = B$, we can use the fact that $A^{-1}A = I$ and $IX = X$ as follows.

$AX = B$	given
$A^{-1}(AX) = A^{-1}B$	multiplication axiom of equality for matrices
$(A^{-1}A)X = A^{-1}B$	associative axiom of multiplication for matrices
$IX = A^{-1}B$	multiplication inverse axiom for matrices
$X = A^{-1}B$	multiplication identity axiom for matrices

In using the multiplication axiom of equality for matrices we must be careful to multiply on the left on both sides of the equation, since, unlike multiplication of real numbers, multiplication of matrices is not commutative. Since $X = A^{-1}B$, we must find A^{-1}.

$$A^{-1} = \frac{1}{-4+2}\begin{pmatrix} -2 & -2 \\ 1 & 2 \end{pmatrix} = \begin{pmatrix} 1 & 1 \\ -\frac{1}{2} & -1 \end{pmatrix}$$

Now we can find X.

$$X = A^{-1}B = \begin{pmatrix} 1 & 1 \\ -\frac{1}{2} & -1 \end{pmatrix}\begin{pmatrix} 2 \\ 3 \end{pmatrix} = \begin{pmatrix} 5 \\ -4 \end{pmatrix}$$

9.3 Exercises

Find the inverse, if it exists, of each of the following matrices.

1. $\begin{pmatrix} 4 & 3 \\ 2 & 1 \end{pmatrix}$

2. $\begin{pmatrix} 1 & 1 \\ -1 & -1 \end{pmatrix}$

3. $\begin{pmatrix} 1 & 0 \\ 0 & 2 \end{pmatrix}$

4. $\begin{pmatrix} -1 & -5 \\ 2 & 3 \end{pmatrix}$

5. $\begin{pmatrix} 3 & 4 \\ 9 & 12 \end{pmatrix}$

6. $\begin{pmatrix} 2 & 7 \\ -2 & 7 \end{pmatrix}$

7. $\begin{pmatrix} 1 & 7 \\ 4 & 2 \end{pmatrix}$

Solve the matrix equation $AX = B$ for X, given matrices A and B as follows.

8. $A = \begin{pmatrix} 1 & 3 \\ -2 & 4 \end{pmatrix}$, $B = \begin{pmatrix} 2 \\ 6 \end{pmatrix}$

11. $A = \begin{pmatrix} 10 & -3 \\ 2 & -1 \end{pmatrix}$, $B = \begin{pmatrix} 4 \\ 8 \end{pmatrix}$

9. $A = \begin{pmatrix} 2 & -2 \\ 1 & 0 \end{pmatrix}$, $B = \begin{pmatrix} -4 \\ 3 \end{pmatrix}$

12. $A = \begin{pmatrix} 5 & 7 \\ 8 & 9 \end{pmatrix}$, $B = \begin{pmatrix} -4 \\ 33 \end{pmatrix}$

10. $A = \begin{pmatrix} 2 & -4 \\ 1 & -2 \end{pmatrix}$, $B = \begin{pmatrix} 8 \\ 5 \end{pmatrix}$

Using $X = \begin{pmatrix} x & y \\ z & w \end{pmatrix}$, and $0 = \begin{pmatrix} 0 & 0 \\ 0 & 0 \end{pmatrix}$, show that the following are true.

13. $IX = XI = X$
14. $X \cdot X^{-1} = X^{-1} \cdot X = I$ if X^{-1} exists.
15. $X \cdot 0 = 0 \cdot X = 0$

16. Using the definitions and properties of matrix algebra, prove that for square matrices A and B, if $AB = 0$, and if A^{-1} exists, then $B = 0$. (Assume 0 is a matrix with the appropriate dimension and with all entries zero.)

9.4 Determinants

For the 2×2 matrix

$$X = \begin{pmatrix} x_{11} & x_{12} \\ x_{21} & x_{22} \end{pmatrix},$$

we have defined the number $\delta = x_{11}x_{22} - x_{21}x_{12}$. We call δ the **determinant** o matrix X, and write

$$|X| = \begin{vmatrix} x_{11} & x_{12} \\ x_{21} & x_{22} \end{vmatrix} = x_{11}x_{22} - x_{21}x_{12},$$

or $$\delta(X) = x_{11}x_{22} - x_{21}x_{12}.$$

In this section, we shall investigate methods that can be used to find determinants for any square matrix.

As we shall see, for each square matrix X, with real number elements, there is a unique real number determinant $\delta(X)$. The set of ordered pairs of the form $(X, \delta(X))$ defines a function with domain the set of square matrices X, and range the set of real numbers $\delta(X)$.

The determinant of a 3×3 matrix, which we shall refer to as a 3×3 determinant, can be defined in a manner similar to that given above. Thus, for the matrix

$$A = \begin{pmatrix} a_{11} & a_{12} & a_{13} \\ a_{21} & a_{22} & a_{23} \\ a_{31} & a_{32} & a_{33} \end{pmatrix},$$

we define the determinant as

$$\delta(A) = \begin{vmatrix} a_{11} & a_{12} & a_{13} \\ a_{21} & a_{22} & a_{23} \\ a_{31} & a_{32} & a_{33} \end{vmatrix} = (a_{11}a_{22}a_{33} + a_{12}a_{23}a_{31} + a_{13}a_{21}a_{32}) \\ - (a_{31}a_{22}a_{13} + a_{32}a_{23}a_{11} + a_{33}a_{21}a_{12}).$$

From this definition of the determinant of a 3×3 matrix, we can rearrange terms and factor to get

$$\begin{vmatrix} a_{11} & a_{12} & a_{13} \\ a_{21} & a_{22} & a_{23} \\ a_{31} & a_{32} & a_{33} \end{vmatrix} = a_{11}(a_{22}a_{33} - a_{32}a_{23}) - a_{21}(a_{12}a_{33} - a_{32}a_{13}) \\ + a_{31}(a_{12}a_{23} - a_{22}a_{13}). \quad (2)$$

Note that each of the quantities in parentheses above represents a 2×2 determinant which is the determinant of that part of the 3×3 array remaining when the row and column of the multiplier are eliminated.

$$a_{11}(a_{22}a_{33} - a_{32}a_{23}) \begin{pmatrix} a_{11} & a_{12} & a_{13} \\ a_{21} & a_{22} & a_{23} \\ a_{31} & a_{32} & a_{33} \end{pmatrix}$$

$$a_{21}(a_{12}a_{33} - a_{32}a_{13}) \begin{pmatrix} a_{11} & a_{12} & a_{13} \\ a_{21} & a_{22} & a_{23} \\ a_{31} & a_{32} & a_{33} \end{pmatrix}$$

$$a_{31}(a_{12}a_{23} - a_{22}a_{13}) \begin{pmatrix} a_{11} & a_{12} & a_{13} \\ a_{21} & a_{22} & a_{23} \\ a_{31} & a_{32} & a_{33} \end{pmatrix}$$

Such a 2×2 determinant is called a **minor** of an element in the 3×3 determinant. Thus, the minors of a_{11}, a_{21}, and a_{31} are as shown in the chart below.

element	minor
a_{11}	$\begin{vmatrix} a_{22} & a_{23} \\ a_{32} & a_{33} \end{vmatrix}$
a_{21}	$\begin{vmatrix} a_{12} & a_{13} \\ a_{32} & a_{33} \end{vmatrix}$
a_{31}	$\begin{vmatrix} a_{12} & a_{13} \\ a_{22} & a_{23} \end{vmatrix}$

We see that the 3×3 determinant can be evaluated by multiplying each element in the first column by its minor and combining the products as indicated in (2) above. This is called the **expansion** of the determinant by minors about the first column.

Example 5 Evaluate the determinant

$$\begin{vmatrix} 1 & 3 & -2 \\ -1 & -2 & -3 \\ 1 & 1 & 2 \end{vmatrix}$$

by expanding by minors about the first column.

Using the procedure above, we have

$$\begin{vmatrix} 1 & 3 & -2 \\ -1 & -2 & -3 \\ 1 & 1 & 2 \end{vmatrix} = 1 \begin{vmatrix} -2 & -3 \\ 1 & 2 \end{vmatrix} - (-1) \begin{vmatrix} 3 & -2 \\ 1 & 2 \end{vmatrix} + 1 \begin{vmatrix} 3 & -2 \\ -2 & -3 \end{vmatrix}$$

$$= 1[(-2)(2) - (1)(-3)] + 1[(3)(2) - (1)(-2)]$$
$$+ 1[(3)(-3) - (-2)(-2)]$$
$$= 1(-1) + 1(8) + 1(-13)$$
$$= -1 + 8 - 13$$
$$= -6.$$

To get equation (2) we could have rearranged terms and factored differently by factoring out the three elements of the second or third columns or of any of the three rows of the array. Therefore, expanding by minors about any row or any column results in the same value for the determinant. To determine the correct signs for the terms of other expansions, the following array of signs is helpful.

$$\begin{matrix} + & - & + \\ - & + & - \\ + & - & + \end{matrix}$$

The signs alternate for each row and column, beginning with $+$ in the first row, first column position. Thus, the array of signs is easy to reproduce at will. If the expansion is to be about the second column, for example, the first term would have a minus sign associated with it, the second a plus sign, and the third a minus sign.

Example 6 Evaluate the determinant of Example 5 by expansion by minors about the second column.

Using the methods described above, we have

$$\begin{vmatrix} 1 & 3 & -2 \\ -1 & -2 & -3 \\ 1 & 1 & 2 \end{vmatrix} = -3\begin{vmatrix} -1 & -3 \\ 1 & 2 \end{vmatrix} + (-2)\begin{vmatrix} 1 & -2 \\ 1 & 2 \end{vmatrix} - 1\begin{vmatrix} 1 & -2 \\ -1 & -3 \end{vmatrix}$$

$$= -3(1) - 2(4) - 1(-5)$$
$$= -3 - 8 + 5$$
$$= -6.$$

This method of expansion by minors can be extended to find determinants of any $n \times n$ matrix, as shown in the next example.

Example 7 Evaluate

$$\begin{vmatrix} -1 & -2 & 3 & 2 \\ 0 & 1 & 4 & -2 \\ 3 & -1 & 4 & 0 \\ 2 & 1 & 0 & 3 \end{vmatrix}.$$

Expanding by minors about the fourth row gives

$$-2\begin{vmatrix} -2 & 3 & 2 \\ 1 & 4 & -2 \\ -1 & 4 & 0 \end{vmatrix} + 1\begin{vmatrix} -1 & 3 & 2 \\ 0 & 4 & -2 \\ 3 & 4 & 0 \end{vmatrix} - 0\begin{vmatrix} -1 & -2 & 2 \\ 0 & 1 & -2 \\ 3 & -1 & 0 \end{vmatrix} + 3\begin{vmatrix} -1 & -2 & 3 \\ 0 & 1 & 4 \\ 3 & -1 & 4 \end{vmatrix}$$

$$= -2(6) + 1(-50) - 0 + 3(-41)$$
$$= -185.$$

9.4 Exercises

Evaluate the following determinants.

1. $\begin{vmatrix} 5 & 8 \\ 2 & -4 \end{vmatrix}$

2. $\begin{vmatrix} -3 & 0 \\ 0 & 9 \end{vmatrix}$

3. $\begin{vmatrix} -1 & -2 \\ 5 & 3 \end{vmatrix}$

4. $\begin{vmatrix} 6 & -4 \\ 0 & -1 \end{vmatrix}$

5. $\begin{vmatrix} 9 & 3 \\ -3 & -1 \end{vmatrix}$

6. $\begin{vmatrix} 0 & 2 \\ 1 & 5 \end{vmatrix}$

7. $\begin{vmatrix} 1 & 2 & 0 \\ -1 & 2 & -1 \\ 0 & 1 & 4 \end{vmatrix}$

8. $\begin{vmatrix} 2 & 1 & -1 \\ 4 & 7 & -2 \\ 2 & 4 & 0 \end{vmatrix}$

9. $\begin{vmatrix} 10 & 2 & 1 \\ -1 & 4 & 3 \\ -3 & 8 & 10 \end{vmatrix}$

10. $\begin{vmatrix} 7 & -1 & 1 \\ 1 & -7 & 2 \\ -2 & 1 & 1 \end{vmatrix}$

11. $\begin{vmatrix} 1 & -2 & 3 \\ 0 & 0 & 0 \\ 1 & 10 & -12 \end{vmatrix}$

14. $\begin{vmatrix} 5 & -3 & 2 \\ -5 & 3 & -2 \\ 1 & 0 & 1 \end{vmatrix}$

12. $\begin{vmatrix} 2 & 3 & 0 \\ 1 & 9 & 0 \\ -1 & -2 & 0 \end{vmatrix}$

15. $\begin{vmatrix} 1 & 1 & 0 & 1 \\ 2 & 1 & 0 & 2 \\ 0 & 1 & -1 & 1 \\ 1 & -1 & 1 & 1 \end{vmatrix}$

13. $\begin{vmatrix} 3 & 3 & -1 \\ 2 & 6 & 0 \\ -6 & -6 & 2 \end{vmatrix}$

16. $\begin{vmatrix} 2 & 7 & 0 & -1 \\ 1 & 0 & 1 & 3 \\ 2 & 4 & -1 & -1 \\ -1 & 1 & 0 & 8 \end{vmatrix}$

9.5 Properties of Determinants

To evaluate the determinant of a matrix by expansion by minors often requires a great deal of computation. For a 3×3 matrix, it is necessary to evaluate the determinant of three 2×2 matrices; for a 4×4, twelve determinants must be evaluated, and, in general, for an $n \times n$ matrix, it can be shown that one must evaluate $n!/2$ determinants.* This work is considerably simplified if a determinant has many zero elements. In this section we present a theorem which is useful in transforming matrices so that their determinants are easier to evaluate.

THEOREM 9.1   For any square matrix X and any real number k:

    (a)   If every element in a row (or column) of X is zero, then $\delta(X) = 0$.

    (b)   If corresponding rows and columns of X are interchanged to form the matrix Y, then $\delta(Y) = \delta(X)$.

    (c)   If any two rows (or columns) of X are interchanged to form the matrix Y, then $\delta(Y) = -\delta(X)$.

    (d)   If every element of a row (or column) of X is multiplied by k to form the matrix Y, then $\delta(Y) = k \cdot \delta(X)$.

    (e)   If k times any row (or column) of X is added to another row (or column) of X to form the matrix Y, then $\delta(Y) = \delta(X)$.

    (f)   If two rows (or columns) of X are identical, then $\delta(X) = 0$.

*$n! = n(n - 1)(n - 2) \cdots (2)(1)$.

We shall not prove Theorem 9.1 in general. Proofs are given below for some parts of the theorem for 3×3 matrices and others are included in the exercises. These proofs indicate the general line of reasoning for the general case.

It is easy to see that Theorem 9.1(a) holds in general, since each term of the expansion by minors about the zero row (or column) would have a zero factor, so that $\delta(X) = 0$. For example,

$$\begin{vmatrix} a & b & c \\ 0 & 0 & 0 \\ d & e & f \end{vmatrix} = -0 \begin{vmatrix} b & c \\ e & f \end{vmatrix} + 0 \begin{vmatrix} a & c \\ d & f \end{vmatrix} - 0 \begin{vmatrix} a & b \\ d & e \end{vmatrix} = 0.$$

To demonstrate Theorem 9.1(b) for a 3×3 matrix, note the following. Let

$$X = \begin{vmatrix} a & b & c \\ d & e & f \\ g & h & i \end{vmatrix} \quad \text{and} \quad Y = \begin{vmatrix} a & d & g \\ b & e & h \\ c & f & i \end{vmatrix},$$

where Y was obtained from X by interchanging the corresponding rows and columns of X. Expanding about row 1 of X, we have

$$\delta(X) = a \begin{vmatrix} e & f \\ h & i \end{vmatrix} - b \begin{vmatrix} d & f \\ g & i \end{vmatrix} + c \begin{vmatrix} d & e \\ g & h \end{vmatrix}$$

$$= a(ei - hf) - b(di - gf) + c(dh - ge).$$

Expanding about column 1 of Y gives

$$\delta(Y) = a \begin{vmatrix} e & h \\ f & i \end{vmatrix} - b \begin{vmatrix} d & g \\ f & i \end{vmatrix} + c \begin{vmatrix} d & g \\ e & h \end{vmatrix}$$

$$= a(ei - fh) - b(di - fg) + c(dh - eg).$$

Thus, $\delta(Y) = \delta(X)$.

The proof of Theorem 9.1(e) for a 3×3 matrix in which the first column of Y is formed by adding k times the second column of X to the first column of X is as follows. If

$$\delta(X) = \begin{vmatrix} a & b & c \\ d & e & f \\ g & h & i \end{vmatrix} \quad \text{and} \quad \delta(Y) = \begin{vmatrix} a + kb & b & c \\ d + ke & e & f \\ g + kh & h & i \end{vmatrix},$$

then $\delta(X) = a(ei - hf) - d(bi - hc) + g(bf - ec)$, where we expanded about column 1 of X. Expanding Y about column 1 gives

$$\delta(Y) = (a+kb)\begin{vmatrix} e & f \\ h & i \end{vmatrix} - (d+ke)\begin{vmatrix} b & c \\ h & i \end{vmatrix} + (g+kh)\begin{vmatrix} b & c \\ e & f \end{vmatrix}$$

$$= (a+kb)(ei-hf) - (d+ke)(bi-hc) + (g+kh)(bf-ec)$$
$$= a(ei-hf) + kb(ei-hf) - d(bi-hc) - ke(bi-hc)$$
$$\quad + g(bf-ec) + kh(bf-ec).$$

By combining terms two, four, and six, and factoring out the k, we have

$$= a(ei-hf) - d(bi-hc) + g(bf-ec)$$
$$\quad + k[bei - bhf - ebi + ehc - hbf - hec]$$
$$= a(ei-hf) - d(bi-hc) + g(bf-ec) + k(0)$$
$$= \delta(X).$$

Theorem 9.1(f) is a consequence of (e) and (a), since, by (e), one of the identical rows (or columns) can be multiplied by -1 and added to the other to produce a zero row (or column) without changing the value of the determinant. Then, by (a) the value of the determinant is zero.

The following examples illustrate the use of the properties of determinants to simplify the evaluation of determinants.

Example 8 Without expanding, show that the value of the following determinant is zero.

$$\begin{vmatrix} 2 & 5 & -1 \\ 1 & -15 & 3 \\ -2 & 10 & -2 \end{vmatrix}$$

Examining the columns of the array, we note that each element in column two is -5 times the corresponding element in column three. Thus, if we use Theorem 9.1(e) and multiply column three by 5, and then add the results to the corresponding elements in column two, we have the equivalent determinant

$$\begin{vmatrix} 2 & 0 & -1 \\ 1 & 0 & 3 \\ -2 & 0 & -2 \end{vmatrix}.$$

By Theorem 9.1(a), the value of this determinant is zero.

Example 9 Find $\delta(A)$ if

$$\delta(A) = \begin{vmatrix} 4 & 2 & 1 & 0 \\ -2 & 4 & -1 & 7 \\ -5 & 2 & 3 & 1 \\ 6 & 4 & -3 & 2 \end{vmatrix}.$$

Our goal is to change row one (we could have selected any row or column) to a row in which every element but one is zero, using Theorem 9.1(e). To begin, we multiply column two by -2 and add the results to column one, replacing column one with this new column.

$$\begin{vmatrix} 0 & 2 & 1 & 0 \\ -10 & 4 & -1 & 7 \\ -9 & 2 & 3 & 1 \\ -2 & 4 & -3 & 2 \end{vmatrix}$$

column one replaced by
-2 times column two,
plus column one

We then replace column two by the results we get when we multiply column three by -2 and add to column two.

$$\begin{vmatrix} 0 & 0 & 1 & 0 \\ -10 & 6 & -1 & 7 \\ -9 & -4 & 3 & 1 \\ -2 & 10 & -3 & 2 \end{vmatrix}$$

column two replaced by
-2 times column three,
plus column two

The first row has only one nonzero number and so we expand about row one.

$$\delta(A) = 1 \begin{vmatrix} -10 & 6 & 7 \\ -9 & -4 & 1 \\ -2 & 10 & 2 \end{vmatrix}$$

Now let us try to change column three to a column with two zeros.

$$\begin{vmatrix} 53 & 34 & 0 \\ -9 & -4 & 1 \\ -2 & 10 & 2 \end{vmatrix}$$

row one replaced by
-7 times row two,
added to row one

$$\begin{vmatrix} 53 & 34 & 0 \\ -9 & -4 & 1 \\ 16 & 18 & 0 \end{vmatrix}$$

row three replaced by
-2 times row two,
added to row three

Now we can easily expand about column three to find the value of $\delta(A)$.

$$\delta(A) = -1 \begin{vmatrix} 53 & 34 \\ 16 & 18 \end{vmatrix} = -1(954 - 544) = -410$$

Note that in applying Theorem 9.1(e) we worked with sums of *rows* to get a *column* with only one nonzero number and with sums of *columns* to get a *row* with one nonzero number.

9.5 Exercises

Without evaluating, use Theorem 9.1 to verify that each of the following determinants is zero.

1. $\begin{vmatrix} 1 & 3 & -1 \\ 0 & 0 & 0 \\ 2 & 0 & 1 \end{vmatrix}$

3. $\begin{vmatrix} 3 & 6 & 6 \\ 2 & 0 & 4 \\ 1 & 4 & 2 \end{vmatrix}$

2. $\begin{vmatrix} -1 & 2 & 4 \\ 4 & -8 & -16 \\ 3 & 0 & 5 \end{vmatrix}$

4. $\begin{vmatrix} 1 & 0 & 0 \\ 1 & 0 & 1 \\ 3 & 0 & 0 \end{vmatrix}$

Use the appropriate parts of Theorem 9.1 to verify each of the following.

5. $\begin{vmatrix} 2 & 3 & 4 \\ 3 & 4 & 5 \\ 4 & 5 & 6 \end{vmatrix} = 0$

6. $\begin{vmatrix} 3 & 4 & -1 \\ 1 & 0 & 2 \\ -2 & 3 & 5 \end{vmatrix} = \begin{vmatrix} 5 & 4 & 3 \\ 1 & 0 & 2 \\ -2 & 3 & 5 \end{vmatrix}$

7. $\begin{vmatrix} 1 & 0 & 5 \\ 3 & 2 & -1 \\ 6 & 0 & 2 \end{vmatrix} = \begin{vmatrix} 1 & 3 & 6 \\ 0 & 2 & 0 \\ 5 & -1 & 2 \end{vmatrix}$

8. $\begin{vmatrix} 3 & 4 & 5 \\ -1 & 1 & 2 \\ 0 & 4 & -3 \end{vmatrix} = - \begin{vmatrix} 4 & 3 & 5 \\ 1 & -1 & 2 \\ 4 & 0 & -3 \end{vmatrix}$

9. $\begin{vmatrix} 2 & 0 & 4 \\ 1 & 3 & 2 \\ -1 & 1 & 0 \end{vmatrix} = \begin{vmatrix} 2 & 0 & 4 \\ 1 & 3 & 2 \\ 1 & 1 & 4 \end{vmatrix}$

10. $-2 \begin{vmatrix} 1 & -1 & 2 \\ 3 & 0 & 4 \\ 2 & 5 & 1 \end{vmatrix} = \begin{vmatrix} -2 & 2 & -4 \\ 3 & 0 & 4 \\ 2 & 5 & 1 \end{vmatrix}$

11. (a) Prove Theorem 9.1(c) for

$$X = \begin{vmatrix} a & b & c \\ d & e & f \\ g & h & i \end{vmatrix} \quad \text{and} \quad Y = \begin{vmatrix} d & e & f \\ a & b & c \\ g & h & i \end{vmatrix}.$$

(b) Prove Theorem 9.1(d) for

$$X = \begin{vmatrix} a & b & c \\ d & e & f \\ g & h & i \end{vmatrix} \quad \text{and} \quad Y = \begin{vmatrix} ka & b & c \\ kd & e & f \\ kg & h & i \end{vmatrix}.$$

12. Evaluate the following determinants after using Theorem 9.1 to introduce zeros as illustrated in Example 9.

(a) $\begin{vmatrix} 1 & 0 & 2 & 2 \\ 2 & 4 & 1 & -1 \\ 1 & -3 & 1 & 0 \\ 1 & 1 & 0 & 1 \end{vmatrix}$ (b) $\begin{vmatrix} 2 & -1 & 1 & 0 \\ 1 & 1 & 0 & 1 \\ 0 & -1 & 1 & 1 \\ 1 & 2 & 1 & 2 \end{vmatrix}$

Determinants can be used to find the area of a triangle given the coordinates of its vertices. Given a triangle PQR with vertices (x_1, y_1), (x_3, y_3) and (x_2, y_2), as shown in the figure, we can introduce segments PM, RN, and QS perpendicular to the x-axis, forming trapezoids PMNR, NSQR, and PMSQ. Recall that the area of a trapezoid is given by $\frac{1}{2}$ the sum of the parallel bases times the altitude. For example, the area of trapezoid PMSQ equals $\frac{1}{2}(y_1 + y_3)(x_3 - x_1)$. The area of triangle PQR can be found by subtracting the area of PMSQ from the sum of the areas of PMNR and RNSQ. Thus, for the area A of triangle PQR, we have

$$A = \tfrac{1}{2}(x_3y_1 - x_1y_3 + x_2y_3 - x_3y_2 + x_1y_2 - x_2y_1).$$

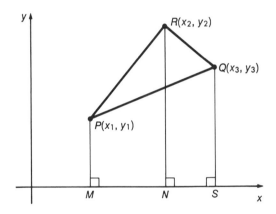

By evaluating the determinant

$$\begin{vmatrix} x_1 & y_1 & 1 \\ x_2 & y_2 & 1 \\ x_3 & y_3 & 1 \end{vmatrix}$$

it can be shown that the area of the triangle is given by $|A|$, where

$$A = \tfrac{1}{2}\begin{vmatrix} x_1 & y_1 & 1 \\ x_2 & y_2 & 1 \\ x_3 & y_3 & 1 \end{vmatrix}.$$

Use the formula given to find the area of the following triangles.

13. $P(0, 1), Q(2, 0), R(1, 3)$
14. $P(2, 5), Q(-1, 3), R(4, 0)$
15. $P(2, -2), Q(0, 0), R(-3, -4)$
16. $P(4, 7), Q(5, -2), R(1, 1)$
17. $P(3, 8), Q(-1, 4), R(0, 1)$
18. $P(-3, -1), Q(4, 2), R(3, -3)$

Chapter Ten

Systems of Equations and Inequalities

Systems of equations or inequalities have become important in many applications. With the advent of the computer, it has become practical to solve large systems with many variables so that new uses for this branch of mathematics have been developed.

A **system of equations** is a set of equations whose solution set is the intersection of the solution sets of the member equations. If some of the members of the system are inequalities, the system is called a **system of inequalities.** It is customary to write a system with only one set brace on the left, or when the context is clear, to omit set braces entirely. For example, the compound sentence $2x + y = 4$ and $x - y = 6$ gives a system which may be written as

$$\begin{cases} 2x + y = 4 \\ x - y = 6 \end{cases} \quad \text{or simply} \quad \begin{aligned} 2x + y &= 4 \\ x - y &= 6. \end{aligned}$$

The solution set of this system is $\{(x, y) \mid 2x + y = 4\} \cap \{(x, y) \mid x - y = 6\}$. In this chapter we shall discuss methods for solving several kinds of systems.

10.1 Linear Systems

A **first degree equation in n unknowns** is any equation of the form

$$a_1 x_1 + a_2 x_2 + \cdots + a_n x_n = k,$$

where $a_1, a_2, \cdots, a_n$ and k are constants, and $x_1, x_2, \cdots, x_n$ are variables. Such equations are also called **linear equations.** Generally, we shall confine our discussion of linear systems to those with two or three variables, although the methods we use can be extended to systems with more variables.

Recall from Chapter 3 that the solution set of a linear equation in two variables is an infinite set of ordered pairs. Since the graph of such an equation is a straight line, there are three possibilities for the solution set of a system of two linear equations in two variables, such as

$$a_1x + b_1y = c_1$$
$$a_2x + b_2y = c_2.$$

1. The two graphs intersect in a single point. The coordinates of this point give the solution of the system. This is the most common case (see Figure 10.1(a)).

2. The graphs are distinct parallel lines. When this is the case, the equations are said to be **inconsistent,** that is, there is no solution common to both equations. The solution set of the linear system is $\varnothing$ (see Figure 10.1(b)).

3. The graphs are the same line. In this case, the equations are said to be **dependent,** since any solution of one equation is also a solution of the other. Here the solution set of the system can be written as either of the two identical sets $\{(x, y) \mid a_1x + b_1y = c_1\}$ or $\{(x, y) \mid a_2x + b_2y = c_2\}$ (see Figure 10.1(c)).

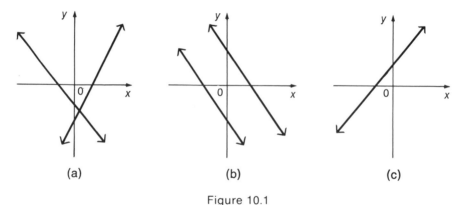

(a)   (b)   (c)

Figure 10.1

A solution of a linear equation $ax + by + cz = k$ with three variables is an **ordered triple** (x, y, z). Thus, the solution set of such an equation is an infinite set of ordered triples. It is shown in geometry that the graph of a linear equation in three variables is a plane. By considering the possible intersections of the planes representing three equations in three unknowns, we see that the solution set of such a system may again be either a single ordered triple (x, y, z), an infinite set of ordered triples, or $\varnothing$.

The most general method of solving a linear system depends on the following theorem, which is stated for equations in three variables. The theorem can be generalized to linear equations with any number of variables.

THEOREM 10.1   The ordered triple (x, y, z) satisfies each of the equations

$$a_1x + b_1y + c_1z = k_1$$
$$a_2x + b_2y + c_2z = k_2,$$

if and only if (x, y, z) satisfies the equation

$$m_1(a_1x + b_1y + c_1z) + m_2(a_2x + b_2y + c_2z)$$
$$= m_1k_1 + m_2k_2, \quad (1)$$

for any nonzero real numbers m_1 and m_2.

To prove Theorem 10.1, we note that k_1 and k_2 can be substituted respectively for $a_1x + b_1y + c_1z$ and $a_2x + b_2y + c_2z$ in equation (1) to give

$$m_1k_1 + m_2k_2 = m_1k_1 + m_2k_2,$$

while from equation (1), setting corresponding terms equal gives

$$k_1 = a_1x + b_1y + c_1z$$
and $\qquad\qquad k_2 = a_2x + b_2y + c_2z.$

To demonstrate how Theorem 10.1 is used, let us solve the system

$$3x - 4y = 1 \qquad\qquad (2)$$

$$2x + 3y = 12. \qquad\qquad (3)$$

We choose the multipliers m_1 and m_2 of Theorem 10.1 so that one of the variables will be eliminated in the expression

$$m_1(3x - 4y) + m_2(2x + 3y) = m_1(1) + m_2(12).$$

Since $a_1 = 3$ and $a_2 = 2$, we can eliminate x in the sum by choosing $m_1 = -2$ and $m_2 = 3$ as follows.

$$-2(3x - 4y) + 3(2x + 3y) = -2(1) + 3(12) \qquad (4)$$
$$-6x + 8y + 6x + 9y = -2 + 36$$
$$17y = 34$$
$$y = 2$$

To find the value of x, we can now either choose new multipliers m_1 and m_2 so as to eliminate y by again applying Theorem 10.1, or we can substitute 2 for y in one of the given equations. By substituting 2 for y in equation (2), we have

$$3x - 4(2) = 1$$
$$3x = 9$$
$$x = 3.$$

Thus, the solution set of the system is $\{(3, 2)\}$, which can be verified by checking in equation (3). For obvious reasons, this method of solution is often called the **elimination method.**

The solution of systems with three or more variables requires repeated use of Theorem 10.1, as illustrated in the following example.

Example 1 Solve the system

$$2x + y - z = 2 \qquad (5)$$

$$x + 3y + 2z = 1 \qquad (6)$$

$$x + y + z = 2. \qquad (7)$$

We need to use several steps to obtain an equation in one variable. First, we must eliminate the same variable from each of two pairs of equations and then eliminate another variable from the two resulting equations. Suppose we wish to eliminate x first. Multiplying equations (5) and (6) by -1 and 2 respectively, we have

$$-2x - y + z = -2$$
$$\underline{2x + 6y + 4z = 2}$$
$$5y + 5z = 0$$

or

$$y + z = 0. \qquad (8)$$

(It is often easier to add vertically rather than horizontally when using Theorem 10.1.) We must now eliminate x again from a different pair of equations, say (6) and (7). To do this, we multiply (6) by -1 and (7) by 1.

$$-x - 3y - 2z = -1$$
$$\underline{x + y + z = 2}$$
$$-2y - z = 1 \qquad (9)$$

Equations (8) and (9) can now be added (both multipliers m_1 and m_2 are 1) to find y.

$$-2y - z = 1$$
$$\underline{y + z = 0}$$
$$-y \qquad = 1$$
$$y = -1$$

By substitution in equation (8), we find z. (We could have used equation (9).)

$$y + z = 0$$
$$-1 + z = 0$$
$$z = 1$$

Now, from equation (7), (we could have used (5) or (6)) using the values for y and z, we find x.

$$x + y + z = 2$$
$$x - 1 + 1 = 2$$
$$x = 2$$

Thus, the solution set of the system is $\{(2, -1, 1)\}$.

10.1 Exercises

Solve the following systems of equations.

1. $2x - 3y = -7$
   $5x + 4y = 17$

2. $4m + 3n = -1$
   $2m + 5n = 3$

3. $5p + 7q = 6$
   $10p - 3q = 46$

4. $12s - 5t = 9$
   $3s - 8t = -18$

5. $6x + 7y = -2$
   $7x - 6y = 26$

6. $2a + 9b = 3$
   $5a + 7b = -8$

7. $\dfrac{x}{2} + \dfrac{y}{3} = 8$

   $\dfrac{2x}{3} + \dfrac{3y}{2} = 17$

8. $\dfrac{x}{5} + 3y = 31$

   $2x - \dfrac{y}{5} = 8$

9. $\dfrac{2}{x} + \dfrac{1}{y} = \dfrac{3}{2}$ $\left(\text{Hint: let } \dfrac{1}{x} = t \text{ and } \dfrac{1}{y} = u.\right)$

   $\dfrac{3}{x} - \dfrac{1}{y} = 1$

10. $\dfrac{1}{x} + \dfrac{3}{y} = \dfrac{16}{5}$

    $\dfrac{5}{x} + \dfrac{4}{y} = 5$

11. $x + 4y - z = 6$
    $2x - y + z = 3$
    $3x + 2y + 3z = 16$

12. $4x - 3y + z = 9$
    $3x + 2y - 2z = 4$
    $x - y + 3z = 5$

13. $5m + n - 3p = -6$
    $2m + 3n + p = 5$
    $-3m - 2n + 4p = 3$

14. $2r - 5s + 4t = -35$
    $5r + 3s - t = 1$
    $r + s + t = 1$

15. $a - 3b - 2c = -3$
    $3a + 2b - c = 12$
    $-a - b + 4c = 3$

16. $2x + 2y + 2z = 6$
    $3x - 3y - 4z = -1$
    $x + y + 3z = 11$

17. Find a and b so that the line with equation $ax + by = 5$ contains the points $(-2, 1)$ and $(-1, -2)$.

18. Find a, b, and c so that the graph of the equation $y = ax^2 + bx + c$ contains the points $(2, 3)$, $(-1, 0)$, and $(-2, 2)$.

19. Betsy invests $10,000, received from an endowment fund, in three ways. With one part, she buys mutual funds which offer a return of 8% per year. The second part, which amounts to twice the first, is used to buy government bonds at 9% per year. She puts the remainder in the bank at 5% annual interest. The first year her investments bring a return of $830. How much did she invest in each way?

20. The Busy Beaver Manufacturing Company makes two products, gadgets and thingamabobs. Both require time on two machines: gadgets, one hour on machine A and two hours on machine B; thingamabobs, three hours on machine A and one hour on machine B. Both machines are operated for 15 hours a day. How many of each product can be produced in a day under these conditions?

10.2 Solution of Linear Systems by Determinants– Cramer's Rule

In the previous section, we discussed the elimination method for solving systems of equations. We can use this method to solve the general system of two equations in two variables,

$$a_1 x + b_1 y = c_1 \tag{10}$$

$$a_2 x + b_2 y = c_2. \tag{11}$$

To eliminate y and solve for x, we multiply equation (10) by b_2 and equation (11) by $-b_1$ and then add.

$$a_1 b_2 x + b_1 b_2 y = c_1 b_2$$
$$\underline{-a_2 b_1 x - b_1 b_2 y = -c_2 b_1}$$
$$(a_1 b_2 - a_2 b_1)x \qquad = c_1 b_2 - c_2 b_1$$

or

$$x = \frac{c_1 b_2 - c_2 b_1}{a_1 b_2 - a_2 b_1}.$$

Similarly, to solve for y, we multiply equation (10) by $-a_2$ and equation (11) by a_1 and add.

$$-a_1 a_2 x - a_2 b_1 y = -a_2 c_1$$
$$\underline{a_1 a_2 x + a_1 b_2 y = a_1 c_2}$$
$$(a_1 b_2 - a_2 b_1)y = a_1 c_2 - a_2 c_1$$

or

$$y = \frac{a_1 c_2 - a_2 c_1}{a_1 b_2 - a_2 b_1}.$$

Both numerators and the common denominator of these values for x and y can be written as determinants, since

$$c_1b_2 - c_2b_1 = \begin{vmatrix} c_1 & b_1 \\ c_2 & b_2 \end{vmatrix},$$

$$a_1c_2 - a_2c_1 = \begin{vmatrix} a_1 & c_1 \\ a_2 & c_2 \end{vmatrix},$$

$$a_1b_2 - a_2b_1 = \begin{vmatrix} a_1 & b_1 \\ a_2 & b_2 \end{vmatrix}.$$

Then we can write the solutions for x and y as

$$x = \frac{\begin{vmatrix} c_1 & b_1 \\ c_2 & b_2 \end{vmatrix}}{\begin{vmatrix} a_1 & b_1 \\ a_2 & b_2 \end{vmatrix}} \quad \text{and} \quad y = \frac{\begin{vmatrix} a_1 & c_1 \\ a_2 & c_2 \end{vmatrix}}{\begin{vmatrix} a_1 & b_1 \\ a_2 & b_2 \end{vmatrix}}.$$

We have just derived the following result, which is known as **Cramer's Rule.**

THEOREM 10.2   Given the system

$$a_1x + b_1y = c_1$$
$$a_2x + b_2y = c_2,$$

with $a_1b_2 - a_2b_1 \neq 0$, then

$$x = \frac{\begin{vmatrix} c_1 & b_1 \\ c_2 & b_2 \end{vmatrix}}{\begin{vmatrix} a_1 & b_1 \\ a_2 & b_2 \end{vmatrix}} \quad \text{and} \quad y = \frac{\begin{vmatrix} a_1 & c_1 \\ a_2 & c_2 \end{vmatrix}}{\begin{vmatrix} a_1 & b_1 \\ a_2 & b_2 \end{vmatrix}}.$$

For convenience, we shall denote the three determinants in the solution as

$$\begin{vmatrix} a_1 & b_1 \\ a_2 & b_2 \end{vmatrix} = D, \quad \begin{vmatrix} c_1 & b_1 \\ c_2 & b_2 \end{vmatrix} = D_x, \quad \text{and} \quad \begin{vmatrix} a_1 & c_1 \\ a_2 & c_2 \end{vmatrix} = D_y.$$

Note that D is the determinant of the matrix of coefficients of the given system; D_x is the determinant of a matrix obtained by replacing the coefficients of x in the coefficient matrix by the respective constants; and D_y is defined similarly.

Example 2   Use Cramer's Rule to solve the system

$$5x + 7y = -1$$
$$6x + 8y = 1.$$

By Cramer's Rule, we know

$$x = \frac{D_x}{D} \quad \text{and} \quad y = \frac{D_y}{D}.$$

Thus we need to evaluate D, D_x, and D_y. We have

$$D = \begin{vmatrix} 5 & 7 \\ 6 & 8 \end{vmatrix} = 5(8) - 6(7) = -2,$$

$$D_x = \begin{vmatrix} -1 & 7 \\ 1 & 8 \end{vmatrix} = (-1)(8) - (1)(7) = -15,$$

$$D_y = \begin{vmatrix} 5 & -1 \\ 6 & 1 \end{vmatrix} = 5(1) - (6)(-1) = 11.$$

Thus we have $x = \frac{-15}{-2}$, and $y = \frac{11}{-2}$. The solution set is $\{(\frac{15}{2}, \frac{-11}{2})\}$, as can be verified by substituting within the given system.

In a similar manner, Cramer's Rule can be generalized to systems of three equations in three variables (or n equations in n variables).

THEOREM 10.3 Given the system

$$a_1 x + b_1 y + c_1 z = d_1$$
$$a_2 x + b_2 y + c_2 z = d_2$$
$$a_3 x + b_3 y + c_3 z = d_3,$$

with

$$D_x = \begin{vmatrix} d_1 & b_1 & c_1 \\ d_2 & b_2 & c_2 \\ d_3 & b_3 & c_3 \end{vmatrix}, \quad D_y = \begin{vmatrix} a_1 & d_1 & c_1 \\ a_2 & d_2 & c_2 \\ a_3 & d_3 & c_3 \end{vmatrix},$$

$$D_z = \begin{vmatrix} a_1 & b_1 & d_1 \\ a_2 & b_2 & d_2 \\ a_3 & b_3 & d_3 \end{vmatrix}, \quad D = \begin{vmatrix} a_1 & b_1 & c_1 \\ a_2 & b_2 & c_2 \\ a_3 & b_3 & c_3 \end{vmatrix} \neq 0,$$

then

$$x = \frac{D_x}{D}, \quad y = \frac{D_y}{D}, \quad \text{and} \quad z = \frac{D_z}{D}.$$

Example 3 Use Cramer's Rule to solve the system

$$x + y - z + 2 = 0$$
$$2x - y + z + 5 = 0$$
$$x - 2y + 3z - 4 = 0.$$

To use Cramer's Rule, the system must be rewritten in the form

$$x + y - z = -2$$
$$2x - y + z = -5$$
$$x - 2y + 3z = 4.$$

To find D, we can expand the coefficient matrix by minors about row one.

$$D = \begin{vmatrix} 1 & 1 & -1 \\ 2 & -1 & 1 \\ 1 & -2 & 3 \end{vmatrix} = 1\begin{vmatrix} -1 & 1 \\ -2 & 3 \end{vmatrix} - 1\begin{vmatrix} 2 & 1 \\ 1 & 3 \end{vmatrix} + (-1)\begin{vmatrix} 2 & -1 \\ 1 & -2 \end{vmatrix}$$
$$= 1(-1) - 1(5) - 1(-3)$$
$$= -3$$

If we replace the coefficients of x by the respective constants and expand about row one, we find D_x.

$$D_x = \begin{vmatrix} -2 & 1 & -1 \\ -5 & -1 & 1 \\ 4 & -2 & 3 \end{vmatrix} = -2\begin{vmatrix} -1 & 1 \\ -2 & 3 \end{vmatrix} - 1\begin{vmatrix} -5 & 1 \\ 4 & 3 \end{vmatrix} + (-1)\begin{vmatrix} -5 & -1 \\ 4 & -2 \end{vmatrix}$$
$$= -2(-1) - 1(-19) - 1(14)$$
$$= 7$$

Replacing the coefficients of y by the respective constants, and expanding about column one gives D_y.

$$D_y = \begin{vmatrix} 1 & -2 & -1 \\ 2 & -5 & 1 \\ 1 & 4 & 3 \end{vmatrix} = 1\begin{vmatrix} -5 & 1 \\ 4 & 3 \end{vmatrix} - 2\begin{vmatrix} -2 & -1 \\ 4 & 3 \end{vmatrix} + 1\begin{vmatrix} -2 & -1 \\ -5 & 1 \end{vmatrix}$$
$$= 1(-19) - 2(-2) + 1(-7)$$
$$= -22$$

Finally, replacing the coefficients of z by the respective constants, and expanding about column two gives D_z.

$$D_z = \begin{vmatrix} 1 & 1 & -2 \\ 2 & -1 & -5 \\ 1 & -2 & 4 \end{vmatrix} = -1\begin{vmatrix} 2 & -5 \\ 1 & 4 \end{vmatrix} + (-1)\begin{vmatrix} 1 & -2 \\ 1 & 4 \end{vmatrix} - (-2)\begin{vmatrix} 1 & -2 \\ 2 & -5 \end{vmatrix}$$
$$= -1(13) - 1(6) + 2(-1)$$
$$= -21$$

Thus we have

$$x = \frac{D_x}{D} = \frac{7}{-3} = -\frac{7}{3}, \; y = \frac{D_y}{D} = \frac{-22}{-3} = \frac{22}{3}, \text{ and } z = \frac{D_z}{D} = \frac{-21}{-3} = 7,$$

so that the solution set is $\{(-\frac{7}{3}, \frac{22}{3}, 7)\}$.

Example 4   Use Cramer's Rule to solve the system

$$2x - 3y + 4z = 10$$
$$6x - 9y + 12z = 24$$
$$x + 2y - 3z = 5.$$

We need to find D, D_x, D_y, and D_z. Here

$$D = \begin{vmatrix} 2 & -3 & 4 \\ 6 & -9 & 12 \\ 1 & 2 & -3 \end{vmatrix} = 2 \begin{vmatrix} -9 & 12 \\ 2 & -3 \end{vmatrix} - 6 \begin{vmatrix} -3 & 4 \\ 2 & -3 \end{vmatrix} + 1 \begin{vmatrix} -3 & 4 \\ -9 & 12 \end{vmatrix}$$
$$= 2(3) - 6(1) + 1(0)$$
$$= 0.$$

As stated in Theorem 10.3, Cramer's Rule does not apply if $D = 0$ and so to check for this possibility, we calculate D first. In this case, the system contains equations which are either dependent or inconsistent. We could now use the elimination method to tell which. Verify that this system is inconsistent.

10.2 Exercises

Use Cramer's Rule to solve the following systems of equations.

1. $2x - 3y = 4$
    $x + 5y = 2$

2. $7x + 3y = 5$
    $2x + 4y = 3$

3. $5x + 2y = 7$
    $6x + y = 8$

4. $x - 9y = 4$
    $3x + 2y = 5$

5. $12x - 8y = 3$
    $15x + 4y = 9$

6. $2x - 10y = 5$
    $7x + 6y = 8$

7. $x + 2y + 3z = 4$
    $4x + 3y + 2z = 1$
    $-x - 2y - 3z = 0$

8. $2x - y + 3z = 1$
    $3x + 2y - z = 4$
    $5x - y + z = 2$

9. $x - 2y + 3z = 4$
    $5x + 7y - z = 2$
    $2x + 2y - 5z = 3$

10. $-3x - 2y - z = 4$
     $4x + y + z = 5$
     $3x - 2y + 2z = 1$

11. $x + 2y = 10$
     $3x + 4z = 7$
     $ - y - z = 1$

12. $5x - 2y = 3$
     $4y + z = 8$
     $x + 2z = 4$

10.3 Solution of Linear Systems by Matrices

Consider the linear system of equations

$$a_1x + b_1y + c_1z = d_1$$
$$a_2x + b_2y + c_2z = d_2$$
$$a_3x + b_3y + c_3z = d_3.$$

We can write the coefficients as a 3×3 matrix,

$$\begin{pmatrix} a_1 & b_1 & c_1 \\ a_2 & b_2 & c_2 \\ a_3 & b_3 & c_3 \end{pmatrix},$$

which we shall call the **coefficient matrix** of the system. If we adjoin to the coefficient matrix the column of constants, we have the **augmented matrix** of the system,

$$\begin{pmatrix} a_1 & b_1 & c_1 & d_1 \\ a_2 & b_2 & c_2 & d_2 \\ a_3 & b_3 & c_3 & d_3 \end{pmatrix}.$$

We can operate with the rows of this augmented matrix just as we would with the equations of a system of equations, since the augmented matrix is actually an abbreviated form of the system. We can perform any transformation of the matrix which will result in an equivalent system. Operations which produce such transformations are stated in the following theorem.

THEOREM 10.4   For any real number k and any augmented matrix of a system of linear equations, the following operations will result in the matrix of an equivalent system.
(a)   Any two rows may be interchanged.
(b)   The elements of any row may be multiplied by the nonzero real number k.
(c)   A row may be changed by adding to the elements of the row k times the elements of another row.

We can use Theorem 10.4 to solve a linear system. For example, to solve

$$x - y + 5z = -6$$
$$3x + 3y - z = 10$$
$$x + 3y + 2z = 5,$$

we begin by writing the augmented matrix of the linear system,

$$\begin{pmatrix} 1 & -1 & 5 & -6 \\ 3 & 3 & -1 & 10 \\ 1 & 3 & 2 & 5 \end{pmatrix}.$$

We wish to use Theorem 10.4 to transform this matrix into one in which the value of the variables will be easy to see. That is, since each column in the matrix represents the coefficients of one variable, we wish to transform the matrix so that the augmented matrix is of the form

$$\begin{pmatrix} 1 & 0 & 0 & m \\ 0 & 1 & 0 & n \\ 0 & 0 & 1 & p \end{pmatrix}$$

(where m, n, and p are real numbers) from which, by rewriting the matrix as a linear system, we have

$$\begin{aligned} x & \quad\quad = m \\ & y \quad\; = n \\ & \quad z = p. \end{aligned}$$

In performing this transformation, it is best to work by columns beginning in each column with the element which is to become 1. In our example,

$$\begin{pmatrix} 1 & -1 & 5 & -6 \\ 3 & 3 & -1 & 10 \\ 1 & 3 & 2 & 5 \end{pmatrix},$$

we already have 1 in the desired position of the first column, so we begin by using Theorem 10.4(c) to change the second row so that the first element is zero. We do this by multiplying each element in row one by -3 and adding the result to the corresponding elements in row two to get the matrix

$$\begin{pmatrix} 1 & -1 & 5 & -6 \\ 0 & 6 & -16 & 28 \\ 1 & 3 & 2 & 5 \end{pmatrix}.$$

Now, to change the first element in row three to zero, we use Theorem 10.4(c) and multiply row one by -1 and add the results to row three.

$$\begin{pmatrix} 1 & -1 & 5 & -6 \\ 0 & 6 & -16 & 28 \\ 0 & 4 & -3 & 11 \end{pmatrix}$$

The same procedure is followed to transform columns two and three with one additional step necessary to transform the desired element in each column to 1, which should be done first.

$$\begin{pmatrix} 1 & -1 & 5 & -6 \\ 0 & 1 & \dfrac{-8}{3} & \dfrac{14}{3} \\ 0 & 4 & -3 & 11 \end{pmatrix} \qquad \begin{array}{l} \text{row two multiplied by } \tfrac{1}{6} \\ \text{(Theorem 10.4(b))} \end{array}$$

$$\begin{pmatrix} 1 & 0 & \dfrac{7}{3} & \dfrac{-4}{3} \\ 0 & 1 & \dfrac{-8}{3} & \dfrac{14}{3} \\ 0 & 4 & -3 & 11 \end{pmatrix}$$

row one added to row two
(Theorem 10.4(c))

$$\begin{pmatrix} 1 & 0 & \dfrac{7}{3} & \dfrac{-4}{3} \\ 0 & 1 & \dfrac{-8}{3} & \dfrac{14}{3} \\ 0 & 0 & \dfrac{23}{3} & \dfrac{-23}{3} \end{pmatrix}$$

-4 times row two added to row three
(Theorem 10.4(c))

$$\begin{pmatrix} 1 & 0 & \dfrac{7}{3} & \dfrac{-4}{3} \\ 0 & 1 & \dfrac{-8}{3} & \dfrac{14}{3} \\ 0 & 0 & 1 & -1 \end{pmatrix}$$

row three multiplied by $\frac{3}{23}$
(Theorem 10.4(b))

$$\begin{pmatrix} 1 & 0 & 0 & 1 \\ 0 & 1 & \dfrac{-8}{3} & \dfrac{14}{3} \\ 0 & 0 & 1 & -1 \end{pmatrix}$$

$-\frac{7}{3}$ times row three added to row one
(Theorem 10.4(c))

$$\begin{pmatrix} 1 & 0 & 0 & 1 \\ 0 & 1 & 0 & 2 \\ 0 & 0 & 1 & -1 \end{pmatrix}$$

$\frac{8}{3}$ times row three added to row two
(Theorem 10.4(c))

The linear system associated with this final matrix is

$$\begin{aligned} x\quad\quad &= 1 \\ y\quad &= 2 \\ z &= -1, \end{aligned}$$

and the solution set is $\{(1,\ 2,\ -1)\}$.

Example 5 Use Theorem 10.4 to solve the linear system

$$\begin{aligned} 3x - 4y &= 1 \\ 5x + 2y &= 19. \end{aligned}$$

The augmented matrix is

$$\begin{pmatrix} 3 & -4 & 1 \\ 5 & 2 & 19 \end{pmatrix}.$$

We want the final matrix to be of the form

$$\begin{pmatrix} 1 & 0 & k \\ 0 & 1 & j \end{pmatrix} \quad \text{for } k \text{ and } j \text{ real numbers.}$$

The transformations are performed as follows.

$$\begin{pmatrix} 1 & \dfrac{-4}{3} & \dfrac{1}{3} \\ 5 & 2 & 19 \end{pmatrix} \qquad \text{row one multiplied by } \tfrac{1}{3}$$

$$\begin{pmatrix} 1 & \dfrac{-4}{3} & \dfrac{1}{3} \\ 0 & \dfrac{26}{3} & \dfrac{52}{3} \end{pmatrix} \qquad -5 \text{ times row one added to row two}$$

$$\begin{pmatrix} 1 & \dfrac{-4}{3} & \dfrac{1}{3} \\ 0 & 1 & 2 \end{pmatrix} \qquad \text{row two multiplied by } \tfrac{3}{26}$$

$$\begin{pmatrix} 1 & 0 & 3 \\ 0 & 1 & 2 \end{pmatrix} \qquad \tfrac{4}{3} \text{ times row two added to row one}$$

From the last matrix we get the system

$$\begin{aligned} x \quad\;\; &= 3 \\ y &= 2, \end{aligned}$$

which gives the solution set $\{(3, 2)\}$. Note that the solution can be read directly from the third column of the final matrix.

The method of solving linear systems described in this section is called the *Gaussian Method*. Since it is easily adapted for use on computers, it is more efficient than Cramer's Rule for solving large systems of equations.

10.3 Exercises

Use matrices to solve the following systems of equations.

1. $\begin{aligned} x + y &= 10 \\ 2x - 5y &= 5 \end{aligned}$

2. $\begin{aligned} 3x - 2y &= 4 \\ 3x + y &= -2 \end{aligned}$

3. $\begin{aligned} 2x - 3y &= 10 \\ 2x + 2y &= 5 \end{aligned}$

4. $\begin{aligned} 6x + y &= 5 \\ 5x + y &= 3 \end{aligned}$

5. $\begin{aligned} 2x - 5y &= 10 \\ 3x + y &= 15 \end{aligned}$

6. $\begin{aligned} 4x - y &= 3 \\ -2x + 3y &= 1 \end{aligned}$

7. $\begin{aligned} x + y - z &= 6 \\ 2x - y + z &= -9 \\ x - 2y + 3z &= 1 \end{aligned}$

8. $\begin{aligned} x + 3y - 6z &= 7 \\ 2x - y + z &= 1 \\ x + 2y + 2z &= -1 \end{aligned}$

9. $\begin{aligned} -x + y \quad\;\; &= -1 \\ y - z &= 6 \\ x \quad\;\; + z &= -1 \end{aligned}$

10.

$$x + y \quad\quad = 1$$
$$2x \quad\quad - z = 0$$
$$y + 2z = -2$$

11.

$$2x - y + 3z = 0$$
$$x + 2y - z = 5$$
$$2y + z = 1$$

12.

$$4x + 2y - 3z = 6$$
$$x - 4y + z = -4$$
$$-x \quad\quad + 2z = 2$$

13.

$$3x + 2y \quad\quad - w = 0$$
$$2x \quad\quad + z + 2w = 5$$
$$x + 2y - z \quad\quad = -2$$
$$2x - y + z + w = 2$$

14.

$$x + 3y - 2z - w = 9$$
$$4x + y + z + 2w = 2$$
$$-3x - y + z - w = -5$$
$$x - y - 3z - 2w = 2$$

10.4 Solutions of Linear Systems of Equations by Inverses

Another way to use matrices to solve linear systems is to write the system as a matrix equation $AX = B$, where X is a matrix of the variables of the system. As shown in the previous chapter, multiplying both sides of the matrix equation by A^{-1} gives

$$X = A^{-1}B.$$

Thus, to find X, we must find A^{-1} and then the product $A^{-1}B$. Setting equal the corresponding elements of X and $A^{-1}B$ gives the solution of the system.

Example 6 Use the inverse of the coefficient matrix to solve the linear system

$$2x - 3y = 4$$
$$x + 5y = 2.$$

To represent the system as a matrix equation, we use one matrix for the coefficients, one for the variables, and one for the constants, as follows.

$$A = \begin{pmatrix} 2 & -3 \\ 1 & 5 \end{pmatrix}, \quad X = \begin{pmatrix} x \\ y \end{pmatrix} \quad \text{and} \quad B = \begin{pmatrix} 4 \\ 2 \end{pmatrix}$$

The system can then be written in matrix form as the equation $AX = B$, since

$$AX = \begin{pmatrix} 2 & -3 \\ 1 & 5 \end{pmatrix}\begin{pmatrix} x \\ y \end{pmatrix} = \begin{pmatrix} 2x - 3y \\ x + 5y \end{pmatrix} = \begin{pmatrix} 4 \\ 2 \end{pmatrix} = B.$$

To solve the system, we first find A^{-1} as follows.

$$A^{-1} = \frac{1}{13}\begin{pmatrix} 5 & 3 \\ -1 & 2 \end{pmatrix} = \begin{pmatrix} \dfrac{5}{13} & \dfrac{3}{13} \\ \dfrac{-1}{13} & \dfrac{2}{13} \end{pmatrix}$$

Next, we find the product $A^{-1}B$.

$$A^{-1}B = \begin{pmatrix} \dfrac{5}{13} & \dfrac{3}{13} \\ \dfrac{-1}{13} & \dfrac{2}{13} \end{pmatrix} \begin{pmatrix} 4 \\ 2 \end{pmatrix} = \begin{pmatrix} 2 \\ 0 \end{pmatrix}$$

Since $X = A^{-1}B$,

$$X = \begin{pmatrix} x \\ y \end{pmatrix} = \begin{pmatrix} 2 \\ 0 \end{pmatrix}$$

Thus, the solution set of the system is $\{(2, 0)\}$.

Example 7 Use the inverse of the coefficient matrix to solve the system

$$2x - 5y = 20$$
$$3x + 4y = 7.$$

The matrices we need are

$$A = \begin{pmatrix} 2 & -5 \\ 3 & 4 \end{pmatrix}, \qquad X = \begin{pmatrix} x \\ y \end{pmatrix}, \qquad \text{and} \qquad B = \begin{pmatrix} 20 \\ 7 \end{pmatrix}.$$

We begin by finding A^{-1}.

$$A^{-1} = \frac{1}{23} \begin{pmatrix} 4 & 5 \\ -3 & 2 \end{pmatrix} = \begin{pmatrix} \dfrac{4}{23} & \dfrac{5}{23} \\ -\dfrac{3}{23} & \dfrac{2}{23} \end{pmatrix}$$

We now find $A^{-1}B$.

$$A^{-1}B = \begin{pmatrix} \dfrac{4}{23} & \dfrac{5}{23} \\ -\dfrac{3}{23} & \dfrac{2}{23} \end{pmatrix} \begin{pmatrix} 20 \\ 7 \end{pmatrix} = \begin{pmatrix} 5 \\ -2 \end{pmatrix}$$

Then, since $X = A^{-1}B$, we have

$$X = \begin{pmatrix} x \\ y \end{pmatrix} = \begin{pmatrix} 5 \\ -2 \end{pmatrix},$$

and the solution of the system is $x = 5$, $y = -2$.

10.4 Exercises

Use the inverse of the coefficient matrix to solve the following systems of linear equations.

1. $7x + 3y = 5$
    $2x + 4y = 3$

2. $5x + 2y = 7$
    $6x + y = 8$

3. $x - 9y = 4$
    $3x + 2y = 5$

4. $12x - 8y = 3$
    $15x + 4y = 9$

5. $2x - 10y = 5$
   $7x + 6y = 8$

6. $5x + 3y = 5$
   $-4x + y = 13$

7. $x + y = 10$
   $2x - 5y = 5$

8. $4x - y = 3$
   $-2x + 3y = 1$

9. $2x - 3y = 10$
   $2x + 2y = 5$

10. $3x - 2y = 4$
    $3x + y = -2$

10.5 Nonlinear Systems of Equations

A **nonlinear system of equations** is one in which at least one of the equations is not a first degree equation. Some nonlinear systems can be solved by the elimination method, using a theorem which is a generalization of Theorem 10.1.

THEOREM 10.5 If R and S are algebraic expressions in x and y, then (x, y) satisfies both $R = k_1$ and $S = k_2$ (k_1 and k_2 real numbers) if and only if (x, y) also satisfies the equation

$$m_1 R + m_2 S = m_1 k_1 + m_2 k_2,$$

for nonzero real numbers m_1 and m_2.

The proof parallels the proof of Theorem 10.1. To illustrate the use of Theorem 10.5, consider the system

$$x^2 + y^2 = 4 \tag{12}$$

$$2x^2 - y^2 = 8. \tag{13}$$

We can use Theorem 10.5 with $m_1 = m_2 = 1$ to get

$$\begin{aligned}
(x^2 + y^2) + (2x^2 - y^2) &= 4 + 8 \\
3x^2 &= 12 \\
x^2 &= 4 \\
x &= 2 \text{ or } x = -2.
\end{aligned}$$

Then, by substitution into equation (12), we can find the corresponding value of y.

$$\begin{array}{ccc}
2^2 + y^2 = 4 & \text{or} & (-2)^2 + y^2 = 4 \\
y^2 = 0 & & y^2 = 0 \\
y = 0 & & y = 0.
\end{array}$$

From this result, we see that the solution set of the system is $\{(2, 0), (-2, 0)\}$, as shown in the graph of Figure 10.2 on page 246.

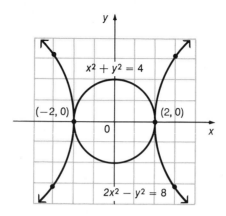

Figure 10.2

Another method for solving systems of equations, the **substitution method,** is particularly useful in solving a nonlinear system which includes a linear equation. For example, in the system

$$3x^2 - 2y = 5 \tag{14}$$

$$x + 3y = -4, \tag{15}$$

equation (15) can be solved for x and the resulting expression substituted into equation (14) as shown below.

$$x + 3y = -4 \tag{15}$$
$$x = -3y - 4$$

Substituting this value of x into equation (14) gives

$$3x^2 - 2y = 5 \tag{14}$$
$$3(-3y - 4)^2 - 2y = 5$$
$$3(9y^2 + 24y + 16) - 2y = 5$$
$$27y^2 + 70y + 43 = 0.$$

Using the quadratic formula, we have

$$y = \frac{-70 \pm \sqrt{4900 - 4644}}{54}$$

$$y = -1 \text{ or } y = \frac{-43}{27}.$$

Substituting these values for y into equation (15), we can find the corresponding values for x.

$$x + 3(-1) = -4 \quad \text{or} \quad x + 3\left(\frac{-43}{27}\right) = -4$$

$$x = -1 \qquad\qquad x = \frac{7}{9}$$

The solution set is $\{(-1, -1), (\frac{7}{9}, -\frac{43}{27})\}$, as shown on the graph of Figure 10.3.

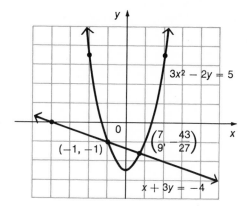

Figure 10.3

Sometimes a combination of the elimination method and the substitution method is effective in solving a system, as illustrated in Example 8.

Example 8   Solve the system

$$x^2 + 3xy + y^2 = 22 \tag{16}$$

$$x^2 - xy + y^2 = 6. \tag{17}$$

By Theorem 10.5, using $m_1 = 1$ and $m_2 = -1$, we have

$$
\begin{aligned}
x^2 + 3xy + y^2 &= 22 \\
-x^2 + xy - y^2 &= -6 \\
\hline
4xy &= 16.
\end{aligned} \tag{18}
$$

Now we solve equation (18) for either x or y and substitute the resulting expression into one of the given equations. Let us solve for y.

$$y = \frac{4}{x} \qquad (x \neq 0) \tag{19}$$

(Note that if $x = 0$ there is no value of y which satisfies the system.) Substituting for y in equation (17) and simplifying gives

$$x^2 - x\left(\frac{4}{x}\right) + \left(\frac{4}{x}\right)^2 = 6$$

$$x^2 - 4 + \frac{16}{x^2} = 6$$

$$x^4 - 4x^2 + 16 = 6x^2$$

$$x^4 - 10x^2 + 16 = 0$$

$$(x^2 - 2)(x^2 - 8) = 0$$

$$x^2 = 2 \text{ or } x^2 = 8$$

$$x = \sqrt{2} \text{ or } x = -\sqrt{2} \text{ or } x = 2\sqrt{2} \text{ or } x = -2\sqrt{2}.$$

We substitute these values of x into equation (19) to find the corresponding values for y.

$$\text{If } x = \sqrt{2}, \ y = \frac{4}{\sqrt{2}} = 2\sqrt{2}.$$

$$\text{If } x = -\sqrt{2}, \ y = -\frac{4}{\sqrt{2}} = -2\sqrt{2}.$$

$$\text{If } x = 2\sqrt{2}, \ y = \frac{4}{2\sqrt{2}} = \sqrt{2}.$$

$$\text{If } x = -2\sqrt{2}, \ y = -\frac{4}{2\sqrt{2}} = -\sqrt{2}.$$

From this we have the solution set of the system,

$$\{(\sqrt{2}, 2\sqrt{2}), (-\sqrt{2}, -2\sqrt{2}), (2\sqrt{2}, \sqrt{2}), (-2\sqrt{2}, -\sqrt{2})\}.$$

Example 9 Solve the system

$$x^2 + y^2 = 16 \tag{20}$$

$$|x| + y = 4. \tag{21}$$

The substitution method is required here. From the definition of absolute value, equation (21) can be rewritten as follows.

$$|x| = 4 - y \tag{21}$$

$$x = 4 - y \text{ or } x = -(4 - y) = y - 4. \tag{22}$$

(Since $|x| \geq 0$ for all real x, here we must have $4 - y \geq 0$, or $4 \geq y$.) Substituting from each part of the compound sentence into equation (20) we get the same equation.

$$(4 - y)^2 + y^2 = 16 \quad \text{or} \quad (y - 4)^2 + y^2 = 16$$
$$(16 - 8y + y^2) + y^2 = 16$$
$$2y^2 - 8y = 0$$
$$2y(y - 4) = 0$$
$$y = 0 \text{ or } y = 4.$$

From equation (22) we have

$$\text{If } y = 0, \ x = 4 - 0 \text{ or } x = 0 - 4$$
$$x = 4 \text{ or } x = -4$$
$$\text{If } y = 4, \ x = 4 - 4 \text{ or } x = 4 - 4$$
$$x = 0.$$

Thus, the solution set is $\{(4, 0), (-4, 0), (0, 4)\}$, as shown in Figure 10.4.

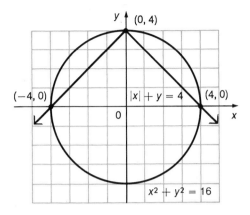

Figure 10.4

10.5 Exercises

Solve the following systems of equations.

1. $x^2 + y^2 = 13$
   $2x + y = 1$

2. $2x^2 - y^2 = 7$
   $y = 3x - 7$

3. $3x^2 + 2y^2 = 5$
   $x - y = -2$

4. $x^2 + y^2 = 5$
   $-3x + 4y = 2$

5. $5x^2 - y^2 = 9$
   $x^2 + y^2 = 45$

6. $x^2 + 3y^2 = 91$
   $2x^2 - y^2 = 119$

7. $2x^2 + 2y^2 = 20$
   $3x^2 + 3y^2 = 30$

8. $x^2 + y^2 = 4$
   $5x^2 + 5y^2 = 28$

9. $9x^2 + 4y^2 = 1$
   $x^2 + y^2 = 1$

10. $2x^2 - 3y^2 = 8$
    $6x^2 + 5y^2 = 24$

11. $x^2 + 2xy - y^2 = 14$
    $x^2 - y^2 = -16$

12. $3x^2 + xy + 3y^2 = 7$
    $x^2 + y^2 = 2$

13. $x^2 + 4y^2 = 25$
    $3xy = 18$

14. $5x^2 - 2y^2 = 6$
    $2xy = 4$

15. $x^2 - xy + y^2 = 5$
    $2x^2 + xy - y^2 = 10$

16. $3x^2 + 2xy - y^2 = 9$
    $x^2 - xy + y^2 = 9$

17. $x^2 + 2xy - y^2 + y = 1$
    $3x + y = 6$

18. $x^2 - 4x + y + xy = -10$
    $2x - y = 10$

19. $y = |x - 1|$
    $y = x^2 - 4$

20. $2x^2 - y^2 = 4$
    $|x| = |y|$

10.6 Systems of Inequalities

In Chapters 3 and 4, we saw that the solution set of an inequality in two variables is usually an infinite set whose graph is one or more regions of the coordinate plane. The solution set of a system of inequalities, such as

$$4y < -3x + 12$$
$$x^2 < 2y,$$

is the intersection of the solution sets of its members, or

$$\{(x, y) \mid 4y < -3x + 12\} \cap \{(x, y) \mid x^2 < 2y\}.$$

Since this description of the solution set is not very informative, it is customary to find the graph of the solution set. To do this, we graph both inequalities on the same coordinate axes and identify, by shading, the region common to both graphs, as shown in Figure 10.5. Since the points on the boundary lines are not in the solution sets, these lines are dashed.

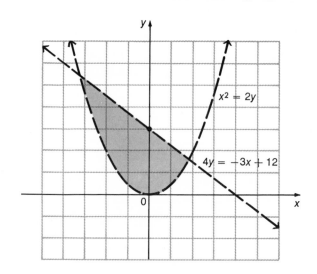

Figure 10.5

Example 10   Graph the solution set of the system

$$y \geq 2^x$$
$$9x^2 + 4y^2 \leq 36$$
$$2x + y < 1.$$

The graph is obtained by graphing the three inequalities on the same axes and shading the region common to all three as shown in Figure 10.6. Note that two boundary lines are solid and one is dashed. The points where the dashed line intersects the solid line are shown as open circles since they do not belong to the solution set. The solution set is written as

$$\{(x, y)\,|\,y \geq 2^x\} \cap \{(x, y)\,|\,9x^2 + 4y^2 \leq 36\} \cap \{(x, y)\,|\,2x + y < 1\},$$

although in systems such as this a graph is more readily understood.

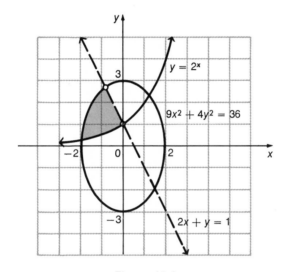

Figure 10.6

Example 11   Graph the solution set of the system

$$|x| \leq 3$$
$$y \leq 0$$
$$y \geq |x| + 1.$$

The graph is shown in Figure 10.7 on page 252. From the graph, we see that the solution sets of $y \leq 0$ and $y \geq |x| + 1$ have no elements in common. Hence, the solution set of the system is $\varnothing$

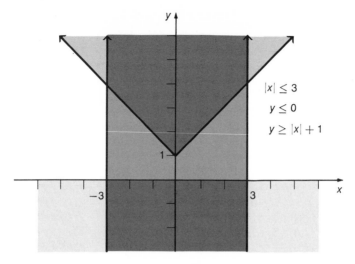

$$|x| \leq 3$$
$$y \leq 0$$
$$y \geq |x| + 1$$

Figure 10.7

10.6 Exercises

Graph the solution set of each of the following systems of inequalities.

1. $x + 2y \leq 4$
   $y \geq x^2 - 1$

2. $x^2 + y^2 \leq 9$
   $-x \geq y^2$

3. $x^2 - y^2 < 1$
   $-1 < y < 1$

4. $2x^2 - y^2 > 4$
   $2y^2 - x^2 > 4$

5. $y \geq x^2 + 4x + 4$
   $y < -x^2$

6. $x^2 + 4y^2 \leq 16$
   $x + y < -1$

7. $x^2 + y^2 \leq 4$
   $x^2 + 4y^2 \geq 16$

8. $y \geq x^2 - 4$
   $y < 2$

9. $4x^2 + 9y^2 > 1$
   $x^2 - y^2 \geq 1$
   $-4 \leq x \leq 4$

10. $2x - 3y < 6$
    $x + y \leq 3$
    $x \geq -2$

11. $y \geq 3^x$
    $y \leq |5x| + 3$
    $|x| \leq 2$

12. $y \leq \log x$
    $|x - 8| \leq 3$
    $y \geq -1$

Complex Numbers

So far in this text we have dealt exclusively with real numbers. However, the set of reals does not include enough numbers for our needs. For example, there is no real number solution of $x^2 + 1 = 0$. In this chapter we discuss a larger set of numbers having the reals as a subset, that is, the set of complex numbers.

11.1 Operations on Complex Numbers

Virtually all our work so far in this text has been with real numbers. However, we know that not all polynomial functions have real number zeros ($x^2 + 1$ does not, for example). To find zeros for every polynomial function, we need a new set of numbers, called the set of complex numbers.

Let us define the number i as follows.

▶ $$i = \sqrt{-1}$$

That is, we define i to be a number whose square is $-1 (i^2 = -1)$. We define $\sqrt{-a}$, $a > 0$, by saying

▶ $$\sqrt{-a} = i\sqrt{a} \qquad (a > 0)$$

Thus, $\sqrt{-16} = 4i$, $\sqrt{-75} = 5\sqrt{3}i$, and so on.

We shall call $a + bi$, a and b real numbers, a **complex number.** If $b \neq 0$, the complex number is also called an **imaginary number.** If $a = 0$ and $b \neq 0$, then the complex number is a nonzero multiple of i and is called a **pure imaginary number.** If $b = 0$, the complex number is of the form $a + 0i$ or just a, a real number, so that every real number is a complex number. Hence, both the real numbers and the pure imaginary numbers are subsets of the set of complex numbers. When a complex number is written in the form $a + bi$, where a and b are real numbers, the number is said to be written in **standard form.**

Let us define **equality** for complex numbers as

▶ $\qquad a + bi = c + di$ if and only if $a = c$ and $b = d$.

Thus, $2 + mi = k + 3i$ if and only if $2 = k$ and $m = 3$.

The complex numbers satisfy the field axioms. Some of these are verified later in this section, with the remainder left for the exercises. Assuming these properties, we can write

▶ $\qquad (a + bi) + (c + di) = (a + c) + (b + d)i,$

which defines the **sum** of two complex numbers. In a similar way,

$$
\begin{aligned}
(a + bi)(c + di) &= ac + adi + bic + bidi \\
&= ac + adi + cbi + bdi^2 \\
&= ac + (ad + cb)i + bd(-1)
\end{aligned}
$$

▶ $\qquad (a + bi)(c + di) = (ac - bd) + (ad + bc)i,$

which defines the **product** of two complex numbers.

Example 1 Find (a) the sum of $3 - 4i$ and $-2 + 6i$, (b) the product of $2 - 3i$ and $3 + 4i$.

(a) $(3 - 4i) + (-2 + 6i) = [3 + (-2)] + (-4 + 6)i = 1 + 2i$

(b) To use the definition to evaluate $(2 - 3i)(3 + 4i)$, we let $a = 2$, $b = -3$, $c = 3$, and $d = 4$. This gives

$$
\begin{aligned}
(2 - 3i)(3 + 4i) &= [2 \cdot 3 - (-3)4] + [2 \cdot 4 + (-3)3]i \\
&= (6 + 12) + (8 - 9)i \\
&= 18 - i.
\end{aligned}
$$

An alternate way to work Example 1(b), is to treat $2 - 3i$ and $3 + 4i$ as binomials, and multiply them as we did binomials, to get

$$
\begin{aligned}
(2 - 3i)(3 + 4i) &= 2 \cdot 3 + 2(4i) + (-3i)(3) + (-3i)(4i) \\
&= 6 + 8i - 9i - 12i^2 \\
&= 6 - i + 12 \\
&= 18 - i.
\end{aligned}
$$

Since $(a + bi) + (0 + 0i) = (a + 0) + (b + 0)i = a + bi$, for all complex numbers $a + bi$, we call $0 + 0i$ the **additive identity** for complex numbers. And since

$$(a + bi) + (-a - bi) = [a + (-a)] + [b + (-b)]i$$
$$= 0 + 0i,$$

we call $-a - bi$ the **negative** or **additive inverse** of $a + bi$.

Using the definition of additive inverse, we define **subtraction** of complex numbers as follows.

▶ $$(a + bi) - (c + di) = (a + bi) + (-c - di)$$

Since $$(a + bi)(1 + 0i) = (a \cdot 1 - b \cdot 0) + (b \cdot 1 + a \cdot 0)i$$
$$= a + bi,$$

we call $1 + 0i$ the **multiplicative identity** for complex numbers. What about a multiplicative inverse for $a + bi$? If we rewrite

$$\frac{1}{a + bi}$$

in the standard form for complex numbers (by multiplying numerator and denominator by $a - bi$), we have

$$\frac{1(a - bi)}{(a + bi)(a - bi)} = \frac{a - bi}{a^2 + b^2}$$

$$= \frac{a}{a^2 + b^2} - \frac{b}{a^2 + b^2}i.$$

To check that this is indeed the multiplicative inverse of $a + bi$, we multiply

$$(a + bi)\left[\frac{a}{a^2 + b^2} - \frac{b}{a^2 + b^2}i\right] = \frac{a(a + bi) - b(a + bi)i}{a^2 + b^2}$$

$$= \frac{a^2 + abi - abi + b^2}{a^2 + b^2}$$

$$= 1.$$

Hence, the **multiplicative inverse** of $a + bi$ (a and b not both equal to 0) is

▶ $$\frac{a}{a^2 + b^2} - \frac{b}{a^2 + b^2}i.$$

As with real numbers, we write the quotient of two complex numbers $a + bi$ and $c + di$ as

$$\frac{a + bi}{c + di} = (a + bi)\left(\frac{1}{c + di}\right).$$

Since $\dfrac{1}{c+di}$ is the multiplicative inverse of $c+di$, we can write

$$\frac{a+bi}{c+di} = (a+bi)\left(\frac{c-di}{c^2+d^2}\right)$$

$$= \frac{(a+bi)(c-di)}{c^2+d^2}$$

▶ $$\frac{a+bi}{c+di} = \frac{ac+bd}{c^2+d^2} + \frac{bc-ad}{c^2+d^2}i.$$

This last equality is called the **quotient** of the complex numbers $a+bi$ and $c+di$.

In practice, to divide two complex numbers such as $3-2i$ and $-4+3i$, it is customary to multiply both numerator and denominator of

$$\frac{3-2i}{-4+3i}$$

by $-4-3i$, the **conjugate** of $-4+3i$. Doing this, we have

$$\frac{(3-2i)(-4-3i)}{(-4+3i)(-4-3i)} = \frac{-18-i}{25} = \frac{-18}{25} - \frac{1}{25}i.$$

We have seen that we can define addition and multiplication of complex numbers so that the associative, commutative, and distributive properties hold. In the exercises, you are asked to show that the complex numbers are closed under addition and multiplication. The set of complex numbers has identity elements for both addition and multiplication. Every complex number has an additive inverse, and every complex number (except $0+0i$) has a multiplicative inverse. Because complex numbers satisfy all these properties, we say that complex numbers form a field, and we speak of the **field of complex numbers.** We saw in Chapter 1 that the set of real numbers, together with the operations of addition and multiplication, forms a field. However, the field of complex numbers, unlike the field of real numbers, cannot be ordered, as shown in Exercise 50 of this section.

11.1 Exercises

Perform the following operations and express all results in standard form.

1. $(3+2i)+(4-3i)$
2. $(4-i)+(2+5i)$
3. $(6-4i)-(3+2i)$
4. $(5-2i)-(5+3i)$
5. $(-2+3i)-(-4+2i)$
6. $(-3+5i)-(-4+3i)$

7. $(2+i)(3-2i)$
8. $(-2+3i)(4-2i)$
9. $(2+4i)(-1+3i)$
10. $(1+3i)(2-5i)$
11. $(5+2i)(5-3i)$
12. $(-3+2i)^2$

13. $(2 + i)^2$

14. $(\sqrt{6} - i)(\sqrt{6} + i)$

15. $(2 - i)(2 + i)$

16. $(5 + 4i)(5 - 4i)$

17. $i(3 - 4i)(3 + 4i)$

18. $i(2 + 5i)(2 - 5i)$

19. $i(3 - 4i)^2$

20. $i(2 + 6i)^2$

21. $\dfrac{1 + i}{1 - i}$

22. $\dfrac{2 - i}{2 + i}$

23. $\dfrac{4 - 3i}{4 + 3i}$

24. $\dfrac{5 + 6i}{5 - 6i}$

25. $\dfrac{4 + i}{6 + 2i}$

26. $\dfrac{3 - 2i}{5 + 3i}$

27. $\dfrac{5 - 2i}{6 - i}$

28. $\dfrac{3 - 4i}{2 - 5i}$

29. $\dfrac{1 - 3i}{1 + i}$

30. $\dfrac{-3 + 4i}{2 - i}$

Simplify (Hint: first calculate i^4.)

31. i^6

32. i^7

33. i^8

34. i^{19}

35. i^{85}

36. i^{78}

Simplify.

37. $\sqrt{-9}$

38. $\sqrt{-25}$

39. $\sqrt{-18}$

40. $\sqrt{-45}$

41. $\sqrt{-150}$

42. $\sqrt{-180}$

Prove that complex numbers satisfy the following properties.

43. Commutative property for addition.

44. Commutative property for multiplication.

45. Associative property for addition.

46. Associative property for multiplication.

47. Distributive property of multiplication over addition.

48. Closure property of addition.

49. Closure property of multiplication.

50. We know that the order relation for real numbers, $<$, satisfies three properties:

(1)   Trichotomy: if a and b are real numbers, then either $a < b$, $a = b$, or $a > b$.

(2)   If a, b, and c are real numbers and if $a < b$, then $a + c < b + c$.

(3)   If a and b are real numbers such that $0 < a$ and $0 < b$, then $0 < ab$.

We shall now show that if a and b are complex numbers it is not possible to give a definition of $a < b$ in such a way that the three properties above still hold. We shall prove this by assuming that a valid meaning can be given to $a < b$ for a and b complex numbers, and then we shall show that such an assumption leads to an absurdity. Fill in the blanks in the following proof with one of the three properties from above.

THEOREM   It is not possible to define an order relation $<$ for complex numbers in such a way that the three properties above still hold.

Proof

$i \neq 0$	$i = \sqrt{-1} \neq 0$
Either $0 < i$ or $i < 0$	_____
Let us assume $0 < i$	assumption
$0 < i^2$	Use $a = b = i$ in property ____
$0 < -1$	$i^2 = -1$

(Note: $0 < -1$ seems to be a contradiction. However, we have only assumed that $<$ can be somehow defined for complex numbers; we have not assumed that $<$ would then have the same meaning it does for real numbers.)

$1 < 0$	add 1 to both sides of the previous step by property ____
$0 < 1$	use $a = b = -1$ in property ____
$1 < 0$ and $0 < 1$	contradicts property ____

The assumption that $0 < i$ has led to a contradiction. In the same way, it can be shown that if we assume $i < 0$ we get another contradiction. Since $i \neq 0$, and since $i < 0$ and $0 < i$ lead to contradictions, we see it is not possible to define $<$ for complex numbers in such a way that the three properties still hold.

11.2   Trigonometric Form of Complex Numbers

It is not possible to graph complex numbers on the xy-coordinate system used throughout this text. Thus, to graph a complex number such as $2 - 3i$ we must modify the coordinate system. One way to do this is to call the horizontal axis the **real axis** and the vertical axis the **imaginary axis.** Then complex numbers can be graphed as shown in Figure 11.1. Each complex number graphed in this way determines a unique directed line segment, the segment from the origin to the point representing the complex number. Such directed line segments (like OP of Figure 11.1) are called **vectors.**

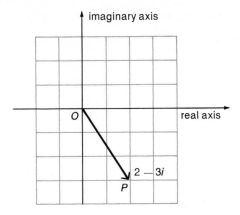

Figure 11.1

We know from the previous section that the sum of the two complex numbers $4 + i$ and $1 + 3i$ is given by $(4 + i) + (1 + 3i) = 5 + 4i$. As shown in Figure 11.2, we can also use a graphical method to add these two complex numbers. The sum of two complex numbers is represented by the vector which is the diagonal of the parallelogram having as adjacent sides the vectors determined by the two given complex numbers. This diagonal is called the **resultant** of the two vectors.

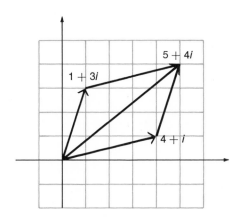

Figure 11.2

Figure 11.3 shows a complex number $x + yi$ which determines a vector OP. Let $r(r \geq 0)$ represent the length of OP, and let θ be the smallest positive angle (measured in a counterclockwise direction) between the positive real axis and OP. Point P could be located uniquely if we knew only r and θ, so that we can use r and θ as coordinates of point P. We call (r, θ) the **polar coordinates** of point P. The following relationships between r, θ, x, and y can be verified from Figure 11.3.

▶ $x = r \cos \theta$ ▶ $r = \sqrt{x^2 + y^2}$

▶ $y = r \sin \theta$ ▶ $\tan \theta = y/x$

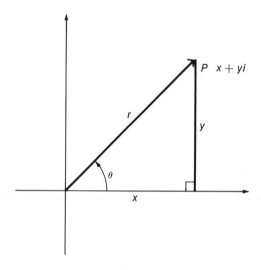

Figure 11.3

We shall discuss polar coordinates in more detail in Chapter 13.

Example 2   Find the polar coordinates of the complex number $1 + i$.

Here $x = 1$ and $y = 1$. Thus, $r = \sqrt{1^2 + 1^2} = \sqrt{2}$, and $\tan \theta = \frac{1}{1} = 1$, so that $\theta = 45°$. Hence the polar coordinates of $1 + i$ are $(\sqrt{2}, 45°)$, as shown in Figure 11.4.

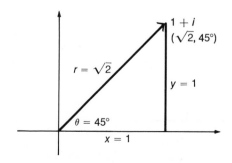

Figure 11.4

We know $x = r \cos \theta$ and $y = r \sin \theta$. We can use these facts to write

$$x + yi = r \cos \theta + (r \sin \theta)i \quad \text{or}$$

▶

$$x + yi = r(\cos \theta + i \sin \theta).$$

The expression $r(\cos \theta + i \sin \theta)$ is called the **polar** or **trigonometric form** of the complex number $x + yi$. The number r is called the **modulus** or **absolute value** of $x + yi$, while θ is called the **argument** of $x + yi$. Using the work from Example 2 above, we can express $1 + i$ in trigonometric form by writing

$$1 + i = \sqrt{2}(\cos 45° + i \sin 45°).$$

Example 3 Express $2(\cos 300° + i \sin 300°)$ in standard form.
We know $\cos 300° = 1/2$, while $\sin 300° = -\sqrt{3}/2$. Hence

$$2(\cos 300° + i \sin 300°) = 2\left(\frac{1}{2} - i\frac{\sqrt{3}}{2}\right)$$

$$= 1 - i\sqrt{3}.$$

11.2 Exercises

Graph each of the following complex numbers.

1. $-2 + 3i$
2. $-4 + 5i$
3. $8 - 5i$
4. $6 - 5i$
5. $2 - 2\sqrt{3}i$

6. $4\sqrt{2} + 4\sqrt{2}i$
7. $-4i$
8. $3i$
9. -8
10. 2

Find the resultant of each of the following pairs of complex numbers.

11. $2 - 3i, -1 + 4i$
12. $-4 - 5i, 2 + i$
13. $-5 + 6i, 3 - 4i$

14. $8 - 5i, -6 + 3i$
15. $-2, 4i$
16. $5, -4i$

Find the polar coordinates of each of the following complex numbers.

17. $2 - 2i$
18. $-1 - i$
19. $3i$

20. -4
21. $1 + i\sqrt{3}$
22. $-2 + 2i\sqrt{3}$

Write in the form $x + yi$ the complex numbers whose polar coordinates are given. Graph each complex number.

23. $(2, 45°)$
24. $(3, 60°)$
25. $(1, 135°)$

26. $(2, 300°)$
27. $(3, 90°)$
28. $(2, 180°)$

Convert the following complex numbers to the form $x + yi$.

29. $2(\cos 45° + i \sin 45°)$
30. $4(\cos 60° + i \sin 60°)$
31. $10(\cos 90° + i \sin 90°)$
32. $8(\cos 270° + i \sin 270°)$
33. $4(\cos 240° + i \sin 240°)$
34. $2(\cos 330° + i \sin 330°)$

Write each of the following in trigonometric form.

35. $3 - 3i$
36. $-2 + 2i\sqrt{3}$

37. $-3 - 3i\sqrt{3}$
38. $1 + i\sqrt{3}$

11.3   De Moivre's Theorem

We can find the product of the two complex numbers $1 + i\sqrt{3}$ and $-2\sqrt{3} + 2i$ as shown in Section 11.1:

$$(1 + i\sqrt{3})(-2\sqrt{3} + 2i) = -2\sqrt{3} + 2i - 2i(3) + 2i^2\sqrt{3}$$
$$= -2\sqrt{3} + 2i - 6i - 2\sqrt{3}$$
$$= -4\sqrt{3} - 4i.$$

We could also obtain this same product by first converting the complex numbers $1 + i\sqrt{3}$ and $-2\sqrt{3} + 2i$ to trigonometric form. Using the method explained in the previous section, we have

$$1 + i\sqrt{3} = 2(\cos 60° + i \sin 60°)$$

and $$-2\sqrt{3} + 2i = 4(\cos 150° + i \sin 150°).$$

If we now multiply the trigonometric forms together, and use the trigonometric identities for the cosine and the sine of the sum of two angles, we have

$$[2(\cos 60° + i \sin 60°)][4(\cos 150° + i \sin 150°)]$$
$$= 2 \cdot 4(\cos 60° \cdot \cos 150° + i \sin 60° \cdot \cos 150°$$
$$+ i \cos 60° \cdot \sin 150° + i^2 \sin 60° \cdot \sin 150°)$$
$$= 8[(\cos 60° \cdot \cos 150° - \sin 60° \cdot \sin 150°)$$
$$+ i(\sin 60° \cdot \cos 150° + \cos 60° \cdot \sin 150°)]$$
$$= 8[\cos(60° + 150°) + i \sin(60° + 150°)]$$
$$= 8(\cos 210° + i \sin 210°).$$

Note that the modulus of the product, 8, is equal to the *product* of the moduli of the factors, $2 \cdot 4$, while the argument of the product, $210°$, is the *sum* of the arguments of the factors, $60° + 150°$. In general, we have the following result.

THEOREM 11.1   If $r_1(\cos \theta_1 + i \sin \theta_1)$ and $r_2(\cos \theta_2 + i \sin \theta_2)$ are any two complex numbers written in trigonometric form, then

$$[r_1(\cos \theta_1 + i \sin \theta_1)][r_2(\cos \theta_2 + i \sin \theta_2)]$$
$$= r_1 r_2[\cos(\theta_1 + \theta_2) + i \sin(\theta_1 + \theta_2)].$$

Example 4   Find the product of $3(\cos 45° + i \sin 45°)$ and $2(\cos 135° + i \sin 135°)$.

Using Theorem 11.1, we have

$$[3(\cos 45° + i \sin 45°)][2(\cos 135° + i \sin 135°)]$$
$$= 3 \cdot 2[\cos(45° + 135°) + i \sin(45° + 135°)]$$
$$= 6(\cos 180° + i \sin 180°),$$

which can be expressed as $6(-1) = -6$. Thus, the two complex numbers of this example are complex factors of -6.

We can use the results of Theorem 11.1 to find the square of a complex number. We have

$$[r(\cos \theta + i \sin \theta)]^2 = [r(\cos \theta + i \sin \theta)][r(\cos \theta + i \sin \theta)]$$
$$= r \cdot r[\cos(\theta + \theta) + i \sin(\theta + \theta)]$$
$$= r^2(\cos 2\theta + i \sin 2\theta).$$

In the same way, we can show

$$[r(\cos \theta + i \sin \theta)]^3 = r^3(\cos 3\theta + i \sin 3\theta).$$

These results suggest the plausibility of the following theorem.

THEOREM 11.2   De MOIVRE'S THEOREM
If $r(\cos \theta + i \sin \theta)$ is a complex number expressed in trigonometric form, and if n is any real number, then

$$[r(\cos \theta + i \sin \theta)]^n = r^n(\cos n\theta + i \sin n\theta).$$

De Moivre's theorem can be proved for all positive integer values of n by the method of mathematical induction, which we discuss in Section 14.7. A more general proof requires methods we shall not discuss.

Example 5   Find $(1 + i\sqrt{3})^8$.

We can use De Moivre's theorem if we first convert $1 + i\sqrt{3}$ into trigonometric form, as shown below.

$$1 + i\sqrt{3} = 2(\cos 60° + i \sin 60°)$$

Now we can apply De Moivre's theorem.

$$(1 + i\sqrt{3})^8 = [2(\cos 60° + i \sin 60°)]^8$$
$$= 2^8[\cos(8 \cdot 60°) + i \sin(8 \cdot 60°)]$$
$$= 256(\cos 480° + i \sin 480°)$$
$$= 256(\cos 120° + i \sin 120°)$$
$$= 256\left(-\frac{1}{2} + i\frac{\sqrt{3}}{2}\right)$$
$$= -128 + 128\sqrt{3}i$$

11.3 Exercises

Find each of the following products. Write each product in the form $x + yi$.

1. $[2(\cos 30° + i \sin 30°)][3(\cos 60° + i \sin 60°)]$
2. $[3(\cos 60° + i \sin 60°)][4(\cos 150° + i \sin 150°)]$
3. $[2(\cos 135° + i \sin 135°)][2(\cos 225° + i \sin 225°)]$
4. $[4(\cos 60° + i \sin 60°)][6(\cos 330° + i \sin 330°)]$
5. $[6(\cos 240° + i \sin 240°)][8(\cos 300° + i \sin 300°)]$
6. $[8(\cos 210° + i \sin 210°)][2(\cos 330° + i \sin 330°)]$

Find each of the following powers. Express all answers in the form $x + yi$.

7. $[2(\cos 60° + i \sin 60°)]^3$
8. $[3(\cos 120° + i \sin 120°)]^4$
9. $(\cos 45° + i \sin 45°)^8$
10. $[2(\cos 120° + i \sin 120°)]^3$
11. $[5(\cos 100° + i \sin 100°)]^3$
12. $[3(\cos 40° + i \sin 40°)]^3$

13. $(\sqrt{3} + i)^5$
14. $(2\sqrt{2} - 2\sqrt{2}i)^6$
15. $(2 - 2\sqrt{3}i)^4$
16. $\left(\dfrac{\sqrt{2}}{2} - \dfrac{\sqrt{2}}{2}i\right)^8$

The quotient of two complex numbers $r_1(\cos \theta_1 + i \sin \theta_1)$ *and* $r_2(\cos \theta_2 + i \sin \theta_2)$ *can be shown to equal*

$$\frac{r_1(\cos \theta_1 + i \sin \theta_1)}{r_2(\cos \theta_2 + i \sin \theta_2)} = \frac{r_1}{r_2}[\cos(\theta_1 - \theta_2) + i \sin(\theta_1 - \theta_2)],$$

where $r_2(\cos \theta_2 + i \sin \theta_2) \neq 0$. *Find each of the following quotients. Write each answer in the form* $x + yi$.

17. $\dfrac{3(\cos 60° + i \sin 60°)}{(\cos 30° + i \sin 30°)}$

18. $\dfrac{9(\cos 135° + i \sin 135°)}{3(\cos 45° + i \sin 45°)}$

19. $\dfrac{16(\cos 300° + i \sin 300°)}{8(\cos 60° + i \sin 60°)}$

20. $\dfrac{24(\cos 150° + i \sin 150°)}{2(\cos 30° + i \sin 30°)}$

21. $\dfrac{2 - 2\sqrt{3}i}{1 - \sqrt{3}i}$

22. $\dfrac{4 + 4\sqrt{2}i}{2 - 2\sqrt{2}i}$

11.4 Roots of Complex Numbers

The complex number $a + bi$ is an **nth root** of the complex number $x + yi$ if

$$(a + bi)^n = x + yi.$$

We can use deMoivre's theorem to find nth roots of complex numbers. For example, to find the cube roots of the complex number $8(\cos 135° + i \sin 135°)$ we look for a complex number, say $r(\cos \alpha + i \sin \alpha)$,* that will satisfy

$$[r(\cos \alpha + i \sin \alpha)]^3 = 8(\cos 135° + i \sin 135°).$$

By deMoivre's theorem, this equation becomes

$$r^3(\cos 3\alpha + i \sin 3\alpha) = 8(\cos 135° + i \sin 135°).$$

One way to satisfy this equation is to set $r^3 = 8$ and also $\cos 3\alpha + i \sin 3\alpha = \cos 135° + i \sin 135°$. The first of these conditions implies that $r = 2$, and the second implies

$$\cos 3\alpha = \cos 135° \qquad \text{and} \qquad \sin 3\alpha = \sin 135°$$

or
$$\cos 3\alpha = \frac{-\sqrt{2}}{2} \qquad \text{and} \qquad \sin 3\alpha = \frac{\sqrt{2}}{2}.$$

This last pair of equations can only be satisfied if

$$3\alpha = 135° + 360° \cdot k, \qquad k \text{ any integer,}$$

or
$$\alpha = \frac{135° + 360° \cdot k}{3}, \qquad k \text{ any integer.}$$

If we let $k = 0, 1, 2, 3, \ldots$, we obtain

$$\alpha = 45°, 165°, 285°, 405°, \ldots.$$

The values of α after the first three are redundant since they produce the same values of $\sin \alpha$ and $\cos \alpha$ that were obtained before. Hence, we get all of the cube roots (3 of them) by letting $k = 0, 1,$ and 2. When $k = 0$, we obtain

$$2(\cos 45° + i \sin 45°).$$

When $k = 1$, we have

$$2(\cos 165° + i \sin 165°),$$

and when $k = 2$, we have

$$2(\cos 285° + i \sin 285°).$$

Hence, these are the three cube roots of $8(\cos 135° + i \sin 135°)$.

We generalize this result in the following theorem.

* α is the Greek letter "alpha."

THEOREM 11.3   If n is any positive integer, and r is a positive real number, then the complex number $r(\cos\theta + i\sin\theta)$ has exactly n distinct nth roots, given by

$$r^{1/n}(\cos\alpha + i\sin\alpha),$$

where

$$\alpha = \frac{\theta + 360° \cdot k}{n}, \qquad k = 0, 1, 2, \ldots, n-1.$$

Example 6   Find all 4th roots of $-8 + 8\sqrt{3}i$.
    First we write $-8 + 8\sqrt{3}i$ in trigonometric form as

$$-8 + 8\sqrt{3}i = 16(\cos 120° + i\sin 120°).$$

Here $r = 16$ and $\theta = 120°$. The 4th roots of this number have modulus $16^{1/4} = 2$ and arguments given as follows.

$$\text{If } k = 0, \qquad \frac{120° + 360° \cdot 0}{4} = 30°,$$

$$\text{if } k = 1, \qquad \frac{120° + 360° \cdot 1}{4} = 120°,$$

$$\text{if } k = 2, \qquad \frac{120° + 360° \cdot 2}{4} = 210°,$$

$$\text{if } k = 3, \qquad \frac{120° + 360° \cdot 3}{4} = 300°.$$

Using these angles, we can write the 4th roots as

$$2(\cos 30° + i\sin 30°)$$
$$2(\cos 120° + i\sin 120°)$$
$$2(\cos 210° + i\sin 210°)$$
$$2(\cos 300° + i\sin 300°).$$

We can also write these four roots in the form $x + yi$ as $\sqrt{3} + i$, $-\sqrt{3} + i$, $-\sqrt{3} - i$, and $\sqrt{3} - i$.

Example 7   Find all 5th roots of 1.
    We can write 1 in trigonometric form as

$$1 = 1 + 0i = 1(\cos 0° + i\sin 0°).$$

The modulus of the 5th roots is $1^{1/5} = 1$, while the arguments are given by

$$\frac{0° + 360° \cdot k}{5}, \qquad k = 0, 1, 2, 3, \text{ or } 4.$$

Using these arguments we can write the 5th roots as

$$1(\cos 0° + i \sin 0°)$$
$$1(\cos 72° + i \sin 72°)$$
$$1(\cos 144° + i \sin 144°)$$
$$1(\cos 216° + i \sin 216°)$$
$$1(\cos 288° + i \sin 288°).$$

Note that the first of these roots equals 1, but that the others cannot easily be expressed in the form $x + yi$. The five 5th roots all lie on a unit circle, and are equally spaced around it, as shown in Figure 11.5.

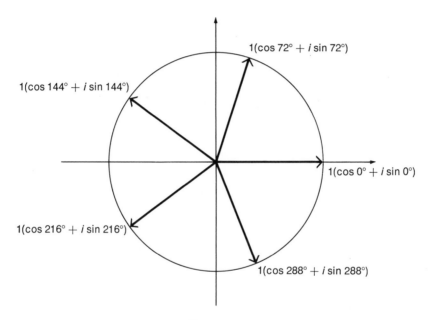

Figure 11.5

11.4 Exercises

Find and graph all cube roots of each of the following complex numbers.

1. 1
2. i
3. $-8i$

4. 27
5. $1 + \sqrt{3}i$
6. $2 - 2\sqrt{3}i$

Find and graph all the following roots of 1.

7. 2nd
8. 4th

9. 6th
10. 8th

Find and graph all the following roots of i.

11. 2nd   12. 4th

Find all solutions of each of the following equations.

13. $x^3 - 1 = 0$ (Hint: write this equation as $x^3 = 1$.)

14. $x^3 + 27 = 0$

15. $x^4 + 1 = 0$

16. $x^4 + 16 = 0$

17. $x^4 - i = 0$

18. $x^5 - i = 0$

19. $x^3 - (4 + 4\sqrt{3}i) = 0$

20. $x^4 - (8 + 8\sqrt{3}i) = 0$

Chapter Twelve

Polynomial and Rational Functions

A **polynomial function** of degree n is a function of the form

$$P(x) = a_n x^n + a_{n-1} x^{n-1} + \cdots + a_1 x + a_0, \qquad (a_n \neq 0).$$

We call a_n the **leading coefficient** of $P(x)$. We have discussed two special types of polynomial functions, those of degree 1 (linear) and those of degree 2 (quadratic). We shall discuss polynomial functions in general in this chapter.

12.1 Complex Zeros of Quadratic Polynomials

In Chapter 4 we developed the quadratic formula for finding the zeros of a quadratic function. (Recall: a quadratic polynomial is of degree 2, and a zero of a function $f(x)$ is a value of x that makes $f(x) = 0$.) The properties used to prove the quadratic formula in Chapter 4 are equally valid for complex numbers, so we can now state a more general form of the quadratic formula.

THEOREM 12.1   QUADRATIC FORMULA

The zeros of $y = ax^2 + bx + c$ (a, b, and c complex numbers, $a \neq 0$), are given by

$$x = \frac{-b \pm \sqrt{b^2 - 4ac}}{2a}.$$

For example, to find the zeros of $y = x^2 + x + 1$, we note that $a = 1$, $b = 1$, and $c = 1$. Using the quadratic formula, we have

$$x = \frac{-1 \pm \sqrt{1-4}}{2}$$

$$= \frac{-1 \pm \sqrt{-3}}{2}.$$

Thus, the zeros of the function above can be written as the complex numbers

$$\frac{-1 + i\sqrt{3}}{2} \quad \text{and} \quad \frac{-1 - i\sqrt{3}}{2}.$$

Example 1   Find the zeros of $y = x^2 + 8x + 25$.

Here we have $a = 1$, $b = 8$, and $c = 25$. Using the quadratic formula, we can write

$$x = \frac{-8 \pm \sqrt{64 - 100}}{2}$$

$$= \frac{-8 \pm 6i}{2}$$

$$x = -4 + 3i \text{ or } x = -4 - 3i.$$

As mentioned in the quadratic formula in Theorem 12.1, the coefficients of the polynomial must be complex numbers, but not necessarily real numbers. For example, to find the zeros of $y = 2ix^2 - ix + 3i$ we first note that $a = 2i$, $b = -i$, and $c = 3i$. This gives

$$x = \frac{i \pm \sqrt{(-i)^2 - 4(2i)(3i)}}{4i}$$

$$= \frac{i \pm \sqrt{-1 - 24i^2}}{4i}$$

$$= \frac{i \pm \sqrt{-1 + 24}}{4i}$$

$$= \frac{i \pm \sqrt{23}}{4i}$$

$$= \frac{(i \pm \sqrt{23})i}{4i \cdot i}$$

$$= \frac{-1 \pm i\sqrt{23}}{-4}$$

$$= \frac{1}{4} \pm \frac{i\sqrt{23}}{4}.$$

12.1 Exercises

Find all zeros of each of the following.

1. $y = x^2 + 9$
2. $y = x^2 + 16$
3. $y = 4m^2 + 25$
4. $y = 9p^2 + 49$
5. $y = p^2 - 4p + 8$
6. $y = r^2 + 8r + 25$
7. $y = 4z^2 - 4z + 5$
8. $y = 9g^2 - 12g + 13$
9. $y = k^2 + 2k + 3$
10. $y = n^2 - 3n + 4$

11. $y = 2m^2 - m + 1$
12. $y = 2x^2 - 3x + 2$
13. $y = 3k^2 - k + 2$
14. $y = 5z^2 - 3z + 1$
15. $y = x^2 + ix + 1$
16. $y = 3m^2 - 2m + i$
17. $y = ip^2 - 2p + i$
18. $y = 2ix^2 - 3x + 3$
19. $y = (1 + i)x^2 - x + (1 - i)$
20. $y = (2 + i)r^2 - 3r + (2 - i)$

Write quadratic equations having the following numbers as zeros. (Hint: $y = (x - r_1)(x - r_2)$ has r_1 and r_2 as solutions.)

21. i and $-i$
22. $1 + i$ and $1 - i$
23. $2 + 3i$ and $2 - 3i$
24. $4 - i$ amd $4 + i$
25. $\sqrt{3}i$ and $-\sqrt{3}i$
26. $\sqrt{6}i$ and $-\sqrt{6}i$

27. $1 + \sqrt{2}i$ and $1 - \sqrt{2}i$
28. $\sqrt{3} + 2i$ and $\sqrt{3} - 2i$
29. $1 + i$ and $2 - 3i$
30. $4 + 2i$ and $1 - 3i$
31. $2 + 4i$ and $3 - 2i$
32. $5 + i$ and $2 + 3i$

12.2 The Factor Theorem

In this section we begin a discussion of the zeros of general polynomial functions having coefficients from the field of complex numbers (polynomials over the complex numbers). Recall from Section 2.2 that the process of synthetic division is useful when dividing a polynomial by a binomial of the form $x - k$, where k is any real number. For example, this process can be used to divide $3x^3 - 4x^2 - 5x - 25$ by $x - 3$, as follows.

$$
\begin{array}{r|rrrr}
3 & 3 & -4 & -5 & -25 \\
  &   & 9 & 15 & 30 \\
\hline
  & 3 & 5 & 10 & 5
\end{array}
$$

Hence: $3x^3 - 4x^2 - 5x - 25 = (3x^2 + 5x + 10)(x - 3) + 5$.

We can now extend this method to cases where k is any complex number, and to polynomials having complex coefficients. For example, to divide $x^3 + ix^2 + 4x + i$ by $x + 2i$, we can write

$$
\begin{array}{r|rrrr}
-2i & 1 & i & 4 & i \\
    &   & -2i & -2 & -4i \\
\hline
    & 1 & -i & 2 & -3i.
\end{array}
$$

From this, we can write

$$x^3 + ix^2 + 4x + i = (x^2 - ix + 2)(x + 2i) - 3i.$$

These examples suggest the division algorithm for polynomials, given in the following theorem.

THEOREM 12.2   DIVISION ALGORITHM

If $P(x)$ is a polynomial over the complex numbers, and k is any complex number, then there exists a unique polynomial $Q(x)$ over the complex numbers and a unique complex number r such that

$$P(x) = (x - k)Q(x) + r.$$

It is possible to obtain more information about the complex number r (the remainder) of the division algorithm. We know that if we divide $P(x)$ by $x - k$, we can write

$$P(x) = (x - k)Q(x) + r,$$

for some polynomial $Q(x)$ and complex number r. Since this last equality is true for all complex values of x, it is true for $x = k$. Thus, we can write

$$P(k) = (k - k)Q(k) + r$$
$$P(k) = r.$$

Hence we have proved the following theorem.

THEOREM 12.3   REMAINDER THEOREM

If $P(x)$ is a polynomial over the field of complex numbers, and if k is any complex number, then the remainder when $P(x)$ is divided by $x - k$ is equal to $P(k)$.

Thus, instead of evaluating $P(k)$ by substituting k for x in the polynomial $P(x)$, we can find $P(k)$ by dividing $P(x)$ by $x - k$. Then $P(k)$ equals the remainder obtained in the division.

Example 2 Let $P(x) = -x^4 + 3x^2 - 4x - 5$, and find $P(-2)$.

Using the remainder theorem and synthetic division, we have

$$
\begin{array}{r|rrrrr}
-2 & -1 & 0 & 3 & -4 & -5 \\
   &   & 2 & -4 & 2 & 4 \\
\hline
   & -1 & 2 & -1 & -2 & -1.
\end{array}
$$

Hence, $P(-2) = -1$, the remainder obtained when $P(x)$ is divided by $x - (-2) = x + 2$.

By the remainder theorem, if $P(k) = 0$, then $x - k$ is a factor of $P(x)$, and conversely, if $x - k$ is a factor of $P(x)$, then $P(k)$ must equal 0. Thus, we have proved the following theorem.

THEOREM 12.4   FACTOR THEOREM
If $P(x)$ is a polynomial over the complex numbers, and if k is any complex number, then $x - k$ is a factor of $P(x)$ if and only if $P(k) = 0$.

Example 3   Is $x - i$ a factor of $P(x) = 3x^3 + (-4 - 3i)x^2 + (5 + 4i)x - 5i$?
The only way $x - i$ can be a factor of $P(x)$ is for $P(i)$ to be 0. To see if this is the case, we use synthetic division.

$$
\begin{array}{r|rrrr}
i & 3 & -4-3i & 5+4i & -5i \\
  &   & 3i & -4i & 5i \\
\hline
  & 3 & -4 & 5 & 0
\end{array}
$$

Since the remainder is 0, this means $P(i) = 0$, and hence $x - i$ is a factor of $P(x)$.

By the results above, any time we have a zero of $P(x)$ we also have found a factor of $P(x)$. That is, if k is a zero of $P(x)$, then $x - k$ is a factor of $P(x)$, and conversely.

12.2 Exercises

Use synthetic division to decide whether or not the given number is a zero of the given polynomial.

1. $2; x^2 + 2x - 8$
2. $-1; m^2 + 4m - 5$
3. $1 - i; y^2 - 2y + 2$
4. $3 - 2i; r^2 - 6r + 13$
5. $2 + i; k^2 + 3k + 4$
6. $1 - 2i; z^2 - 3z + 5$
7. $2; g^3 - 3g^2 + 4g - 4$

8. $-3; m^3 + 2m^2 - m + 6$
9. $4; 2r^3 - 6r^2 - 9r + 4$
10. $-6; 2y^3 + 9y^2 - 16y + 12$
11. $i; x^3 + 2ix^2 + 2x + i$
12. $-i; p^3 - ip^2 + 3p + 5i$
13. $1 + i; p^3 + 3p^2 - p + 1$
14. $2 - i; 2r^3 - r^2 + 3r - 5$

Evaluate $P(x)$ for the given polynomial and the given value of x.

15. $x = 3; P(x) = x^2 - 4x + 5$
16. $x = -2; P(x) = x^2 + 5x + 6$
17. $x = 2 + i; P(x) = x^2 - 5x + 1$
18. $x = 3 - 2i; P(x) = x^2 - x + 3$
19. $x = 1 - i; P(x) = x^3 + x^2 - x + 1$
20. $x = 2 - 3i; P(x) = x^3 + 2x^2 + x - 5$

Determine the value of k that makes the first polynomial divisible by the second.

21. $3x^3 - 12x^2 - 11x + k; x - 5$
22. $4x^3 + 6x^2 - 5x + k; x + 2$
23. $2x^4 + 5x^3 - 2x^2 + kx + 3; x + 3$
24. $5x^4 + 16x^3 - 15x^2 + kx + 16; x + 4$

12.3 Complex Zeros of General Polynomial Functions

We have seen that if a polynomial $P(x)$ can be divided by $x - k$, then $P(x) = 0$, and conversely. The next theorem shows that every polynomial function of degree greater than or equal to 1 has a zero, and thus that every such polynomial can be factored.

THEOREM 12.5   THE FUNDAMENTAL THEOREM OF ALGEBRA
If $P(x)$ is a polynomial of degree at least 1 over the complex numbers, there exists at least one complex number k such that $P(k) = 0$.

This theorem was first proved by the mathematician Gauss in his doctoral thesis in 1799 when he was 20. Although many proofs of this result have been given, none of them involve only the ideas of this text, and hence no proof is included here. From the fundamental theorem and the factor theorem, we have the following result.

THEOREM 12.6   If $P(x)$ is a polynomial of degree n $(n \geq 1)$ over the complex numbers, there exists a complex number k and a polynomial $Q(x)$ over the complex numbers such that

$$P(x) = (x - k)Q(x).$$

We could also use the factor theorem and the fundamental theorem to factor the polynomial $Q(x)$ of Theorem 12.6. If we continue this process a total of n times, we get the result of the next theorem.

THEOREM 12.7   If $P(x)$ is a polynomial of degree n $(n \geq 1)$ over the complex numbers, then there exist n complex numbers k_1, k_2, $\cdots$, k_n, which need not be distinct, such that

$$P(x) = a(x - k_1)(x - k_2) \cdots (x - k_n),$$

where a is the leading coefficient in $P(x)$.

Using this result, we can prove that a polynomial of degree $n(n \geq 1)$ has at most n distinct zeros. To prove this, we use an indirect proof. Thus, suppose a polynomial $P(x)$ of degree n $(n \geq 1)$ has $n + 1$ distinct zeros. Let us label the distinct zeros $k_1, k_2, k_3, \cdots, k_n,$ and k. By Theorem 12.7, we can use the n zeros $k_1, k_2, \cdots, k_n$ to write

$$P(x) = a(x - k_1)(x - k_2) \cdots (x - k_n),$$

where a is the leading coefficient in $P(x)$. If we now let $x = k$, we have

$$P(k) = a(k - k_1)(k - k_2) \cdots (k - k_n). \tag{1}$$

We have assumed that k is also a zero of $P(x)$, so that $P(k) = 0$. Using this fact, and equation (1), we have

$$0 = a(k - k_1)(k - k_2) \cdots (k - k_n). \tag{2}$$

We assumed $k_1, k_2, \cdots, k_n,$ and k were $n + 1$ *distinct* zeros of $P(x)$. Hence, $k - k_1 \neq 0,\ k - k_2 \neq 0,\ \cdots,\ k - k_n \neq 0$. By the definition of $P(x)$, $a \neq 0$. Thus, the product

$$a(k - k_1)(k - k_2) \cdots (k - k_n) \neq 0. \tag{3}$$

(Here we assume without proof that the product of n nonzero complex numbers is nonzero.) Using (2) and (3), we have

$$0 = a(k - k_1)(k - k_2) \cdots (k - k_n) \neq 0.$$

This is a contradiction. Hence, our original assumption is invalid, and we have the following result.

THEOREM 12.8   If $P(x)$ is a polynomial of degree n $(n \geq 1)$ over the complex numbers, then there exist at most n distinct complex zeros of $P(x)$.

Note that there exist *at most* n distinct zeros. For example, the polynomial $P(x) = x^3 + 3x^2 + 3x + 1 = (x + 1)^3$ is of degree 3 but has only one zero, $x = -1$.

Using the results of the previous section, we can show that both $2 + i$ and $2 - i$ are zeros of $P(x) = x^3 - x^2 - 7x + 15$. It is not just coincidence that both $2 + i$ and its conjugate $2 - i$ are zeros of this polynomial. We can prove that if $a + bi$ is a zero of a polynomial function over the reals, then so is $a - bi$.

To see this, we need the following properties of complex conjugates. Let $z = a + bi$, and write $\bar{z}$ for the conjugate of z, $\bar{z} = a - bi$. We leave the proof of the following equalities for the exercises.

▶ For any complex numbers c and d,
   (a) $\overline{c + d} = \bar{c} + \bar{d}$
   (b) $\overline{cd} = \bar{c} \cdot \bar{d}$
   (c) $\overline{c^n} = (\bar{c})^n.$

Now consider the polynomial over the *reals*

$$P(x) = a_n x^n + a_{n-1} x^{n-1} + \cdots + a_1 x + a_0,$$

where $a_n \neq 0$. If $z = a + bi$ is a zero of $P(x)$, we can write

$$P(z) = a_n z^n + a_{n-1} z^{n-1} + \cdots + a_1 z + a_0 = 0.$$

If we take the conjugate of both sides of this last equation, we have

$$\overline{a_n z^n + a_{n-1} z^{n-1} + \cdots + a_1 z + a_0} = \bar{0}.$$

By the properties of complex conjugates mentioned above, this becomes

$$\overline{a_n z^n} + \overline{a_{n-1} z^{n-1}} + \cdots + \overline{a_1 z} + \overline{a_0} = \bar{0}$$
$$a_n \cdot \overline{z^n} + a_{n-1} \cdot \overline{z^{n-1}} + \cdots + a_1 \cdot \bar{z} + a_0 = 0$$

(Here we use property (b), and the fact that for a real number a, $\bar{a} = a$.)

$$a_n (\bar{z})^n + a_{n-1} (\bar{z})^{n-1} + \cdots + a_1 (\bar{z}) + a_0 = 0.$$

Hence, $\bar{z}$ is also a zero of $P(x)$, and we have proved the following result.

THEOREM 12.9   If $P(x)$ is a polynomial over the reals, and if $a + bi$ is a zero of $P(x)$, then so is $a - bi$.

Note the importance of the requirement that the polynomial have real coefficients. For example, $P(x) = x - (1 + i)$ has $1 + i$ as a zero, but the conjugate $1 - i$ is not a zero.

This last result is important in helping predict the number of real zeros of a polynomial over the reals. A polynomial of odd degree $n(n \geq 1)$ must have at least one real zero (since we have just seen that complex zeros occur in pairs). On the other hand, a polynomial of even degree $n(n \geq 2)$ need have no real zeros, but on the other hand, may have up to n real zeros.

For example, suppose we want to find all zeros of $x^4 - 7x^3 + 18x^2 - 22x + 12$, given that $x = 1 - i$ is a zero. Since $x = 1 - i$ is a zero, we know by Theorem 12.9 that the conjugate $x = 1 + i$ is also a zero. Thus, we should first divide the original polynomial by $1 - i$ and then divide the quotient by $1 + i$, as follows.

$$\begin{array}{r|rrrrr}
1-i & 1 & -7 & 18 & -22 & 12 \\
 & & 1-i & -7+5i & 16-6i & -12 \\
\hline
 & 1 & -6-i & 11+5i & -6-6i & 0
\end{array}$$

$$\begin{array}{r|rrrr}
1+i & 1 & -6-i & 11+5i & -6-6i \\
 & & 1+i & -5-5i & 6+6i \\
\hline
 & 1 & -5 & 6 & 0
\end{array}$$

Now we must find the zeros of the quadratic polynomial $x^2 - 5x + 6$. By factoring, we find the solutions to be 2 and 3, so that the four zeros of the original polynomial are $1 - i$, $1 + i$, 2, and 3.

12.3 Exercises

For each of the following, find a polynomial of lowest degree with real coefficients having the given zeros.

1. $5 + i$ and $5 - i$
2. $3 - 2i$ and $3 + 2i$
3. 2, $1 - i$ and $1 + i$
4. -3, $2 - i$, and $2 + i$
5. $1 + \sqrt{2}$, $1 - \sqrt{2}$, and 1
6. $1 - \sqrt{3}$, $1 + \sqrt{3}$, and -2

7. $2 + i$, $2 - i$, 3, and -1
8. $3 + 2i$, $3 - 2i$, -1, and 2
9. 2, and $3 + i$
10. -1 and $4 - 2i$
11. $1 + \sqrt{2}$ and $1 - i$
12. $\sqrt{3} + 2$ and $2 + 3i$

For each of the following polynomials, one zero is given. Find all others.

13. $x^3 - x^2 - 4x - 6$; $x = 3$
14. $x^3 - 5x^2 + 17x - 13$; $x = 1$
15. $x^3 + x^2 - 4x - 24$; $x = -2 + 2i$
16. $x^3 + x^2 - 20x - 50$; $x = -3 + i$
17. $2x^3 - 2x^2 - x - 6$; $x = 2$
18. $2x^3 - 5x^2 + 6x - 2$; $x = 1 + i$
19. $x^4 + 5x^2 + 4$; $x = -i$
20. $x^4 + 10x^3 + 27x^2 + 10x + 26$; $x = i$
21. $x^4 - 3x^3 + 6x^2 + 2x - 60$; $x = 1 + 3i$
22. $x^4 - 6x^3 - x^2 + 86x + 170$; $x = 5 + 3i$

Prove each of the following statements for any complex numbers c and d.

23. $\overline{c + d} = \overline{c} + \overline{d}$
24. $\overline{cd} = \overline{c} \cdot \overline{d}$
25. $\overline{(c^n)} = (\overline{c})^n$

12.4 Rational Zeros of Polynomial Functions

We know by the fundamental theorem of algebra that every polynomial of degree at least 1 has a zero. However, the fundamental theorem merely asserts that such a zero exists, but gives no help at all in identifying zeros. For second degree polynomials we can use the quadratic formula. Similar but more complicated formulas exist for finding zeros of third and fourth degree polynomial functions. It was proved by the Norwegian mathematician Abel, at age 22, that it is not possible to produce a formula that will find zeros of fifth degree or higher polynomials.

In practice, however, it is ordinarily sufficient to find only rational zeros, or decimal approximations of any irrational zeros. In this section we shall develop a necessary condition for a rational number to be a zero of a polynomial function, and then in the next section we shall discuss approximations of irrational zeros.

Now consider a polynomial function $P(x)$ with integer coefficients. The next theorem gives a useful test for determining whether or not a given rational number is a possible zero of $P(x)$.

THEOREM 12.10   Let $P(x) = a_n x^n + a_{n-1} x^{n-1} + \cdots + a_1 x + a_0$ $(a_n \neq 0)$ be a polynomial with integer coefficients. If p/q is a rational number written in lowest terms with the property that $P(p/q) = 0$, then p is a factor of a_0 and q is a factor of a_n.

Since p/q is a zero of $P(x)$, we can write

$$a_n \left(\frac{p}{q}\right)^n + a_{n-1}\left(\frac{p}{q}\right)^{n-1} + \cdots + a_1\left(\frac{p}{q}\right) + a_0 = 0.$$

This can also be written as

$$a_n \cdot \frac{p^n}{q^n} + a_{n-1} \cdot \frac{p^{n-1}}{q^{n-1}} + \cdots + a_1 \cdot \frac{p}{q} + a_0 = 0.$$

If we multiply both sides of this last result by q^n, and add $-a_0 q^n$ to both sides, we have

$$a_n p^n + a_{n-1} p^{n-1} q + \cdots + a_1 p q^{n-1} = -a_0 q^n.$$

Factoring out p gives

$$p(a_n p^{n-1} + a_{n-1} p^{n-2} q + \cdots + a_1 q^{n-1}) = -a_0 q^n.$$

By this last result, p must be a factor of $-a_0 q^n$ (why?). Since we have assumed that p/q is written in lowest terms, p and q have no common factor other than 1. Hence, p is not a factor of q^n. Thus, we must conclude that p is a factor of a_0. In a similar way, we can show that q is a factor of a_n.

Example 4 Find all rational zeros of $P(x) = 2x^4 - 11x^3 + 14x^2 - 11x + 12$.

If p/q is to be a rational zero of $P(x)$, we know by Theorem 12.10 that p must be a factor of $a_0 = 12$, and q must be a factor of $a_4 = 2$. Hence, p must be $\pm 1, \pm 2, \pm 3, \pm 4, \pm 6,$ or ± 12, while q must be ± 1 or ± 2. From this, we see that any rational zero of $P(x)$ will come from the list

$$\pm 1, \pm \tfrac{1}{2}, \pm 2, \pm 3, \pm \tfrac{3}{2}, \pm 4, \pm 6, \text{ or } \pm 12.$$

Note that we have no assurance that any of these numbers are zeros. However, if $P(x)$ has any rational zeros at all, they will be in the list above. We can check these proposed zeros by synthetic division. By trial and error we can find that $x = 4$ is a zero.

$$
\begin{array}{r|rrrrr}
4 & 2 & -11 & 14 & -11 & 12 \\
  &   & 8 & -12 & 8 & -12 \\
\hline
  & 2 & -3 & 2 & -3 & 0
\end{array}
$$

As a fringe benefit of this calculation, we now only need to look for zeros of the simpler polynomial $Q(x) = 2x^3 - 3x^2 + 2x - 3$. By Theorem 12.10 any rational zero of $Q(x)$ will have a numerator of ± 3 or ± 1, with a denominator of ± 1 or ± 2. Hence, any rational zeros of $Q(x)$ will come from the list

$$\pm 3, \pm \tfrac{3}{2}, \pm 1, \pm \tfrac{1}{2}.$$

Again we use synthetic division and trial and error to find that $x = \tfrac{3}{2}$ is a zero.

$$
\begin{array}{r|rrrr}
\tfrac{3}{2} & 2 & -3 & 2 & -3 \\
 &   & 3 & 0 & 3 \\
\hline
 & 2 & 0 & 2 & 0
\end{array}
$$

The quotient is $2x^2 + 2$, which, by the quadratic formula, has i and $-i$ as zeros. Hence, the solution set of the polynomial function

$$P(x) = 2x^4 - 11x^3 + 14x^2 - 11x + 12$$

is given by $\{4, \tfrac{3}{2}, i, -i\}$.

To find any rational zeros of a polynomial with rational coefficients, first multiply the polynomial by a number that will clear it of all fractional coefficients. Then use Theorem 12.10.

12.4 Exercises

Find all rational zeros of the following polynomials.

1. $x^3 - 2x^2 - 13x - 10$
2. $x^3 + 5x^2 + 2x - 8$
3. $x^3 + 6x^2 - x - 30$
4. $x^3 - x^2 - 10x - 8$
5. $x^3 + 9x^2 - 14x - 24$
6. $x^3 + 3x^2 - 4x - 12$

7. $x^4 + 9x^3 + 21x^2 - x - 30$
8. $x^4 + 4x^3 - 7x^2 - 34x - 24$
9. $6x^3 + 17x^2 - 31x - 12$
10. $15x^3 + 61x^2 + 2x - 8$
11. $12x^3 + 20x^2 - x - 6$
12. $12x^3 + 40x^2 + 41x + 12$
13. $2x^3 + 7x^2 + 12x - 8$
14. $2x^3 + 20x^2 + 68x - 40$
15. $x^4 + 4x^3 + 3x^2 - 10x + 50$
16. $x^4 - 2x^3 + x^2 + 18$
17. Show that $x^2 - 2$ has no rational zeros, so that $\sqrt{2}$ must be irrational.

12.5 Approximate Zeros of Polynomial Functions

We know that every polynomial function of degree at least 1 has a zero. However, we do not know whether or not the function has real zeros; and even if it does have real zeros, we often have no way to find them. In this section we investigate methods of approximating any real zeros a polynomial function may have.

Much of our work in locating real zeros uses the following result, which follows from the fact that graphs of polynomial functions are continuous, with no gaps or sudden jumps.

THEOREM 12.11   If $P(x)$ is a polynomial function over the reals, a and b are real numbers such that $a < b$ and such that $P(a)$ and $P(b)$ are opposite in sign, then there exists a real number c, $a < c < b$, such that $P(c) = 0$.

The next theorem gives a good method for narrowing the search for real zeros.

THEOREM 12.12   If $P(x) = a_n x^n + a_{n-1} x^{n-1} + \cdots + a_1 x + a_0$ is a polynomial over the reals of degree $n \geq 1$, if $a_n > 0$, and if $P(x)$ is divided synthetically by $x - c$, then we have:

(a)   if $c > 0$ and all numbers in the bottom row of the synthetic division are nonnegative, then c is an upper bound for the zeros of $P(x)$.

(b)   if $c < 0$, and if the numbers in the bottom row of the synthetic division alternate in sign (with 0 considered positive or negative, as needed), then c is a lower bound for the zeros of $P(x)$.

Example 5 Approximate the real zeros of $P(x) = x^4 - 6x^3 + 8x^2 + 2x - 1$.

Let us begin to look for zeros by trying $c = -2$. To do this, we divide $P(x)$ by $x + 2$.

$$
\begin{array}{r|rrrrr}
-2 & 1 & -6 & 8 & 2 & -1 \\
   &   & -2 & 16 & -48 & 92 \\
\hline
   & 1 & -8 & 24 & -46 & 91
\end{array}
$$

Since the numbers in the bottom row alternate in sign, and since $-2 < 0$, we know by Theorem 12.12(b) that $x = -2$ is a lower bound for the zeros of $P(x)$. If we divide $P(x)$ by $x + 1$, we get

$$
\begin{array}{r|rrrrr}
-1 & 1 & -6 & 8 & 2 & -1 \\
   &   & -1 & 7 & -15 & 13 \\
\hline
   & 1 & -7 & 15 & -13 & 12.
\end{array}
$$

By this result, $x = -1$ is also a lower bound for any real zeros of $P(x)$. Note that $P(-1) = 12 > 0$, while $P(0) = -1 < 0$. Hence, there is at least one real number zero between $x = -1$ and $x = 0$.

Let us try $c = -.5$. If we divide $P(x)$ by $x + .5$, we have

$$
\begin{array}{r|rrrrr}
-.5 & 1 & -6 & 8 & 2 & -1 \\
    &   & -.5 & 3.25 & -5.625 & 1.8125 \\
\hline
    & 1 & -6.5 & 11.25 & -3.625 & .8125.
\end{array}
$$

Since $P(-.5) > 0$ and $P(0) < 0$, the real number zero is between $-.5$ and 0.

Now try $c = -.4$.

$$
\begin{array}{r|rrrrr}
-.4 & 1 & -6 & 8 & 2 & -1 \\
    &   & -.4 & 2.56 & -4.224 & .8896 \\
\hline
    & 1 & -6.4 & 10.56 & -2.224 & -.1104
\end{array}
$$

Note that $P(-.5) > 0$, while $P(-.4) < 0$. Since $P(-.4)$ is much closer to zero than $P(-.5)$, we are probably safe in saying that, to one decimal place of accuracy, one real number zero of $P(x)$ is $x = -.4$. Further decimal places of accuracy could be obtained by continuing this process. In the same way, we can show that $P(x)$ has three more real zeros; to one decimal of accuracy these three other zeros are $.3$, 2.4, and 3.7.

Descartes' rule of signs, stated in the next theorem, gives a useful, practical test for determining the number of positive or negative real zeros of a given polynomial.

THEOREM 12.13   DESCARTES' RULE OF SIGNS

Let $P(x)$ be a polynomial over the reals.

(a)   The number of positive real zeros of $P(x)$ is either equal to the number of variations in sign occurring in the coefficients of $P(x)$, or else is less than the number of variations by a positive even integer.

(b)   The number of negative real zeros of $P(x)$ either equals the number of variations in sign of $P(-x)$, or else is less than the number of variations by a positive even integer.

For the purposes of this theorem, $(x - 1)^4$ is said to have four positive zeros, each equal to 1. As an example, $P(x) = x^4 - 6x^3 + 8x^2 + 2x - 1$ has three variations in sign:

$$x^4 - 6x^3 + 8x^2 + 2x - 1.$$

$$\underbrace{\quad}_{1} \underbrace{\quad}_{2} \underbrace{\quad}_{3}$$

Thus, by Descartes' rule of signs, $P(x)$ has either 3 or $3 - 2 = 1$ positive real zeros. We found above that $P(x)$ has three positive real zeros. Since $P(x)$ is of degree 4, and has 3 positive real zeros, we know it must have 1 negative real zero, which also corresponds to the result above. We could also verify this with part (b) of Descartes' rule of signs. Note that

$$P(-x) = (-x)^4 - 6(-x)^3 + 8(-x)^2 + 2(-x) - 1$$
$$= x^4 + 6x^3 + 8x^2 - 2x - 1$$

has only one variation in sign. Hence, $P(x)$ has only one negative real zero, which again corresponds to our result from above.

As a further example, note that

$$Q(x) = x^5 + 5x^4 + 3x^2 + 2x + 1$$

has no variations in sign and hence has no positive real zeros. Here

$$Q(-x) = -x^5 + 5x^4 + 3x^2 - 2x + 1,$$

which has three variations in sign. Hence, $Q(x)$ has either 3 or 1 negative real zeros.

12.5 Exercises

Show that each of the following polynomials has a real zero between the numbers given.

1. $x^3 + 3x^2 - 2x - 6$; 1 and 2
2. $x^3 + x^2 - 5x - 5$; 2 and 3
3. $2x^3 - 8x^2 + x + 16$; 2 and 2.5
4. $3x^3 + 7x^2 - 4$; $\frac{1}{2}$ and 1

Show that the real zeros of each of the following polynomials have the indicated bounds.

5. $x^4 - x^3 + 3x^2 - 8x + 8$; no real zeros greater than 2
6. $2x^5 - x^4 + 2x^3 - 2x^2 + 4x - 4$; no real zero greater than 1
7. $x^4 + x^3 - x^2 + 3$; no real zero less than -2
8. $x^5 + 2x^3 - 2x^2 + 5x + 5$; no real zero less than -1

Find all real zeros of each of the following polynomials. Approximate each zero as a decimal to the nearest tenth.

9. $x^3 + 3x^2 - 2x - 6$
10. $x^3 + x^2 - 5x - 5$
11. $x^3 - 4x^2 - 5x + 14$
12. $x^3 + 9x^2 + 34x + 13$
13. $x^3 + 6x - 13$
14. $x^3 - 3x^2 + 4x - 5$
15. $x^4 - 8x^3 + 17x^2 - 2x - 14$
16. $x^4 - 4x^3 - x^2 + 8x - 2$

12.6 Graphing Polynomial Functions

We have graphed many linear functions (straight lines) and we have graphed many quadratic functions (parabolas). Now we shall investigate the graphs of more general polynomial functions. We shall find that point plotting is more important here than with the other graphs mentioned.

As an example, let us graph the function $P(x) = 8x^3 - 12x^2 + 2x + 1$. By Descartes' rule of signs, $P(x)$ has 2 or 0 positive real zeros, and 1 negative real zero. We need to find some ordered pairs belonging to the graph. We can use synthetic division to evaluate, say, $P(3)$.

$$
\begin{array}{r|rrrr}
3 & 8 & -12 & 2 & 1 \\
  &   & 24 & 36 & 114 \\
\hline
  & 8 & 12 & 38 & 115 \\
\end{array}
$$

Hence, $P(3) = 115$, and $(3, 115)$ belongs to the graph. We can also find $P(-1)$.

$$
\begin{array}{r|rrrr}
-1 & 8 & -12 & 2 & 1 \\
   &   & -8 & 20 & -22 \\
\hline
   & 8 & -20 & 22 & -21 \\
\end{array}
$$

From this result, $(-1, -21)$ belongs to the graph. Since we need several such points in order to sketch the graph, it is helpful to use the shortened form of synthetic division shown below, in which we find values of $P(x)$ corresponding to six different values of x. The numbers printed in bold face below are the numbers we obtained in the bottom row of the synthetic divisions above.

x				$P(x)$	ordered pair
	8	-12	2	1	
3	**8**	**12**	**38**	**115**	$(3, 115)$
2	8	4	10	21	$(2, 21)$
1	8	-4	-2	-1	$(1, -1)$
0	8	-12	2	1	$(0, 1)$
-1	**8**	**-20**	**22**	**-21**	$(-1, -21)$
-2	8	-28	58	-115	$(-2, -115)$

By Theorem 12.11, there is a zero between 0 and 1, and between -1 and 0, as well as between 1 and 2. To get the graph, we plot the points of the chart above, and then draw a continuous curve through them, as in Figure 12.1.

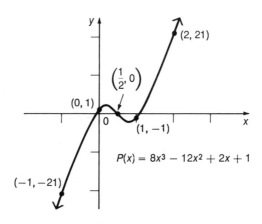

Figure 12.1

Example 6 Graph $P(x) = 3x^4 - 14x^3 + 54x - 3$.

To find points to plot we can again use the abbreviated form of synthetic division presented above.

x					$P(x)$	ordered pair
	3	-14	0	54	-3	
5	3	1	5	79	392	$(5, 392)$
4	3	-2	-8	22	85	$(4, 85)$
3	3	-5	-15	9	24	$(3, 24)$
2	3	-8	-16	22	41	$(2, 41)$
1	3	-11	-11	43	40	$(1, 40)$
0	3	-14	0	54	-3	$(0, -3)$
-1	3	-17	17	37	-40	$(-1, -40)$
-2	3	-20	40	-26	49	$(-2, 49)$
-3	3	-23	69	-153	456	$(-3, 456)$

Since the row in the chart for $x = 5$ contains all positive numbers, the polynomial has no zero greater than 5. Also, since the row for $x = -2$ has numbers which alternate in sign, there is no zero less than -2, by Theorem 12.12. We see here that the polynomial has zeros between 0 and 1, and between -2 and -1. If we plot the points found above and draw a continuous curve through them, we get the graph of Figure 12.2.

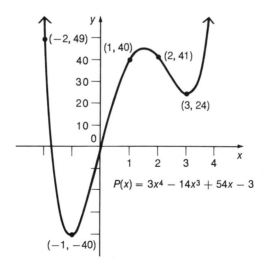

$$P(x) = 3x^4 - 14x^3 + 54x - 3$$

Figure 12.2

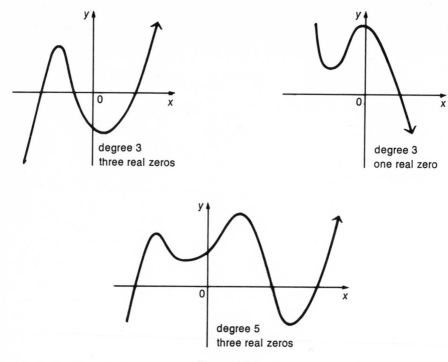

Figure 12.3

In general, the domain of a polynomial function is the set of all real numbers. The range of a polynomial function of odd degree is also the set of all real numbers. Some typical graphs of polynomial functions of odd degree are shown in Figure 12.3. These graphs illustrate that every polynomial function of odd degree has at least one real zero.

Polynomial functions of even degree have a range that takes the form $\{y \mid y \leq k\}$ or $\{y \mid y \geq k\}$ for some real number k. Figure 12.4 shows two typical graphs of polynomial functions of even degree.

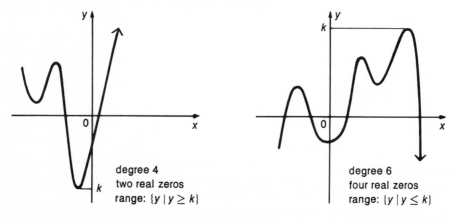

Figure 12.4

12.6 Exercises

Graph each of the following.

1. $P(x) = (x + 1)^3$
2. $P(x) = x^3 + 1$
3. $P(x) = x^3 - 7x - 6$
4. $P(x) = x^3 + x^2 - 4x - 4$
5. $P(x) = x^4 - 5x^2 + 6$
6. $P(x) = x^3 - 3x^2 - x + 3$

7. $P(x) = 6x^3 + 11x^2 - x - 6$
8. $P(x) = x^4 - 2x^2 - 8$
9. $P(x) = x^4 + x^3 - 2$
10. $P(x) = 6x^4 - x^3 - 23x^2 - 4x + 12$
11. $P(x) = 8x^4 - 2x^3 - 47x^2 - 52x - 15$

12.7 Rational Functions

A function of the form

$$\left\{ (x, y) \mid y = \frac{P(x)}{Q(x)}, \ P(x), Q(x) \text{ polynomials}, Q(x) \neq 0 \right\}$$

is called a **rational function.** We shall graph several such functions in this section.

The function

$$y = \frac{2}{1 + x}$$

is undefined for $x = -1$. Hence, the graph of this function will not intersect the vertical line $x = -1$. Since x can equal any number except -1, we can let x approach -1 as closely as we wish. From the following table, note that as x gets closer and closer to -1, $1 + x$ gets closer and closer to 0, and $\frac{2}{(1 + x)}$ gets larger and larger. The vertical line $x = -1$ that is approached by the curve is called a **vertical asymptote.**

x	$1 + x$	$\dfrac{2}{1 + x}$
$-.5$	$.5$	4
$-.8$	$.2$	10
$-.9$	$.1$	20
$-.99$	$.01$	200

Note also that as $|x|$ gets larger and larger, $\frac{2}{1 + x}$ gets closer and closer to 0. Hence, the graph has the x-axis as a **horizontal asymptote.** By using these asymptotes and plotting the intercepts and a few points, we get the graph of Figure 12.5, page 288.

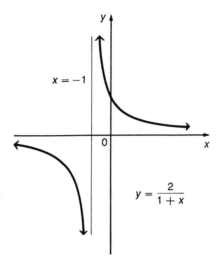

Figure 12.5

Example 7 Graph $y = \dfrac{1}{(x-1)(x+4)}$.

Note that there are vertical asymptotes at $x = 1$ and $x = -4$. As $|x|$ gets larger and larger, $|(x-1)(x+4)|$ also gets larger and larger. Thus, y gets closer and closer to 0, making the x-axis a horizontal asymptote. The vertical asymptotes divide the x-axis into three regions. It is often convenient to consider each region separately. If we find the intercepts and plot a few points, we get the result shown in Figure 12.6.

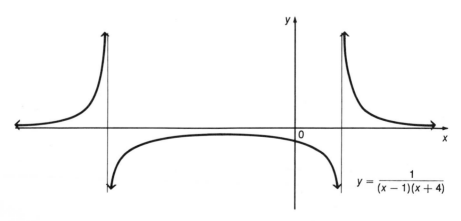

Figure 12.6

To graph

$$y = \frac{x^2 + 1}{x - 2},$$

we can divide $x - 2$ into $x^2 + 1$ using synthetic division.

$$\begin{array}{r|rrr} 2 & 1 & 0 & 1 \\ & & 2 & 4 \\ \hline & 1 & 2 & 5 \end{array}$$

This gives

$$y = x + 2 + \frac{5}{x - 2}.$$

As $|x|$ gets larger and larger, $\dfrac{5}{x - 2}$ gets closer and closer to 0. Hence, the graph approaches the **oblique asymptote** $y = x + 2$. Using this fact, and noticing that $x = 2$ is a vertical asymptote, we get the graph of Figure 12.7.

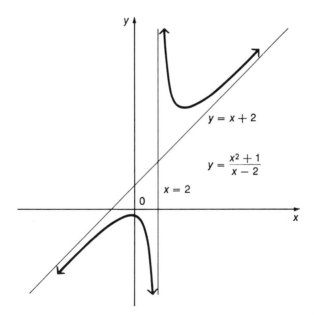

Figure 12.7

12.7 Exercises

Graph each of the following.

1. $y = \dfrac{1}{1 - x}$

2. $y = \dfrac{-2}{x}$

3. $y = \dfrac{2}{3 + 2x}$

4. $y = \dfrac{4}{5 + 3x}$

5. $y = \dfrac{1}{(x + 1)(x - 3)}$

6. $y = \dfrac{1}{(x - 2)(x + 4)}$

7. $y = \dfrac{3}{(x + 4)^2}$

8. $y = \dfrac{2}{(x - 3)^2}$

9. $y = \dfrac{3x}{x - 1}$

10. $y = \dfrac{4x}{1 - 3x}$

11. $y = \dfrac{x + 1}{x - 4}$

12. $y = \dfrac{x - 3}{x + 5}$

13. $y = \dfrac{x}{x^2 - 9}$

14. $y = \dfrac{x^2 + 1}{x + 3}$

15. $y = \dfrac{x^2 - 5}{x + 2}$

16. $y = \dfrac{x^2 - 3x + 2}{x - 3}$

Chapter Thirteen

Analytic Geometry

Some analytic geometry has already been discussed in this text. We studied the analytic geometry of straight lines in Chapter 3, and we shall extend that discussion somewhat in Section 1 of this chapter. We discussed conic sections in Chapter 3, and we shall extend that discussion in this chapter. We shall also discuss another coordinate system, polar coordinates, which can be used to simplify certain types of graphing.

13.1 Straight Lines

Figure 13.1, page 292, shows a line which makes an angle α with the positive x-axis. If α is the smallest such positive angle, it is called the **angle of inclination** of the line. By this definition, we have $0 \leq \alpha < 180°$. We have defined the slope of a line as the change in y divided by the change in x. Using this, and the definition of the tangent of an angle from Chapter 6, we have

▶
$$m = \tan \alpha,$$

where m is the slope of the line, and α is the angle of inclination of the line.

Example 1   Find the angle of inclination of the line through (5, 8) and (8, 11).

The tangent of the angle of inclination equals the slope of the line. Thus, if α is the angle of inclination, we have

$$\tan \alpha = \frac{11 - 8}{8 - 5} = \frac{3}{3} = 1.$$

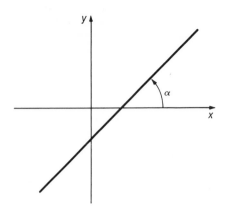

Figure 13.1

Since $\tan \alpha = 1$, and since $0 \leq \alpha < 180°$, we see

$$\alpha = 45°.$$

Example 2   Find the angle of inclination of the line through $(-4, 3)$ and $(2, -5)$.

Here we have

$$\tan \alpha = \frac{-5 - 3}{2 - (-4)} = \frac{-8}{6} = \frac{-4}{3}.$$

Hence, $\alpha \approx 127°$.

We can use the fact that the slope of a line equals the tangent of the angle of inclination to prove some results about straight lines. For example, if two lines are parallel (but not vertical) they will have the same angle of inclination and hence the same slope. Also, if two lines have the same slope, then they will have the same angle of inclination, and be parallel. (Here we use the fact that for any angle of inclination α, $0 \leq \alpha < 180°$.) Thus we have proved the following result.

THEOREM 13.1   Two nonvertical lines are parallel if and only if they have the same slope.

As shown in Figure 13.2, the angles of inclination of two perpendicular lines differ by 90°. Thus, if α_1 and α_2 are the angles of inclination of two perpendicular lines, we have

$$\alpha_2 = \alpha_1 + 90°.$$

Hence we can write

$$\tan \alpha_2 = \tan(\alpha_1 + 90°).$$

Using the fact that $\tan(\alpha_1 + 90°) = -\cot \alpha_1$, we have

$$\tan \alpha_2 = -\cot \alpha_1$$

$$\tan \alpha_2 = -\frac{1}{\tan \alpha_1}.$$

From this, we see that the corresponding slopes m_1 and m_2 satisfy the relationship

$$m_2 = -\frac{1}{m_1}.$$

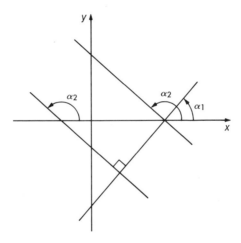

Figure 13.2

Hence, two perpendicular lines have slopes which are negative reciprocals of each other. By repeating these steps in reverse order, and again using the fact that angles of inclination α satisfy the relationship $0 \leq \alpha < 180°$, we can conclude that if two slopes are negative reciprocals of each other, then the corresponding lines are perpendicular. This proves the following theorem.

THEOREM 13.2   Two lines (neither of which is vertical) are perpendicular if and only if the slopes of the lines are negative reciprocals of each other.

Example 3 Find the equation of the line through $(3, -5)$ and perpendicular to $3x + y = -5$.

The slope of the desired line is the negative reciprocal of the slope of $3x + y = -5$. Recall that we can find the slope of $3x + y = -5$ by writing it in the slope-intercept form, $y = mx + b$, where m is the slope and b is the y-intercept. We have

$$3x + y = -5$$
$$y = -3x - 5.$$

The slope here is -3, and thus the slope of the desired perpendicular line is $-1/(-3) = 1/3$. We also know that the desired line goes through $(3, -5)$. We can find its equation by using the point-slope form of the equation of a line: a line with slope m going through (x_1, y_1) has equation $y - y_1 = m(x - x_1)$. In this example we have

$$y - (-5) = \frac{1}{3}(x - 3)$$
$$3y + 15 = x - 3$$
$$3y + 18 = x.$$

13.1 Exercises

Find the angle of inclination of the lines through the following pairs of points.

1. $(1, 2)$, $(4, 5)$
2. $(-2, 8)$, $(1, 11)$
3. $(1, 2)$, $(2, 2 + \sqrt{3})$
4. $(-3, 5)$, $(-3 + \sqrt{3}, 6)$

5. $(2, -1)$, $(3, -4)$
6. $(-3, 2)$, $(-5, 4)$
7. $(2, -9)$, $(-5, 6)$
8. $(3, 7)$, $(-7, 2)$

Determine the slope of all lines parallel to each of the following lines.

9. a line having slope $m = 2$
10. a line having slope $m = -3/4$
11. $3y + 2x = 5$
12. $4y - 5x = 3$

13. $y = 4$
14. $x + 3 = 0$
15. through $(-5, 8)$ and $(3, -4)$
16. through $(6, -2)$ and $(9, -5)$

Determine the slope of all lines perpendicular to each of the following lines.

17. a line having slope $m = 3$
18. a line having slope $m = -5/8$
19. $4x + 2y = 3$
20. $-3x + 7y = 8$

21. $y = -3$
22. $x = 4$
23. through $(7, -2)$ and $(8, -3)$
24. through $(-3, 6)$ and $(5, -4)$

Find the equation of each of the following lines.

25. through $(8, -2)$, parallel to $x + 4y = -3$
26. through $(-4, 3)$, parallel to $3x - y = 1$
27. through $(-1, 3)$, perpendicular to $3x + 2y = 4$

28. through $(2, -4)$, perpendicular to $-4y + 2x = 5$
29. Find k so that the line through the points $(-2, 3)$ and $(4, k)$ is
    (a) parallel to a line having slope $m = 1/2$.
    (b) perpendicular to a line having slope $m = -3/4$.
30. Find k so that the line through $(4, -1)$ and $(k, 2)$ is
    (a) parallel to $3y + 2x = 6$.
    (b) perpendicular to $2y - 5x = 1$.

13.2 Analytic Proofs

In this section we shall use methods of analytic geometry to prove certain results from plane geometry. We shall find that many of these results can be proved more easily by analytic proofs than by the synthetic proofs of most plane geometry courses. Many analytic proofs depend on the distance formula, which we discussed earlier, and the midpoint formula, as stated below.

THEOREM 13.3   The midpoint of the segment with endpoints (x_1, y_1) and (x_2, y_2) is

$$\left(\frac{x_1 + x_2}{2}, \frac{y_1 + y_2}{2} \right).$$

Proof   To prove this result, let A have coordinates (x_1, y_1), let B have coordinates (x_2, y_2), and let B' be the midpoint of segment AB, as shown in Figure 13.3, page 296. Let AC be a horizontal line, and let $B'C'$ and BC each be perpendicular to AC. Triangles $AB'C'$ and ABC are similar, so that

$$\frac{AB'}{AB} = \frac{AC'}{AC}.$$

Since B' is the midpoint of AB,

$$\frac{AB'}{AB} = \frac{1}{2}.$$

Thus

$$\frac{AC'}{AC} = \frac{1}{2},$$

or

$$AC' = \frac{1}{2}AC.$$

From Figure 13.3, we have $AC = x_2 - x_1$. The x-coordinate of points C' and B' is thus

$$x_1 + \frac{1}{2}AC = x_1 + \frac{1}{2}(x_2 - x_1) = \frac{1}{2}(x_1 + x_2).$$

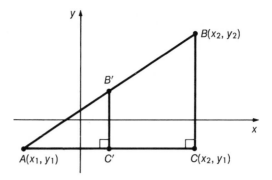

Figure 13.3

In the same way, the y-coordinate of point B' is

$$\frac{1}{2}(y_1 + y_2),$$

which proves the theorem.

We can use this theorem to prove the following result from plane geometry: the line segment joining the midpoints of two sides of a triangle is parallel to the third side and has half its length. To prove this by analytic methods, we place a general triangle on a coordinate system. We can do this in a way most convenient to the proof. Here it would be helpful to place one vertex at the origin $(0, 0)$, with one side of the triangle along the x-axis, as shown in Figure 13.4. We can label another vertex as $(a, 0)$, with the third vertex labeled (b, c). By the midpoint formula, the midpoint of AB is

$$\left(\frac{b + 0}{2}, \frac{c + 0}{2}\right) = \left(\frac{b}{2}, \frac{c}{2}\right),$$

while the midpoint of BC is

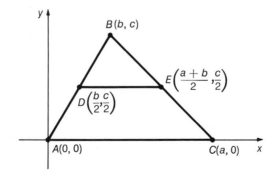

Figure 13.4

$$\left(\frac{a+b}{2}, \frac{0+c}{2}\right) = \left(\frac{a+b}{2}, \frac{c}{2}\right).$$

The slope of DE can be found from the two endpoints, $\left(\frac{b}{2}, \frac{c}{2}\right)$ and $\left(\frac{a+b}{2}, \frac{c}{2}\right)$.

$$\text{slope of } DE = \frac{\dfrac{c}{2} - \dfrac{c}{2}}{\dfrac{(a+b)}{2} - \dfrac{b}{2}} = 0$$

Hence, DE is horizontal and thus parallel to AC. The length of DE can be found by the distance formula. We have

$$\sqrt{\left(\frac{a+b}{2} - \frac{b}{2}\right)^2 + \left(\frac{c}{2} - \frac{c}{2}\right)^2} = \sqrt{\left(\frac{a}{2}\right)^2} = \frac{a}{2},$$

which is half the length of AC. Hence, DE is parallel to AC and half as long as AC.

Example 4 Prove that the diagonals of a parallelogram bisect each other.

Figure 13.5 shows parallelogram $ABCD$ placed on a coordinate system with A at the origin and side AD along the x-axis. We assign coordinates (b, c) to B and $(a, 0)$ to D. Since DC is parallel to AB and the same length as AB, we write the coordinates of C as $(a + b, c)$. To show that the diagonals bisect each other, we show that they have the same midpoint. By the midpoint formula, we have

$$\text{midpoint of } AC = \left(\frac{a+b+0}{2}, \frac{c+0}{2}\right) = \left(\frac{a+b}{2}, \frac{c}{2}\right)$$

$$\text{midpoint of } BD = \left(\frac{a+b}{2}, \frac{c+0}{2}\right) = \left(\frac{a+b}{2}, \frac{c}{2}\right).$$

Thus, AC and BD bisect each other.

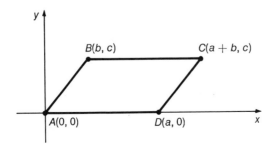

Figure 13.5

13.2 Exercises

Find the midpoints of the line segments having the following endpoints.

1. $(8, 11)$, $(16, -9)$
2. $(-6, 5)$, $(8, -3)$
3. $(7, 5)$, $(8, 9)$
4. $(-5, 8)$, $(8, -5)$
5. $(-1, 5)$, $(2, -4)$
6. $(3, 8)$, $(-5, -9)$

Find the other endpoint of the segments having endpoints and midpoints as given.

7. endpoint $(-3, 6)$, midpoint $(5, 8)$
8. endpoint $(2, -8)$, midpoint $(3, -5)$
9. endpoint $(6, -1)$, midpoint $(-2, 5)$
10. endpoint $(-5, 3)$, midpoint $(-7, 6)$

Give analytic proofs of each of the following.

11. The midpoint of the hypotenuse of a right triangle is equally distant from all three vertices.
12. The diagonals of a rectangle are equal.
13. The diagonals of a square are perpendicular.
14. The diagonals of an isosceles trapezoid are equal.
15. The line segments joining consecutive midpoints of the sides of a quadrilateral form a parallelogram.
16. The sum of the squares of the lengths of the diagonals of a parallelogram is equal to the sum of the squares of the lengths of the four sides.
17. If two medians of a triangle are equal, then the triangle is isosceles.
18. If the diagonals of a parallelogram are equal, the parallelogram is a rectangle.
19. The perpendicular bisectors of the sides of a triangle meet in a point.

13.3   Two Useful Theorems

In this section we shall develop two useful formulas for straight lines—the formula for the distance from a point to a line and the formula for the angle between two lines. To find the distance from the point (x_1, y_1) to the line $y = mx + b(m \neq 0)$, as shown in Figure 13.6, first recall that distance from a point to a line is measured along the perpendicular drawn from the point to the line. We shall find the slope of this perpendicular line, and then its equation. Then we can find (x_2, y_2), the point of intersection of the two lines. Finally, we can use the distance formula to find the distance from (x_1, y_1) to (x_2, y_2). This process requires a considerable amount of algebra, as we shall see.

We must first find the slope of the line through (x_1, y_1) and (x_2, y_2).

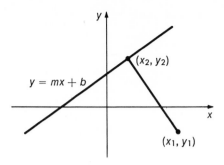

Figure 13.6

Since this line is to be perpendicular to $y = mx + b$, its slope must be the negative reciprocal of the slope of $y = mx + b$, or $\dfrac{-1}{m}$. Thus, the line through (x_1, y_1) and (x_2, y_2) has an equation given by the point-slope form of the equation of a line as

$$y - y_1 = -\frac{1}{m}(x - x_1)$$

$$my - my_1 = x_1 - x. \tag{1}$$

To find the coordinates of the point of intersection (x_2, y_2), we find the simultaneous solution of equation (1) and $y = mx + b$. Substituting $y = mx + b$ into (1) gives

$$m(mx + b) - my_1 = x_1 - x$$

$$m^2 x + bm - y_1 m = x_1 - x$$

$$(m^2 + 1)x = x_1 - bm + y_1 m.$$

Thus, $$x_2 = x = \frac{x_1 - bm + y_1 m}{m^2 + 1}. \tag{2}$$

This gives the x-coordinate of the point of intersection. Verify that the y-coordinate is given by

$$y_2 = \frac{mx_1 + y_1 m^2 + b}{m^2 + 1}. \tag{3}$$

Now we use the distance formula to find the distance, d, between (x_1, y_1) and (x_2, y_2). Using (2) and (3) above, we have

$$d^2 = \left(\frac{x_1 - bm + y_1 m - x_1 m^2 - x_1}{m^2 + 1}\right)^2 + \left(\frac{mx_1 + y_1 m^2 + b - y_1 m^2 - y_1}{m^2 + 1}\right)^2$$

$$= \frac{(-b + y_1 - x_1 m)^2 m^2}{(m^2 + 1)^2} + \frac{(b + mx_1 - y_1)^2}{(m^2 + 1)^2}$$

$$= \frac{(-b + y_1 - x_1 m)^2 (m^2 + 1)}{(m^2 + 1)^2}$$

$$d^2 = \frac{(-b + y_1 - x_1 m)^2}{(m^2 + 1)}.$$

Since distance is nonnegative, we can now write

$$d = \frac{|-b + y_1 - x_1 m|}{\sqrt{m^2 + 1}} \quad \text{or} \quad \frac{|y_1 - mx_1 - b|}{\sqrt{m^2 + 1}}.$$

This result is summarized in the next theorem.

THEOREM 13.4   The distance between the point (x_1, y_1) and the line $y = mx + b$ is given by

$$\frac{|y_1 - mx_1 - b|}{\sqrt{m^2 + 1}}.$$

Note the similarity of the numerator to the slope-intercept form of the equation of a line, $y = mx + b$, or $y - mx - b = 0$.

It can be shown that if the equation of the line is given by $Ax + By + C = 0$, the result of Theorem 13.4 can be expressed as follows.

THEOREM 13.5   DISTANCE FROM A POINT TO A LINE
The distance between the point (x_1, y_1) and the line $Ax + By + C = 0$ is given by

$$\frac{|Ax_1 + By_1 + C|}{\sqrt{A^2 + B^2}}.$$

Figure 13.7 shows two nonparallel lines, one of slope m_1 and the other of slope m_2. We know that the slope of a line also gives the tangent of the angle of inclination, or

$$m_1 = \tan \theta_1 \qquad\qquad (4)$$

$$m_2 = \tan \theta_2. \qquad\qquad (5)$$

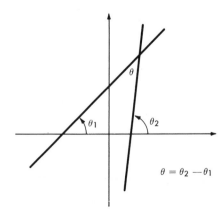

Figure 13.7

Let θ be the acute angle between the two lines. Verify that $\theta = \theta_2 - \theta_1$. Using the formula for $\tan(x + y)$ from Chapter 7, we have

$$\tan(\theta_2 - \theta_1) = \tan\theta = \frac{\tan\theta_2 - \tan\theta_1}{1 + \tan\theta_2 \cdot \tan\theta_1}. \tag{6}$$

We can now substitute the results from (4) and (5) into equation (6), obtaining

$$\tan\theta = \frac{m_2 - m_1}{1 + m_1 \cdot m_2}. \tag{7}$$

If we had let θ be the obtuse angle in Figure 13.7, we would have obtained a formula of the same form as (7), but having opposite sign. Since an acute angle has a positive tangent, while an obtuse angle has a negative tangent, the sign of (7) determines which case we have. If the value of (7) is positive, θ is the acute angle between the lines, while if the value of (7) is negative, θ is the obtuse angle between the two lines. (Since $\tan 90°$ is undefined, (7) will not hold for perpendicular lines.) This can all be summarized as in the next theorem.

THEOREM 13.6   ANGLE BETWEEN TWO LINES

Let $y = m_1 x + b_1$ and $y = m_2 x + b_2$ be any two non-parallel, nonvertical, nonperpendicular lines. Then θ, the angle between the lines, is given by

$$\tan\theta = \frac{m_2 - m_1}{1 + m_1 m_2}.$$

If one of the two lines is vertical, the formula becomes

$$\tan\theta = \frac{1}{m},$$

where m is the slope of the nonvertical line.

Example 5   Find the angle between the lines $3x - 2y = 5$ and $4x = y - 5$.

The slope of $3x - 2y = 5$ is $3/2$, and the slope of $4x = y - 5$ is 4. Let $m_2 = 4$, with $m_1 = 3/2$. If θ is the angle between the two lines, we have

$$\tan\theta = \frac{m_2 - m_1}{1 + m_1 m_2}$$

$$= \frac{4 - \frac{3}{2}}{1 + 4 \cdot \frac{3}{2}}$$

$$= \frac{5}{14}$$

$$\tan\theta \approx .357$$

$$\theta \approx 20°.$$

13.3 Exercises

Find the distances from the given points to the line $3x - 4y = -2$.

1. $(-5, 3)$
2. $(2, -4)$
3. $(-4, -2)$
4. $(-3, -6)$

Find the acute angles (to the nearest degree) between the following pairs of lines.

5. $x = 2y, x + y = 0$
6. $x - 3y = 4, x = 2y$
7. $3x - 2y = 1, 4x + y = 5$
8. $2x - 3y = -1, 2x + 3y = 4$

Find all angles of the triangles having vertices as given.

9. $(2, 4), (2, 6), (4, 4)$
10. $(-3, -4), (-3, -7), (-6, -4)$
11. $(3, -2), (5, -2), (5, -2 + 2\sqrt{3})$
12. $(4, -8), (4, -10), (3, -6)$
13. The inclination of a line is $60°$. Find the slope and inclination of any line making an angle of $45°$ with the first line.
14. The inclination of a line is $135°$. Find the slope and inclination of any line making an angle of $60°$ with the first line.

The area of a triangle can be found by the formula $A = \frac{1}{2}ab \sin C$, where a and b are the lengths of two of the sides, and C is the included angle. Find the areas of the triangles having vertices as listed.

15. $(2, -6), (2, -3), (5, -4)$
16. $(-5, 8), (-3, 7), (-2, -1)$

The figure below shows a line with equation $x = -p$, the point $(p, 0)$, and a curve with the property that any point on the curve is equidistant from the given line and the given point.

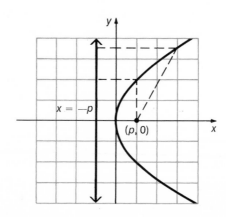

17. Use the distance formula to find the distance from any point (x, y) of the curve to the point $(p, 0)$.

18. Use the formula for the distance from a point to a line to find the distance from (x, y) to the line $x = -p$.

19. Equate the results from Exercises 17 and 18 and simplify.

20. Verify that the curve is a parabola.

The point $(p, 0)$ is called the focus of a parabola with equation $y^2 = 4px$. Find the focus of each of the following parabolas.

21. $x = 16y^2$
22. $x = -32y^2$
23. $x = y^2$
24. $2x = -y^2$
25. $y = 4x^2$ (Hint: $(0, p)$ is the focus of a parabola with equation $x^2 = 4py$.)
26. $y = -18x^2$

13.4 Translation of Coordinate Axes

When graphing functions it is often easier to graph the function on a coordinate system that has been transformed from the usual one. We shall discuss two such simplifying transformations—in this section we discuss translations of coordinate axes, with rotation of the axes discussed in the next section. Figure 13.8 shows two coordinate systems. One has an x-axis and a y-axis. We shall refer to this system as the xy-system. The other has an x'-axis and a y'-axis, and will be called the $x'y'$-system. Because the x'-axis is parallel to the x-axis, and the y'-axis is parallel to the y-axis, the $x'y'$-system is called a **translation** of the xy-system. The origin of the $x'y'$-system has coordinates $(0, 0)$ with respect to the $x'y'$-system, but coordinates (h, k) with respect to the xy-system. To find formulas converting coordinates from one system to another, consider point P of Figure 13.8, page 304. Point P has coordinates (x, y) with respect to the xy-system, and (x', y') with respect to the $x'y'$-system. In Figure 13.8, we have $DP = x$, $DC = h$, and $CP = x'$. Since

$$DP = PC + CD,$$

we have $$x = x' + h.$$

In the same way, $$y = y' + k.$$

While the discussion above tacitly assumes that the origin of the $x'y'$-system is in the first quadrant of the xy-system, the results obtained are true in general, as summarized in the following theorem.

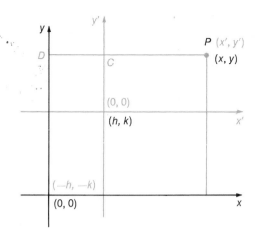

Figure 13.8

THEOREM 13.7   If an xy-coordinate system is translated so that the origin of the $x'y'$-system has coordinates $(h,\ k)$ with respect to the xy-system, and if $(x,\ y)$ and $(x',\ y')$ are the corresponding coordinates of the same point, then

$$x = x' + h \text{ and } y = y' + k,$$

or $\qquad x' = x - h \text{ and } y' = y - k.$

Example 6   Graph $\dfrac{(x-5)^2}{9} + \dfrac{(y+3)^2}{25} = 1.$ $\qquad\qquad$ (1)

Let $x' = x - 5$ and $y' = y + 3$. Then equation (1) becomes

$$\frac{x'^2}{9} + \frac{y'^2}{25} = 1$$

with respect to the new $x'y'$-system. This equation represents an ellipse with x' intercepts 3 and -3, and y' intercepts 5 and -5. Since $x' = x - 5$ and $y' = y + 3 = y - (-3)$, the origin of the $x'y'$-system has coordinates $(h,\ k) = (5,\ -3)$ in the xy-system. The resulting graph is shown in Figure 13.9.

Example 7   Graph $2x + y^2 - 8y = -26$.

We can complete the square on y.

$$y^2 - 8y + 16 = -2x - 26 + 16$$
$$(y-4)^2 = -2x - 10$$
$$(y-4)^2 = -2(x+5)$$

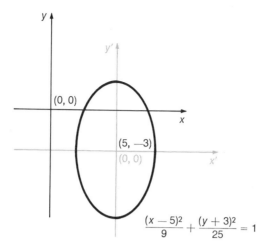

Figure 13.9

Here we can let $y' = y - 4$ and $x' = x + 5$. The equation thus becomes

$$y'^2 = -2x',$$

which is the equation of a parabola with vertex at the origin of the $x'y'$-system and opening to the left. (Parabolas of this type are discussed in Exercises 17–26 of Section 13.3.) The origin of the $x'y'$-system has coordinates $(h, k) = (-5, 4)$ in the xy-system. The graph of the parabola is shown in Figure 13.10.

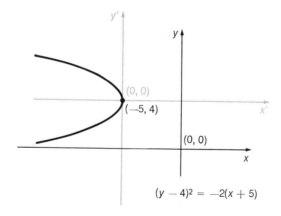

Figure 13.10

Example 8   Graph $4x^2 + 24x - y^2 - 12y - 4 = 0$.

If we complete the square on both x and y, we have

$$4(x^2 + 6x \qquad) - (y^2 + 12y \qquad) = 4$$
$$4(x^2 + 6x + 9) - (y^2 + 12y + 36) = 4 + 36 - 36$$
$$4(x + 3)^2 - (y + 6)^2 = 4$$
$$(x + 3)^2 - \frac{(y + 6)^2}{4} = 1.$$

To simplify the graphing, we can translate the axes by letting $x' = x + 3$ and $y' = y + 6$. This yields

$$x'^2 - \frac{y'^2}{4} = 1,$$

a hyperbola with x' intercepts of 1 and -1. Here the origin of the $x'y'$ system has xy coordinates $(h, k) = (-3, -6)$. The asymptotes are given by

$$x'^2 - \frac{y'^2}{4} = 0,$$

or

$$x' = \frac{\pm y'}{2}.$$

These results lead to the graph of Figure 13.11.

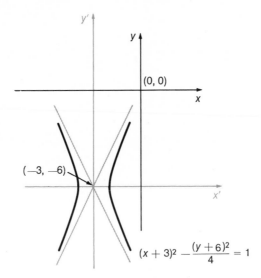

Figure 13.11

Generalizing from the examples presented above, we can say that if the general second degree relation

$$Ax^2 + Cy^2 + Dx + Ey + F = 0$$

has a graph, it is one of the following.

▶   A circle (or a point) if $A = C \neq 0$.
▶   An ellipse (or a point) if $AC > 0$, and $A \neq C$,
▶   A hyperbola (or two intersecting lines) if $AC < 0$,
▶   A parabola (or one line, or two parallel lines) if either $A = 0$ or $C = 0$, but not both.

(Note: the coefficient B is reserved for an xy term, as discussed in the next section.)

13.4 Exercises

Graph each of the following.

1. $(x - 5)^2 = 4y$
2. $(y + 2)^2 = -6x$
3. $(y - 5)^2 = 3(x + 2)$
4. $(x + 1) = -4(y - 3)^2$
5. $\dfrac{(x + 3)^2}{9} + \dfrac{(y - 5)^2}{4} = 1$
6. $\dfrac{(x - 4)^2}{16} + \dfrac{(y + 6)^2}{25} = 1$
7. $\dfrac{(x - 6)^2}{9} - \dfrac{(y + 2)^2}{25} = 1$
8. $\dfrac{(y + 5)^2}{16} - \dfrac{(x - 7)^2}{25} = 1$
9. $x^2 - 2x - 6y = 11$
10. $2x - y^2 - 6y - 11 = 0$
11. $x - 2y^2 - 20y - 47 = 0$
12. $3x^2 - 30x - 2y + 79 = 0$
13. $x^2 - 2x + y^2 + 6y + 1 = 0$
14. $x^2 + 10x + y^2 + 4y + 13 = 0$
15. $x^2 - 6x + y^2 - 10y + 42 = 0$
16. $x^2 + 4x - y^2 + 10y - 20 = 0$
17. $9x^2 - 90x + 25y^2 - 200y + 400 = 0$
18. $x^2 + 6x + 4y^2 - 20y + 105 = 0$
19. $x^2 - 6x - 4y^2 - 32y - 71 = 0$
20. $16x^2 + 96x - 9y^2 - 36y - 36 = 0$
21. $5y^2 - 20y - 3x^2 - 24x - 28 = 0$

13.5 Rotation of Axes

If we begin with an xy-coordinate system having origin 0, and rotate the axes about 0 through an angle θ, we have performed a rotation transformation. We can use trigonometric identities to obtain equations for converting the coordinates of a point from the xy-system to the $x'y'$-system. To do this, let P be any point $(P \neq 0)$, and let P have coordinates (x, y) in the xy-system and (x', y') in the $x'y'$-system. Let $OP = r$, and let α represent the angle made by OP and the x'-axis. As shown in Figure 13.12, page 308,

$$\cos(\theta + \alpha) = \frac{OA}{r} = \frac{x}{r}$$

$$\sin(\theta + \alpha) = \frac{AP}{r} = \frac{y}{r}$$

$$\cos \alpha = \frac{OB}{r} = \frac{x'}{r}$$

$$\sin \alpha = \frac{PB}{r} = \frac{y'}{r}.$$

These four statements can be rewritten as

$$x = r \cos(\theta + \alpha), \qquad y = r \sin(\theta + \alpha),$$
$$x' = r \cos \alpha, \qquad y' = r \sin \alpha.$$

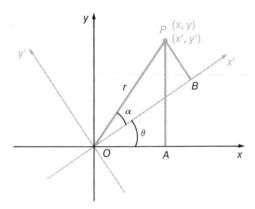

Figure 13.12

Using the trigonometric identity for the cosine of the sum of two angles, we have

$$x = r \cos (\theta + \alpha)$$
$$= r(\cos \theta \cdot \cos \alpha - \sin \theta \cdot \sin \alpha)$$
$$= (r \cos \alpha)\cos \theta - (r \sin \alpha)\sin \theta$$
$$x = x'\cos \theta - y'\sin \theta.$$

In the same way, using the identity for the sine of the sum of two angles, we have

$$y = x'\sin \theta + y'\cos \theta.$$

We have proved the following result.

THEOREM 13.8   If the coordinate axes are rotated about the origin through an angle θ, and if the coordinates of a point P are (x, y) and (x', y') with respect to the xy-system and the $x'y'$-system, then

$$x = x' \cos \theta - y' \sin \theta$$

$$y = x' \sin \theta + y' \cos \theta.$$

Example 9 The equation of a curve is $x^2 + y^2 + 2xy + 2\sqrt{2}x - 2\sqrt{2}y = 0$. Find the resulting equation if the axes are rotated $45°$.

If we let $\theta = 45°$, then using Theorem 13.8 we have

$$x = \frac{\sqrt{2}}{2}x' - \frac{\sqrt{2}}{2}y'$$

$$y = \frac{\sqrt{2}}{2}x' + \frac{\sqrt{2}}{2}y'.$$

Substituting these values in the given equation yields

$$x^2 + y^2 + 2xy + 2\sqrt{2}x - 2\sqrt{2}y = 0$$

$$\left[\frac{\sqrt{2}}{2}x' - \frac{\sqrt{2}}{2}y'\right]^2 + \left[\frac{\sqrt{2}}{2}x' + \frac{\sqrt{2}}{2}y'\right]^2 +$$

$$2\left[\frac{\sqrt{2}}{2}x' - \frac{\sqrt{2}}{2}y'\right] \times \left[\frac{\sqrt{2}}{2}x' + \frac{\sqrt{2}}{2}y'\right] +$$

$$2\sqrt{2}\left[\frac{\sqrt{2}}{2}x' - \frac{\sqrt{2}}{2}y'\right] - 2\sqrt{2}\left[\frac{\sqrt{2}}{2}x' + \frac{\sqrt{2}}{2}y'\right] = 0.$$

Expanding these terms, we have

$$\frac{1}{2}x'^2 - x'y' + \frac{1}{2}y'^2 + \frac{1}{2}x'^2 + x'y' + \frac{1}{2}y'^2 + x'^2 - y'^2$$

$$+ 2x' - 2y' - 2x' - 2y' = 0.$$

Collecting terms gives

$$2x'^2 - 4y' = 0,$$

or finally, $$x'^2 - 2y' = 0,$$

which is the equation of a parabola. The graph of this parabola is shown in Figure 13.13, page 310.

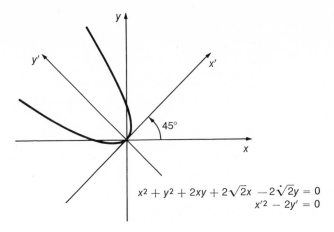

$$x^2 + y^2 + 2xy + 2\sqrt{2}x - 2\sqrt{2}y = 0$$
$$x'^2 - 2y' = 0$$

Figure 13.13

We have learned to graph relations whose equations can be stated in the general form $Ax^2 + Cy^2 + Dx + Ey + F = 0$. In the preceeding example, the rotation of axes eliminated the xy term. Thus, to graph an equation which has an xy term, it is necessary to find an appropriate angle of rotation to eliminate the xy term. The necessary angle of rotation can be determined by using the next theorem. The proof of the theorem is quite lengthy and so we shall not present it here.

THEOREM 13.9   The xy term is removed from the general second degree relation

$$Ax^2 + Bxy + Cy^2 + Dx + Ey + F = 0$$

by a rotation of the axes through an angle θ, $0 < \theta < 90°$, where

$$\cot 2\theta = \frac{A - C}{B}.$$

Theorem 13.9 can be used to find the appropriate angle of rotation, θ. To find the equations of Theorem 13.8, we must then find $\sin \theta$ and $\cos \theta$. The following example illustrates a way to obtain $\sin \theta$ and $\cos \theta$ from $\cot 2\theta$ without first identifying the angle θ.

Example 10   Rotate the axes and graph $52x^2 - 72xy + 73y^2 = 200$.
   Here we have

$$\cot 2\theta = \frac{52 - 73}{-72} = \frac{-21}{-72} = \frac{7}{24}.$$

To find $\sin \theta$ and $\cos \theta$, we use the trigonometric identities

$$\sin \theta = \sqrt{\frac{1 - \cos 2\theta}{2}}, \qquad \cos \theta = \sqrt{\frac{1 + \cos 2\theta}{2}}.$$

By sketching a right triangle, as in Figure 13.14, we see that $\cos 2\theta = 7/25$. (Recall: for the two quadrants for which we are concerned, cosine and cotangent have the same sign.)

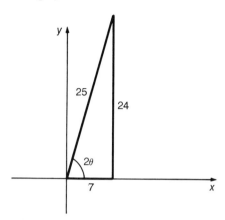

Figure 13.14

We have

$$\sin \theta = \sqrt{\frac{1 - 7/25}{2}} = \sqrt{\frac{9}{25}} = \frac{3}{5}$$

$$\cos \theta = \sqrt{\frac{1 + 7/25}{2}} = \sqrt{\frac{16}{25}} = \frac{4}{5}.$$

Thus the transformation here is given by

$$x = \frac{4}{5}x' - \frac{3}{5}y'$$

$$y = \frac{3}{5}x' + \frac{4}{5}y'.$$

Substituting these values of x and y into the original equation yields

$$52\left(\frac{4}{5}x' - \frac{3}{5}y'\right)^2 - 72\left(\frac{4}{5}x' - \frac{3}{5}y'\right)\left(\frac{3}{5}x' + \frac{4}{5}y'\right) + 73\left(\frac{3}{5}x' + \frac{4}{5}y'\right)^2 = 200.$$

This becomes

$$52\left(\frac{16}{25}x'^2 - \frac{24}{25}x'y' + \frac{9}{25}y'^2\right) - 72\left(\frac{12}{25}x'^2 + \frac{7}{25}x'y' - \frac{12}{25}y'^2\right)$$

$$+ 73\left(\frac{9}{25}x'^2 + \frac{24}{25}x'y' + \frac{16}{25}y'^2\right) = 200.$$

Combining terms, we have

$$25x'^2 + 100y'^2 = 200.$$

Dividing both sides by 200 yields

$$\frac{x'^2}{8} + \frac{y'^2}{2} = 1,$$

which is the equation of an ellipse having intercepts $(0,\ \sqrt{2})$, $(0,\ -\sqrt{2})$, $(2\sqrt{2},\ 0)$, and $(-2\sqrt{2},\ 0)$ with respect to the $x'y'$-system, as shown in Figure 13.15.

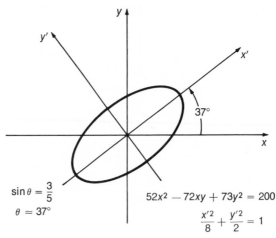

Figure 13.15

The following theorem enables us to determine, from the general equation, the type of graph to expect.

THEOREM 13.10   If the equation

$$Ax^2 + Bxy + Cy^2 + Dx + Ey + F = 0$$

has a graph, then the graph will be
(a)   a circle, ellipse (or point) if $B^2 - 4AC < 0$,
(b)   a parabola (or a line, or two parallel lines) if $B^2 - 4AC = 0$,
(c)   a hyperbola (or two intersecting lines) if $B^2 - 4AC > 0$,
(d)   a straight line if $A = B = C = 0$, and $D \neq 0$ or $E \neq 0$.

13.5 Exercises

Perform the following rotations, using the angle of rotation given.

1. $x^2 - xy + y^2 = 6$, $\theta = 45°$
2. $2x^2 - xy + 2y^2 = 25$, $\theta = 45°$

Remove the xy term from each of the following by performing a suitable rotation. Graph each curve.

3. $3x^2 - 2xy + 3y^2 = 8$
4. $x^2 + xy + y^2 = 3$
5. $x^2 - 4xy + y^2 = -5$
6. $x^2 + 2xy + y^2 + 4\sqrt{2}x - 4\sqrt{2}y = 0$
7. $7x^2 + 6\sqrt{3}xy + 13y^2 = 64$
8. $7x^2 + 2\sqrt{3}xy + 5y^2 = 24$
9. $3x^2 - 2\sqrt{3}xy + y^2 - 2x - 2\sqrt{3}y = 0$
10. $2x^2 + 2\sqrt{3}xy + 4y^2 = 5$

13.6 Polar Coordinates

The rectangular coordinate system that we have used throughout this text is very convenient and useful, but it is not the only coordinate system that is commonly used. We shall investigate another coordinate system in this section. Figure 13.16 shows a point P, which has coordinates (x, y) with respect to the rectangular xy-system. We could also locate point P by drawing a line through P and the origin O, and measuring the distance from O to P along this line. If the line OP makes an angle θ with the positive x-axis, and if the distance OP equals some number r, then we say P has **polar coordinates** (r, θ). If r is positive, then the distance is measured along the terminal side of the angle. If r is negative, the distance is measured in a direction opposite to the terminal side. The point O is called the **pole,** and the positive x-axis is called the **polar axis.** While a point has uniquely determined coordinates (x, y) with respect to a given rectangular coordinate system, it has many different coordinates (r, θ) with respect to a given polar coordinate system. Thus, $(1, 45°)$, $(1, 405°)$, $(-1, 225°)$, and $(-1, -135°)$ all represent the same point.

From Figure 13.16, page 314, we see that

$$\cos \theta = \frac{x}{r} \quad \text{and} \quad \sin \theta = \frac{y}{r},$$

from which

$$x = r \cos \theta \quad \text{and} \quad y = r \sin \theta.$$

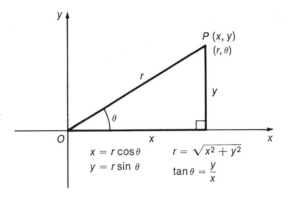

Figure 13.16

If we now square both sides of these equations and add the results, we have

$$x^2 + y^2 = r^2(\cos^2\theta + \sin^2\theta) = r^2(1) = r^2.$$

Also, we can write

$$\tan \theta = \frac{\sin \theta}{\cos \theta} = \frac{\dfrac{y}{r}}{\dfrac{x}{r}} = \frac{y}{x}.$$

Thus, we have proved the following theorem.

THEOREM 13.11   (a)   If a point P has rectangular coordinates (x, y), then the polar coordinates (r, θ) are given by

$$r^2 = x^2 + y^2 \qquad\qquad \tan \theta = \frac{y}{x}.$$

(b)   If a point P has polar coordinates (r, θ) then the rectangular coordinates are given by

$$x = r \cos \theta \qquad\qquad y = r \sin \theta.$$

Graphing in polar coordinates is much the same as graphing in rectangular coordinates—we obtain some representative ordered pairs belonging to the relation and then sketch the graph. For example, to graph $r = 1 + \cos \theta$, we first complete the ordered pairs shown, and then sketch the graph of Figure 13.17. This curve is called a **cardioid,** because of its heart shape.

θ	0	30°	45°	60°	90°	120°	135°	150°	180°	270°	315°
r	2	1.9	1.7	1.5	1	.5	.3	.1	0	1	1.7

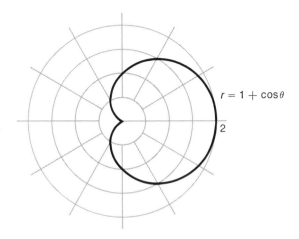

$r = 1 + \cos\theta$

2

Figure 13.17

Example 11 Graph $r = 2$.

Since r always has the value 2, this equation represents a circle of radius 2, as shown in Figure 13.18.

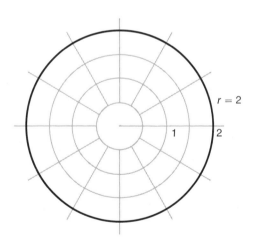

$r = 2$

1 2

Figure 13.18

Example 12 Graph $r^2 = \cos 2\theta$.

First, we complete a table of ordered pairs as shown, and then sketch the graph, as in Figure 13.19, page 316. This curve is called a **lemniscate.**

θ	0	30°	45°	135°	150°	180°
r	± 1	$\pm .7$	0	0	$\pm .7$	± 1

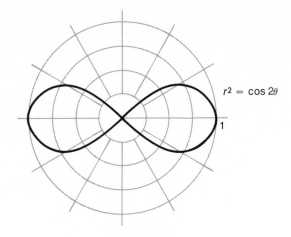

$r^2 = \cos 2\theta$

Figure 13.19

Example 13 Graph $r = \dfrac{4}{1 + \sin \theta}$.

We can again complete a table of ordered pairs, which gives the graph shown in Figure 13.20.

θ	0	30°	45°	60°	90°	120°	135°	150°	180°	210°	225°
r	4	2.7	2.3	2.1	2	2.1	2.3	2.7	4	8	13.65

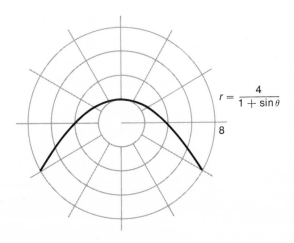

$r = \dfrac{4}{1 + \sin \theta}$

8

Figure 13.20

Example 14 Graph $r = 2\theta$ (θ measured in radians).

Figure 13.21 shows this graph, called a **spiral of Archimedes.**

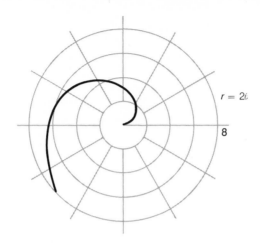

Figure 13.21

We can convert the equation

$$r = \frac{4}{1 + \sin \theta}$$

to rectangular coordinates, using the results of Theorem 13.11. We have

$$r = \frac{4}{1 + \sin \theta}$$
$$r + r \sin \theta = 4$$
$$\sqrt{x^2 + y^2} + y = 4$$
$$\sqrt{x^2 + y^2} = 4 - y$$
$$x^2 + y^2 = (4 - y)^2$$
$$x^2 = -8(y - 2),$$

which is the equation of a parabola, and could be graphed using the methods shown earlier.

13.6 Exercises

Locate each of the following points on a polar coordinate system.

1. $(1, 75°)$
2. $(2, 150°)$
3. $(-4, 60°)$
4. $(-3, -240°)$
5. $(0, 40°)$
6. $(-5, 100°)$
7. $(2, 30°)$
8. $(2, 400°)$

For each of the following equations, find an equivalent equation in rectangular coordinates and graph.

9. $r = 2 \sin \theta$
10. $r = 2 \cos \theta$

11. $r = \dfrac{2}{1 - \cos \theta}$

12. $r = \dfrac{3}{1 - \sin \theta}$

13. $r = 2 \cos \theta - 2 \sin \theta$

14. $r = \dfrac{3}{4 \cos \theta - \sin \theta}$

15. $r = 2 \sec \theta$
16. $r = -5 \csc \theta$
17. $r(\cos \theta + \sin \theta) = 2$
18. $r(2 \cos \theta + \sin \theta) = 2$
19. $r \sin \theta + 2 = 0$
20. $r \sec \theta = 5$

Graph each of the following.

21. $r = 2 + 2 \cos \theta$
22. $r = 2(4 + 3 \cos \theta)$
23. $r = 3 + \cos \theta$ (limaçon)
24. $r = 2 - \cos \theta$ (limaçon)
25. $r = \sin 4\theta$ (eight-leaved rose)
26. $r = 3 \cos 5\theta$ (five-leaved rose)
27. $r^2 = 4 \cos 2\theta$ (lemniscate)
28. $r^2 = 4 \sin 2\theta$ (lemniscate)
29. $r = 4(1 - \cos \theta)$ (cardioid)
30. $r = 3(2 - \cos \theta)$ (cardioid)
31. $r = 5\theta$ (spiral of Archimedes—measure θ in radians)
32. $r = \theta$ (spiral of Archimedes—measure θ in radians)
33. $r\theta = \pi$ (hyperbolic spiral—measure θ in radians)
34. $r^2 = \theta$ (parabolic spiral—measure θ in degrees)
35. $\ln r = \theta$ (logarithmic spiral—measure θ in radians)
36. $r = 2 \sin \theta \tan \theta$ (cissoid)

37. $r = \dfrac{\cos 2\theta}{\cos \theta}$

38. $r = \dfrac{3}{2 + \sin \theta}$

39. $r = \sin \theta \cos^2 \theta$

Chapter Fourteen

Combinatorics

Most of the topics in this chapter are useful in the study of probability. Counting theory–the study of combinations and permutations–deals with methods for counting the number of ways in which certain objectives can be reached. We shall study only the basics of this interesting subject. The binomial theorem, which is important in many branches of mathematics, is a central idea in the study of probability theory.

14.1 Permutations

After making do with your old automobile for several years, you finally decide to replace it with a shiny new one. You drive over to Ned's New Car Emporium to choose the car of your dreams. Once there, you find that you can select from 5 models, each with 14 power options, in your choice of 8 exterior colors and 4 interior colors. How many different new cars are available to you? Problems of this sort are best solved by means of the counting principles which we shall discuss in this section and the next.

Suppose there are 3 roads from town A to town B and 2 roads from town B to town C. For each of the 3 roads from A to B, there are 2 different routes from B to C. Hence, to travel from A to C by way of B there are $3 \cdot 2 = 6$ different ways, as illustrated in Figure 14.1.

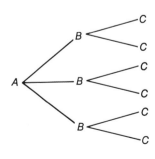

Figure 14.1

Generalizing from examples such as this, we have the **successive events axiom,**

▶ If one event can occur in m ways and then a second event can occur in n ways, then both events can occur in mn ways, provided the outcome of the first event does not influence the outcome of the second.

The successive events axiom can be extended to any number of events, provided that the outcome of no one event influences the outcome of another. Thus we can apply this axiom to determine how many choices we actually have in selecting a new car at Ned's. Applying the axiom we find that

$$5 \cdot 14 \cdot 8 \cdot 4 = 2240$$

different new cars are available from this one dealer.

Example 1 A restaurant offers a choice of 3 different salads, 5 main dishes, and 2 desserts. How many different meals can be selected?

Since the first event (selecting a salad) can occur in 3 ways, the second event (selecting a main dish) in 5 ways, and the third event (selecting a dessert) in 2 ways, there are $3 \cdot 5 \cdot 2$ or 30 different choices.

Example 2 A teacher has 5 different books which he wishes to arrange on his desk. How many different arrangements are possible?

With a little thought, we can use the successive events axiom here. There are 5 choices to be made, one for each space which will hold a book. To select a book for the first space, the teacher has 5 choices, for the second space, 4 choices (one book has already been put in the first space), for the third space, 3 choices, and so on. By the successive events axiom, we see that the number of arrangements is

$$5 \cdot 4 \cdot 3 \cdot 2 \cdot 1 = 120.$$

In using the successive events axiom, we shall frequently encounter such products as $5 \cdot 4 \cdot 3 \cdot 2 \cdot 1$, as we did in Example 2. For convenience in writing these products, we introduce the symbol $n!$ (read "n factorial"), which is defined as follows for any counting number n.

$$n! = n(n-1)(n-2)(n-3) \cdots (2)(1)$$

Thus, we may write $5 \cdot 4 \cdot 3 \cdot 2 \cdot 1$ as 5! Also, $3! = 3 \cdot 2 \cdot 1 = 6$. By this definition of $n!$, we see that $n[(n-1)!] = n!$ for all natural numbers $n \geq 2$. It is convenient to have this relation hold also for $n = 1$, so we define

▶ $$0! = 1.$$

Example 3 Suppose the teacher in Example 2 wishes to place only 3 of the 5 books on his desk. How many arrangements of 3 books are possible?

The teacher again has 5 ways to fill the first space, 4 ways to fill the second space, and 3 ways to fill the third. Since he wants to use only 3 books, there are only 3 spaces to be filled (3 events) instead of 5. Thus, there are

$$5 \cdot 4 \cdot 3 = 60 \text{ arrangements.}$$

The number 60 here is called the number of **permutations** of 5 things taken 3 at a time, written $60 = P(5, 3)$. Generalizing from these examples, we have the following theorems.

THEOREM 14.1   If $P(n, n)$ denotes the number of permutations of n elements taken n at a time, then

$$P(n, n) = n!.$$

THEOREM 14.2   If $P(n, r)$ denotes the number of permutations of n elements taken r at a time, then

$$P(n, r) = \frac{n!}{(n - r)!}.$$

As illustrated by Example 2, Theorem 14.1 is a direct consequence of the successive events axiom. To prove Theorem 14.2, we note that

$$P(n, r) = n(n - 1)(n - 2) \cdots (n - r + 1)$$
$$= \frac{n(n - 1)(n - 2) \cdots (n - r + 1)(n - r) \cdots (2)(1)}{(n - r)(n - r - 1) \cdots (2)(1)}$$
$$= \frac{n!}{(n - r)!}.$$

Example 4   Find the number of permutations of 8 elements taken 3 at a time.
    Using the successive events axiom, we note that there are 3 choices to be made, so that $P(8, 3) = 8 \cdot 7 \cdot 6 = 336$. However, we can instead use the formula given above for $P(n, r)$ as follows.

$$P(8, 3) = \frac{8!}{5!} = \frac{8 \cdot 7 \cdot 6 \cdot 5 \cdot 4 \cdot 3 \cdot 2 \cdot 1}{5 \cdot 4 \cdot 3 \cdot 2 \cdot 1}$$
$$= 8 \cdot 7 \cdot 6$$
$$= 336$$

Example 5   In how many ways can 6 students be seated in a row of 6 desks?
    Here we have

$$P(6, 6) = 6! = 6 \cdot 5 \cdot 4 \cdot 3 \cdot 2 \cdot 1 = 720.$$

14.1 Exercises

1. Evaluate each of the following.
   (a) $P(7, 7)$
   (b) $P(5, 3)$
   (c) $P(6, 5)$
   (d) $P(4, 2)$

2. In how many ways can 5 pictures be arranged in a row? How many arrangements are possible if only 2 of them are used?

3. In how many ways can 6 people be seated in a row of 6 seats?

4. In how many ways can 7 out of 10 people be assigned to 7 seats?

5. In how many ways can 5 players be assigned to the 5 positions on a basketball team, assuming any player can play any position? In how many ways can 10 players be assigned to the 5 positions?

6. A couple has narrowed down their choice of names for a new baby to three first names and five middle names. How many different first and middle name arrangements are possible?

7. How many different homes are available if a builder offers a choice of 5 basic plans, 3 roof styles, and 2 exterior finishes?

8. An auto manufacturer produces 7 models, each available in 6 different colors, with 4 different upholstery fabrics, and 5 interior colors. How many varieties of the auto are available?

9. A concert is to consist of 5 works: 2 modern, 2 classical, and a piano concerto. In how many ways can the program be arranged?

10. If the program in Exercise 9 must be shortened to 3 works, chosen from the 5, how many arrangements are possible?

11. In Exercise 9, how many different programs are possible if the two modern works are to be played first, then the two classical, and then the concerto?

12. How many 4 letter radio station call letters can be made if the first letter must be K or W and no letter may be repeated? How many if repeats are allowed?

13. How many of the 4 letter call letters in Exercise 12 with no repeats end in K?

14. A business school gives courses in typing, shorthand, transcription, business English, technical writing, and accounting. How many ways can a student arrange his program if he takes 3 courses?

15. How many different license plate numbers can be formed using three letters followed by three digits if no repeats are allowed? How many if there are no repeats and either letters or numbers come first?

16. Show that $P(n, n - 1) = P(n, n)$.

14.2 Combinations

In the last section we discussed a method for finding the number of *ordered* subsets of size r which could be formed from a set of n elements ($n \geq r$). If we are not interested in the order within a subset, there will be fewer different subsets. For example, consider the set $\{a, b, c\}$. The number of ordered subsets of 2 elements which can be formed from the given set of 3 elements is

$$P(3, 2) = \frac{3!}{(3 - 2)!} = \frac{3!}{1!} = 3 \cdot 2 \cdot 1 = 6.$$

We can display these 6 subsets.

$$
\begin{array}{ccc}
\{a, b\} & \{a, c\} & \{b, c\} \\
\{b, a\} & \{c, a\} & \{c, b\}
\end{array}
$$

If we are not interested in the order of the elements, $\{a, b\}$ and $\{b, a\}$ are the same subset, $\{a, c\}$ and $\{c, a\}$ are the same, and $\{b, c\}$ and $\{c, b\}$ are the same. We see that there are only 3 subsets if we disregard order. These three sets are called the two-element **combinations** of the set $\{a, b, c\}$. There is only one three-element combination of $\{a, b, c\}$, the set itself, although there are $3!$, or 6, three-element permutations of $\{a, b, c\}$. To find the number of *ordered* subsets of r elements in a set of n elements, we use *permutations*. To find the number of *unordered* subsets, we use *combinations*. Since each combination of r elements, from a set of n elements, forms $r!$ permutations, we can find the number of combinations of n elements taken r at a time by dividing the number of permutations, $P(n, r)$, by $r!$ to get

$$\frac{P(n, r)}{r!} \text{ combinations.}$$

This expression can be rewritten as follows.

$$\frac{P(n, r)}{r!} = \frac{\dfrac{n!}{(n - r)!}}{r!}$$

$$= \frac{n!}{(n - r)!r!}$$

In summary, we have the following theorem.

THEOREM 14.3   If $C(n, r)$ denotes the number of combinations of n elements taken r at a time, then

$$C(n, r) = \frac{n!}{(n - r)!r!}.$$

We can now use Theorem 14.3 to find $C(3, 2)$ which we found to be equal to 3 in the discussion above.

$$C(3, 2) = \frac{3!}{(3-2)!2!} = \frac{3 \cdot 2 \cdot 1}{1 \cdot 2 \cdot 1} = \frac{3}{1} = 3$$

Example 6 How many committees of size three can be formed from a group of 8 people?

A committee is an unordered set, so we want $C(8, 3)$.

$$C(8, 3) = \frac{8!}{5!3!}$$

$$= \frac{8 \cdot 7 \cdot 6 \cdot 5 \cdot 4 \cdot 3 \cdot 2 \cdot 1}{5 \cdot 4 \cdot 3 \cdot 2 \cdot 1 \cdot 3 \cdot 2 \cdot 1}$$

$$= \frac{8 \cdot 7 \cdot 6}{3 \cdot 2 \cdot 1}$$

$$= 56$$

Example 7 From a group of 30 employees, 3 are to be selected to work on a special project.
(a) In how many different ways can the employees be selected?
(b) In how many ways can the group of three be selected if it has been decided that a particular man must work on the project?

(a) Here we wish to know how many three-element combinations can be formed from a set of 30 elements. (We want combinations, and not permutations, since order within the group of 3 is irrelevant.)

$$C(30, 3) = \frac{30!}{27!3!}$$

$$= \frac{30(29)(28)(27) \cdots (3)(2)(1)}{27(26)(25) \cdots (2)(1)(3)(2)(1)}$$

$$= \frac{30 \cdot 29 \cdot 28}{3 \cdot 2 \cdot 1}$$

$$= 4060$$

There are 4060 ways to select the project group.
(b) Since one man has already been selected for the project, the problem is reduced to selecting two more from the remaining 29 employees.

$$C(29, 2) = \frac{29!}{27!2!} = 406$$

In this case, the project group can be selected in 406 different ways.

Example 8 Find $C(n, n)$.

Using Theorem 14.3, with $r = n$, we have

$$C(n, n) = \frac{n!}{(n - n)!n!} = \frac{n!}{0!n!} = \frac{n!}{1 \cdot n!} = 1.$$

This is reasonable since there is only one way of selecting a group of n objects from a set of n objects when order is disregarded.

Example 9 Find $C(n, 1)$.

Again we can use Theorem 14.3.

$$C(n, 1) = \frac{n!}{(n - 1)!1!}$$

$$= \frac{n(n - 1)!}{(n - 1)! \cdot 1}$$

$$= n$$

This result agrees with what we already know: there are n distinct one element subsets of an n element set.

Example 10 Find $C(n, n - 1)$.

Here we have

$$C(n, n - 1) = \frac{n!}{[n - (n - 1)]!(n - 1)!}$$

$$= \frac{n!}{1!(n - 1)!}$$

$$= n.$$

We can use the results of Examples 8–10 as theorems and apply them directly to special cases. Thus, from Example 8, $C(5, 5) = 1$, $C(8, 8) = 1$, and so on. Also, from Examples 9 and 10 we can write

$$C(7, 1) = C(7, 6) = 7,$$
$$C(10, 1) = C(10, 9) = 10,$$

and in general,

$$C(n, 1) = C(n, n - 1) = n.$$

The examples above are all special cases of the general statement expressed in Theorem 14.4, which follows.

THEOREM 14.4   If $C(n, r)$ denotes the number of combinations of n elements taken r at a time, then

$$C(n, r) = C(n, n - r).$$

The proof of this theorem is left for the exercises.

14.2 Exercises

1. Evaluate each of the following.
   (a) $C(6, 5)$
   (b) $C(4, 2)$
   (c) $C(8, 5)$
   (d) $C(10, 2)$

2. In how many ways can an employer select two new employees from a group of four applicants?

3. A club has 30 members. If a committee of 4 is selected in a random manner, how many committees are possible?

4. How many samples of 3 apples can be drawn from a crate of 25 apples?

5. A group of 3 students are to be selected randomly from a group of 12 students to participate in a special class. In how many ways can this be done? In how many ways can the group which will not participate be selected?

6. Hal's Hamburger Heaven sells hamburgers with cheese, relish, lettuce, tomato, mustard or ketchup. How many different hamburgers can be made using any three of the extras?

7. How many different two card hands can be dealt from a deck of 52 cards?

8. How many different 12 card pinochle hands can be dealt from a deck of 48 cards? (A pinochle deck has the 9, 10, J, Q, K, and A of each suit twice for a total of 48 cards.)

9. Five cards are marked with the numbers 1, 2, 3, 4, and 5, then shuffled, and two cards are then drawn. How many different two card combinations are possible?

10. If a bag contains 15 marbles, how many samples of two marbles can be drawn from it? How many samples of four marbles?

11. In Exercise 10, if the bag contains 3 yellow, 4 white, and 8 blue marbles, how many samples of two can be drawn in which both marbles are blue?

12. In Exercise 4, if it is known that there are 5 rotten apples in the crate:
    (a) How many samples of 3 could be drawn in which all 3 are rotten?
    (b) How many samples of 3 could be drawn in which there is one rotten apple and two good apples?

13. A city council is composed of 5 liberals and 4 conservatives. A delegation of 3 is to be selected randomly to attend a convention.
    (a) How many delegations are possible?
    (b) How many delegations could have all liberals?
    (c) How many delegations could have 2 liberals and 1 conservative?
    (d) If one member of the council serves as mayor, how many delegations are possible which include the mayor?

14. A group of 7 workers decide to send a delegation of 2 to the supervisor to discuss their grievances.

    (a) How many different delegations are possible?

    (b) If it is decided that a particular employee must be in the delegation, how many different delegations are possible?

    (c) If there are 2 women and 5 men in the group, how many delegations would include a woman?

15. Show that $C(n, r) = C(n, n - r)$.

14.3 Basic Properties of Probability

The study of probability is becoming increasingly popular partly because of its importance to applications and partly because of an inherent interest in the subject itself. It is human nature to be interested in chance, and probability deals with chance. In fact, the theory of probability was developed to describe games of chance. Modern usage, however, extends the application of probability theory to a much wider range of activities, including many practical applications.

We shall call the set S of all possible outcomes of a given experiment the **sample space** for the experiment. In this text, all sample spaces will be finite. For example, a sample space for the experiment of tossing a coin includes two outcomes, heads (H) and tails (T). We write this in set notation as

$$S = \{H, T\}.$$

Similarly, a sample space for the experiment of rolling a single die is

$$S = \{1, 2, 3, 4, 5, 6\}.$$

(A single die can have either 1, 2, 3, 4, 5, or 6 showing on top after it is rolled.)

Any subset of the sample space is called an **event.** In the experiment with the die, for example, "the number showing is a three" is an event, say E_1, such that $E_1 = \{3\}$. The **probability** of an event E, written $P(E)$, is the ratio of the number of outcomes in the sample space favorable to E compared to the total number of outcomes in the sample space. Thus, if there are s possible outcomes in the sample space S, each equally likely to occur, and n of these are favorable to event E, we have

$$\blacktriangleright \qquad P(E) = \frac{n}{s}.$$

To use this definition to find $P(E_1)$ as given above, we note that the sample space for this experiment is $S = \{1, 2, 3, 4, 5, 6\}$, while the desired event is $E_1 = \{3\}$. Thus,

$$P(E_1) = \frac{1}{6},$$

since there is one element in E_1 and 6 elements in S.

Example 11 A single die is rolled. Write the following events in set notation and give the probability for each event.

(a) E_2; the number showing is even.
(b) E_3: the number showing is greater than 4.
(c) E_4: the number showing is less than 10.
(d) E_5: the number showing is 8.

Using the definitions above, we have the following.

(a) $E_2 = \{2, 4, 6\}$ and $P(E_2) = \dfrac{3}{6} = \dfrac{1}{2}$.

(b) $E_3 = \{5, 6\}$ and $P(E_3) = \dfrac{2}{6} = \dfrac{1}{3}$.

(c) $E_4 = \{1, 2, 3, 4, 5, 6\}$ and $P(E_4) = \dfrac{6}{6} = 1$.

(d) $E_5 = \varnothing$ and $P(E_5) = \dfrac{0}{6} = 0$.

Note that $E_4 = S$. The event E_4 is certain to occur every time the experiment is performed. We see that an event which is certain to occur, such as E_4, has a probability of 1. On the other hand, E_5 is $\varnothing$ and $P(E_5)$ is 0. The probability of an impossible event, such as E_5, is always 0, since none of the outcomes in the sample space favor the event.

The set of outcomes in the sample space which are *not* favorable to an event E is called the **complement** of E, written E'. For example, in the experiment of drawing a single card from a standard deck of 52 cards, let E be the event "the card is an ace." Then E' is the event "the card is not an ace." From the definition of E', we see that for any event E,

$$E \cup E' = S \qquad\qquad E \cap E' = \varnothing.$$

Example 12 In the experiment of drawing a card from a well-shuffled deck, find the probability of events E, (the card is an ace), and E'.

Since there are four aces in the deck of 52 cards, $P(E) = \dfrac{4}{52} = \dfrac{1}{13}$. Of the 52 cards, 48 are not aces, so that we have $P(E') = \dfrac{48}{52} = \dfrac{12}{13}$.

In Example 12, we note that $P(E) + P(E') = \dfrac{1}{13} + \dfrac{12}{13} = 1$. Recall that for any event E, $P(E) = \dfrac{n}{s}$ (where n is the number of outcomes favorable to E, and s is the total number of outcomes in the sample space). Since E' is the set of all outcomes not favorable to E, $P(E') = \dfrac{s - n}{s}$. Noting that

$$\frac{n}{s} + \frac{s-n}{s} = \frac{s}{s} = 1,$$

we see that in general, for any event E,

▶
$$P(E) + P(E') = 1.$$

This fact can be restated as

▶
$$P(E) = 1 - P(E') \quad \text{or} \quad P(E') = 1 - P(E).$$

These two equations suggest an alternate way to compute the probability of an event. For example, if it is known that $P(E) = \frac{1}{10}$, then we have

$$P(E') = 1 - \frac{1}{10} = \frac{9}{10}.$$

Example 13 Two dice are rolled. Find
(a)   the probability that the sum of the numbers showing is 3,
(b)   the probability that the sum is greater than 3.
      The sample space for this experiment has 36 outcomes. These can be displayed as follows.

Die 2

		1	2	3	4	5	6
	1	1, 1	1, 2	1, 3	1, 4	1, 5	1, 6
	2	2, 1	2, 2	2, 3	2, 4	2, 5	2, 6
Die 1	3	3, 1	3, 2	3, 3	3, 4	3, 5	3, 6
	4	4, 1	4, 2	4, 3	4, 4	4, 5	4, 6
	5	5, 1	5, 2	5, 3	5, 4	5, 5	5, 6
	6	6, 1	6, 2	6, 3	6, 4	6, 5	6, 6

(a)   To find the probability of a sum of 3, we need to find the number of outcomes favorable to a sum of 3. From the sample space, we see that there are 2 such outcomes: 1, 2, and 2, 1. Since the number of outcomes in the sample space is 36, the probability we seek is 2/36 or 1/18.
(b)   The sample space contains 33 outcomes in which the sum is greater than 3. Thus the probability of a sum greater than 3 is 33/36 or 11/12. Another approach to this problem is to use the relationship

$$P(E) = 1 - P(E').$$

For the event "the sum is greater than 3," the complementary event is "the sum is 3 or less," which is satisfied by 3 of the 36 outcomes in the sample space. Here, $P(E') = 3/36$, or 1/12, and the required probability, then, is

$$1 - \frac{1}{12} = \frac{11}{12}.$$

Sometimes probability statements are expressed in terms of odds, a comparison of $P(E)$ with $P(E')$. The **odds** in favor of an event E are expressed as the ratio of $P(E)$ to $P(E')$ or as the fraction $P(E)/P(E')$. For example, if the probability of rain can be established as $1/3$, the odds that it will rain are

$$P(\text{rain}) \text{ to } P(\text{no rain}) = \frac{1}{3} \text{ to } \frac{2}{3}$$

$$= \frac{\frac{1}{3}}{\frac{2}{3}}$$

$$= \frac{1}{2} \text{ (or 1 to 2).}$$

On the other hand, the odds that it will not rain are 2 to 1 (or $2/3$ to $1/3$). If we know that the odds in favor of an event are, say, 3 to 5, then we can see that the probability of the event is $3/8$, while the probability of the complement of the event is $5/8$. In general, if the odds favoring event E are m to n, then

▶ $$P(E) = \frac{m}{m+n} \quad \text{and} \quad P(E') = \frac{n}{m+n}.$$

The probabilities that we have considered to this point have been determined by inherent properties of the experiment in question. Such probabilities are called *a priori* and are based on theory as opposed to observation. Empirical probabilities, those based only on observation, are more likely to occur in practical applications. An example is the life tables used by life insurance companies to determine premiums. A portion of such a table is shown below.

From the table, we see that the probability of dying before the age of 10 is .033, while a 70-year old person has a .259 probability of dying before age 80.

Life table

The table gives the probability that a person alive at the beginning of an age interval will die before reaching the beginning of the next age interval.

period	probability of death	period	probability of death
0–9	.033	50–59	.112
10–19	.007	60–69	.245
20–29	.013	70–79	.259
30–39	.019	80–89	.201
40–49	.047	90–99	.064

When using a life table, it is important to note for which group of individuals the table has been prepared. A table valid for people in the United States would not be applicable to people in other countries such as India. Also, a table prepared on the basis of men would not be valid for women, who usually live about 5 years longer than men.

14.3 Exercises

Write sample spaces in which each outcome is equally likely to occur for the following experiments.

1. A two-headed coin is tossed once.
2. Two fair coins are tossed.
3. Three fair coins are tossed.
4. Slips of paper marked with the numbers 1, 2, 3, 4, and 5 are placed in a box. After mixing well, two slips are drawn.
5. An unprepared student takes a three question true-false quiz in which he guesses the answer for all three questions.
6. A die is rolled and then a coin is tossed.

Write the following events in set notation and give the probability of each event.

7. In the experiment of Exercise 2:
   (a) both coins show the same face.
   (b) at least one coin turns up heads.
8. In Exercise 1:
   (a) the result of the toss is heads.
   (b) the result of the toss is tails.
9. In Exercise 4:
   (a) both slips are marked with even numbers.
   (b) both slips are marked with odd numbers.
   (c) both slips are marked with the same number.
   (d) one slip is marked with an odd number, the other with an even number.
10. In Exercise 5:
   (a) the student gets all three answers correct.
   (b) he gets all three answers wrong.
   (c) he gets exactly two answers correct.
   (d) he gets at least one answer correct.
11. A marble is drawn at random from a box containing 3 yellow, 4 white, and 8 blue marbles. Find the following probabilities.
   (a) a yellow marble is drawn.
   (b) a blue marble is drawn.
   (c) a black marble is drawn.

(d) what are the odds in favor of drawing a yellow marble?

(e) what are the odds against drawing a blue marble?

12. A baseball player with a batting average of .300 comes to bat. What are the odds in favor of his getting a hit?

13. In Exercise 4, what are the odds that the sum of the numbers on the two slips of paper is 5?

14. If the odds that it will rain are 4 to 5, what is the probability of rain?

15. If the odds that a candidate will win an election are 3 to 2, what is the probability that the candidate will lose?

16. A card is drawn from a well-shuffled deck of 52 cards. Find the probability that the card is

(a) a 9,

(b) black,

(c) a black 9,

(d) a heart,

(e) the 9 of hearts,

(f) a face card. (The K, Q, J of any suit.)

17. Two dice are rolled. Find the probability that

(a) the sum of the points is 5.

(b) the sum of the points is 12.

(c) the sum of the points is less than 12. (Hint: use $P(E) = 1 - P(E')$.)

(d) the sum of the points is an even number.

(e) both dice show the same number of points.

18. Use the life table to find the probability of dying in the 4th decade; the 7th decade.

14.4 Probabilities of Alternate and Successive Events

We now consider the probability of a compound event which involves an alternative such as $(E \text{ or } F)$ where E and F are simple events. For example, in the experiment of rolling a die, suppose H is the event "the result is a 3," while K is the event "the result is an even number." What is the probability of $(H$ or $K)$, "the result is a 3 or an even number?" The three event sets and their probabilities are

$$H = \{3\}; \qquad\qquad P(H) = \frac{1}{6}.$$

$$K = \{2, 4, 6\}; \qquad\qquad P(K) = \frac{3}{6} = \frac{1}{2}.$$

$$(H \text{ or } K) = \{2, 3, 4, 6\}; \qquad P(H \text{ or } K) = \frac{4}{6} = \frac{2}{3}.$$

Here $P(H) + P(K) = P(H \text{ or } K)$. Before we assume that this relationship is true in general, let us consider another event for this experiment, "the result is a 2," event G. For G and (K or G) we have

$$G = \{2\}; \qquad\qquad P(G) = \frac{1}{6}.$$

$$(K \text{ or } G) = \{2, 4, 6\}; \qquad P(K \text{ or } G) = \frac{3}{6} = \frac{1}{2}.$$

In this case $P(K) + P(G) \neq P(K \text{ or } G)$.

The difference in the two examples above comes from the fact that events H and K cannot occur simultaneously. Such events are called **mutually exclusive events.** We see that $H \cap K = \emptyset$, which is true in general for any two mutually exclusive events. Events K and G, however, can occur simultaneously. Both are satisfied if the result of the roll is a 2, the element in their intersection ($K \cap G = \{2\}$). Generalizing the above discussion, we have the following theorem.

THEOREM 14.5   For any events E and F,

$$P(E \text{ or } F) = P(E \cup F) = P(E) + P(F) - P(E \cap F).$$

Example 14   For the experiment consisting of one roll of a pair of dice, find the probability that the sum of the points showing is at most 4.

We can rewrite "at most 4" as "2 or 3 or 4." (A sum smaller than 2 is meaningless here.) Then, by Theorem 14.5, we can write

$$\begin{aligned} P(\text{at most 4}) &= P(2 \text{ or } 3 \text{ or } 4) \\ &= P(2) + P(3) + P(4), \end{aligned} \qquad (1)$$

since the events represented by "2," "3," and "4" are mutually exclusive. By the definition of probability, $P(2) = 1/36$, $P(3) = 2/36$, and $P(4) = 3/36$. Substituting into equation (1) above we have

$$P(\text{at most 4}) = \frac{1}{36} + \frac{2}{36} + \frac{3}{36} = \frac{6}{36} = \frac{1}{6}.$$

Example 15   One card is drawn from a well-shuffled deck of 52 cards. What is the probability that it is an ace or a spade? a three or a king?

The events "drawing an ace" and "drawing a spade" are not mutually exclusive since it is possible to draw the ace of spades, an outcome favorable to both events.

By Theorem 14.5, we know

$$P(\text{ace or spade}) = P(\text{ace}) + P(\text{spade}) - P(\text{ace and spade})$$

$$= \frac{4}{52} + \frac{13}{52} - \frac{1}{52}$$

$$= \frac{16}{52}$$

$$= \frac{4}{13}.$$

"Drawing a 3" and "drawing a king" are mutually exclusive events because it is impossible to draw one card which is both a 3 and a king. From Theorem 14.5 we have

$$P(3 \text{ or } K) = P(3) + P(K) - P(3 \text{ and } K) = \frac{4}{52} + \frac{4}{52} - 0 = \frac{8}{52} = \frac{2}{13}.$$

An experiment can consist of more than one event. Suppose we wish to find the probability of the two-event outcome, getting a 3 on a roll of a die and then getting heads on the toss of a coin. A successful outcome here requires success in both parts of the experiment. The sample space has 12 outcomes as shown below.

$$S = \begin{Bmatrix} 1, H & 2, H & 3, H & 4, H & 5, H & 6, H \\ 1, T & 2, T & 3, T & 4, T & 5, T & 6, T \end{Bmatrix}$$

Since only one of the twelve outcomes favors the desired event, the probability we seek is 1/12. By the successive events axiom, probability of a composite event which consists of successive parts (or events) can also be found by multiplying the probabilities of the parts. Thus, in the experiment above, the probability of the first event, $P(3)$, multiplied by the probability of the second event, $P(H)$, gives $P(3 \text{ and } H)$.

$$P(3) = \frac{1}{6}, \qquad P(H) = \frac{1}{2}, \qquad P(3 \text{ and } H) = \frac{1}{12} = \frac{1}{6} \cdot \frac{1}{2}$$

From this example, it might appear that we can generalize to a product theorem. Before we do, let us consider another example.

Example 16 Two marbles are drawn (without replacement) from a box containing 1 red, 3 white, and 2 green marbles. Find the probability that the two marbles drawn are both white.

The sample space for this experiment consists of all possible sets of two marbles. There are $C(6, 2)$ ways in which a sample of two marbles can be drawn from the 6 marbles, and $C(3, 2)$ ways in which the sample can include 2 white marbles. (There are 3 white marbles in the box.) We can compute the probability using the definition as follows.

$$P(W \text{ and } W) = \frac{C(3, 2)}{C(6, 2)}$$

$$= \frac{\dfrac{3!}{2!1!}}{\dfrac{6!}{2!4!}}$$

$$= \frac{3!2!4!}{2!1!6!}$$

$$= \frac{3 \cdot 2}{6 \cdot 5}$$

$$= \frac{1}{5}.$$

To compute the probability by multiplication, as we did above, we find that the probability of drawing a white marble on one draw is 3/6 or 1/2, so that

$$P(W) \cdot P(W) = \frac{1}{2} \cdot \frac{1}{2} = \frac{1}{4}.$$

However, $P(W \text{ and } W) = 1/5$ from the work above. The difficulty in this case is that the probability of drawing white for the second marble is *not* 1/2 but 2/5, since drawing a white marble on the first draw reduces the remaining marbles to 5, of which 2 are white. Thus, to use multiplication to compute $P(W \text{ and } W)$, we must compute the probability that the second marble is white taking into account the first draw of a white marble. This gives

$$P(W \text{ and } W) = P(W \text{ first}) \cdot P(W \text{ second}) = \frac{1}{2} \cdot \frac{2}{5} = \frac{1}{5},$$

which agrees with the probability we computed using the definition.

In the experiment discussed above, if the first marble were replaced before the second was drawn, the probability that both draws resulted in a white marble becomes

$$P(W \text{ first}) \cdot P(W \text{ second}) = \frac{1}{2} \cdot \frac{1}{2} = \frac{1}{4} = P(W \text{ and } W).$$

Thus, it is important to consider the relationship between the events in a two-part experiment.

As Example 16 illustrates, when computing the probability of the second event in a two-part experiment, it is important to consider whether or not the prior occurrence of the first event has affected the outcome of the second event. That is, we must find the probability of the second event *given that the first event has already occurred*. This kind of probability is called **conditional probability.** The conditional probability of event F given that event E has occurred is written $P(F \mid E)$. We can now state a product theorem as follows.

THEOREM 14.6   If E and F are two successive events, the probability of the composite event (E and F) is given by

$$P(E \text{ and } F) = P(E) \cdot P(F|E).$$

If $P(F|E)$ equals $P(F)$, the product theorem becomes

$$P(E \text{ and } F) = P(E) \cdot P(F).$$

When this is the case, E and F are said to be **independent events.** Intuitively, this means that the occurrence of E does not affect the occurrence of F. For example, the two events "heads on the first toss of a coin" (H_1) and "heads on the second toss of a coin" (H_2) are independent, which can be shown as follows. We know that $P(H_1) = 1/2$ and $P(H_2) = 1/2$. Also, $P(H_1 \text{ and } H_2) = 1/4$, because only one of the four outcomes in the sample space

$$\{H_1H_2, H_1T_2, T_1H_2, T_1T_2\},$$

satisfies the condition (H_1 and H_2). Thus,

$$P(H_1) \cdot P(H_2) = P(H_1 \text{ and } H_2)$$

as required by the definition of independent events. Events which are not independent are called **dependent events.**

Example 17   Find the probability of getting heads on five successive tosses of a coin.

For each toss of the coin, $P(H) = 1/2$. Each of the tosses is independent of the others, since what happened on the first toss has no affect on the outcome of the second toss, and so on. Using the product theorem, we have

$$P(5 \text{ heads}) = \frac{1}{2} \cdot \frac{1}{2} \cdot \frac{1}{2} \cdot \frac{1}{2} \cdot \frac{1}{2} = \left(\frac{1}{2}\right)^5 = \frac{1}{32}.$$

14.4 Exercises

1. Mrs. Elliott invites 10 relatives to a party: her mother, two uncles, three brothers, and four cousins. If the chances of any one guest arriving first are equally likely, find the following probabilities.
    (a) The first guest is an uncle or a brother.
    (b) The first guest is a brother or cousin.
    (c) The first guest is a brother or her mother.
2. One card is drawn from a standard deck of 52 cards. What is the probability that the card is
    (a) a 9 or a 10?
    (b) red or a 3?
    (c) a heart or black?
    (d) less than a four? (Consider aces as ones.)

3. Two dice are rolled. Find the probability that
   (a) the sum of the points is at least 10.
   (b) the sum of the points is either 7 or at least 10.
   (c) the sum of the points is 2 or the dice both show the same number.
4. A die is rolled three times. What is the probability that
   (a) all three results are even?
   (b) the first result is a 2 and the second is an even number and the third is any number?
5. If the probability of having a boy is 1/2, what is the probability in a three-child family of having
   (a) two boys and then a girl?
   (b) two boys and one girl in any order?
   (c) all three children of the same sex?
6. If two marbles are drawn without replacement from a jar with four black and three white marbles, what is the probability that
   (a) both are white?
   (b) both are black?
   (c) the first is black and the second is white?
   (d) one is black and the other is white?
7. Two cards are drawn from a deck of 52 cards, without replacement. What is the probability that
   (a) both cards are hearts?
   (b) both cards are black?
   (c) both cards are twos?
   (d) the first card is black and the second is a heart?
   (e) the first card is black and the second is a club?
8. If a marble is drawn from a bag containing 3 yellow, 4 white, and 8 blue marbles, what is the probability that
   (a) the marble is either yellow or white?
   (b) it is either yellow or blue?
   (c) it is either red or white?
9. A smooth talking young man has a 1/3 probability of talking a policeman out of giving him a speeding ticket when stopped. The probability that he is stopped during a given weekend is 1/2. Find the probability that
   (a) he will receive no tickets on a given weekend.
   (b) he will receive no tickets on three consecutive weekends.
10. If two fair coins are tossed, find the probability that
    (a) both coins show heads or both show tails.
    (b) only one coin comes up heads or both show heads.
    (c) at least one coin comes up heads or both coins show tails.
11. In a club with 20 senior and 10 junior members, a committee of three is to be randomly selected. Find the probability that

(a) the committee is composed entirely of senior members or entirely of junior members.

(b) the committee has at least one senior member.

12. Slips of paper marked with the digits 1, 2, 3, 4, and 5 are placed in a box and mixed well. If two slips are drawn (without replacement), find the probability that

(a) the first is even and the second is odd.

(b) both are even.

(c) the first is a three and the second is a number greater than three.

(d) both are marked three.

13. If two cards are drawn from a deck of 52 cards, what is the probability that they are both queens or both red?

14. Use the life table on page 221 to find the following probabilities.

(a) the probability of dying by age 20. (Hint: dying by age 20 means dying in either the first or the second decade.)

(b) the probability of dying by age 50.

(c) the probability of living to age 50.

(d) the probability of living to age 80.

(e) For approximately what age is there a 1/2 probability of dying before that age and a 1/2 probability of surviving beyond it?

14.5 The Binomial Theorem

If we write the expression $(x + y)^n$ for positive integer values of n, we get a family of expressions which are important in the study of mathematics generally, and, in particular, in the study of probability theory. For example

$$(x + y)^1 = x + y$$
$$(x + y)^2 = x^2 + 2xy + y^2$$
$$(x + y)^3 = x^3 + 3x^2y + 3xy^2 + y^3$$
$$(x + y)^4 = x^4 + 4x^3y + 6x^2y^2 + 4xy^3 + y^4$$
$$(x + y)^5 = x^5 + 5x^4y + 10x^3y^2 + 10x^2y^3 + 5xy^4 + y^5.$$

From inspection, we see that these expressions follow a pattern. Let us try to identify the pattern so that we can write a general expression for $(x + y)^n$.

First, observe that each expression begins with x raised to the same power as the binomial itself. That is, the expansion of $(x + y)^1$ has a first term of x^1, $(x + y)^2$ has a first term of x^2, $(x + y)^3$ has a first term of x^3, and so on. Also, the last term in each expansion is y to the same power as the binomial. We can see that the expression of $(x + y)^n$ should begin with the term x^n and the end with the term y^n.

Further, we note that the exponents on x decrease by one in each term after the first, while the exponents on y, beginning with y in the second term, increase by one in each succeeding term. Thus, the *variables*, in the expansion of $(x + y)^n$ have the following pattern.

$$x^n, \; x^{n-1}y, \; x^{n-2}y^2, \; x^{n-3}y^3, \; \cdots, \; xy^{n-1}, \; y^n$$

From this pattern, it can be seen that the sum of the exponents on x and y in each term is n. For example, in the third term above, the variable is $x^{n-2}y^2$, and the sum of the exponents, $n - 2$ and 2, is n.

Now let us try to find a pattern for the *coefficients* in the terms of the expansions shown above. In the product

$$(x + y)^5 = (x + y)(x + y)(x + y)(x + y)(x + y), \tag{2}$$

the term x occurs 5 times, once in each factor. To get the first term of the expansion, we form the product of these 5 x's to get x^5. The product x^5 can occur in just one way, by taking an x from each factor in equation (2), so the coefficient of x^5 is 1. We can get the term with x^4y in more than one way. For example, we can take the x's from the first four factors and the y from the last factor, or we may take the x's from the last four factors and the y from the first, and so on. Since there are five factors in equation (2), from which we wish to select four x's for the term x^4y, there are $C(5, 4)$ ways in which this can be done. Thus the term x^4y occurs in $C(5, 4)$ or 5 ways and has coefficient 5. Continuing in this manner, we can use combinations to find the coefficients for each of the terms of the expansion as follows.

$$(x + y)^5 = x^5 + C(5, 4)x^4y + C(5, 3)x^3y^2 + C(5, 2)x^2y^3 + C(5, 1)xy^4 + y^5$$

The coefficient, 1, of the first and last terms could be written $C(5, 5)$ or $C(5, 0)$ to complete the pattern.

In general, then, the coefficient for a term of $(x + y)^n$ in which the variable is $x^{n-r}y^r$ is $C(n, n - r)$. Thus, we have the **binomial theorem,** which gives the **general binomial expansion.**

THEOREM 14.7   For any positive integer n,

$$\begin{aligned}
(x + y)^n = \; & x^n + C(n, n - 1)x^{n-1}y \\
& + C(n, n - 2)x^{n-2}y^2 + C(n, n - 3)x^{n-3}y^3 \\
& + \cdots + C(n, 1)xy^{n-1} + y^n.
\end{aligned}$$

A formal proof of this theorem is given at the end of the next section.

Example 18 Write out the binomial expansion of $(x + y)^9$.

We can use the binomial theorem to write

$$
\begin{aligned}
(x + y)^9 &= x^9 + C(9, 8)x^8y + C(9, 7)x^7y^2 + C(9, 6)x^6y^3 \\
&\quad + C(9, 5)x^5y^4 + C(9, 4)x^4y^5 + C(9, 3)x^3y^6 \\
&\quad + C(9, 2)x^2y^7 + C(9, 1)xy^8 + y^9 \\
&= x^9 + \frac{9!}{8!1!}x^8y + \frac{9!}{7!2!}x^7y^2 + \frac{9!}{6!3!}x^6y^3 + \frac{9!}{5!4!}x^5y^4 \\
&\quad + \frac{9!}{4!5!}x^4y^5 + \frac{9!}{3!6!}x^3y^6 + \frac{9!}{2!7!}x^2y^7 + \frac{9!}{1!8!}xy^8 + y^9 \\
&= x^9 + 9x^8y + 36x^7y^2 + 84x^6y^3 + 126x^5y^4 + 126x^4y^5 \\
&\quad + 84x^3y^6 + 36x^2y^7 + 9xy^8 + y^9.
\end{aligned}
$$

Example 19 Expand $\left(a - \dfrac{b}{2}\right)^5$.

Again we can use the binomial theorem.

$$
\begin{aligned}
\left(a - \frac{b}{2}\right)^5 &= a^5 + C(5, 4)a^4\left(-\frac{b}{2}\right) + C(5, 3)a^3\left(-\frac{b}{2}\right)^2 \\
&\quad + C(5, 2)a^2\left(-\frac{b}{2}\right)^3 + C(5, 1)a\left(-\frac{b}{2}\right)^4 + \left(-\frac{b}{2}\right)^5 \\
&= a^5 + 5a^4\left(-\frac{b}{2}\right) + 10a^3\left(-\frac{b}{2}\right)^2 + 10a^2\left(-\frac{b}{2}\right)^3 \\
&\quad + 5a\left(-\frac{b}{2}\right)^4 + \left(-\frac{b}{2}\right)^5 \\
&= a^5 - \frac{5}{2}a^4b + \frac{5}{2}a^3b^2 - \frac{5}{4}a^2b^3 + \frac{5}{16}ab^4 - \frac{1}{32}b^5.
\end{aligned}
$$

We can also write any single term of a binomial expansion. For example, if we want to write the tenth term of $(x + y)^n$, $(n \geq 9)$, we see that in the tenth term y is raised to the ninth power (since y has the power of 1 in the second term, the power of 2 in the third term, and so on.) Since the exponents on x and y in any term must have a sum of n, the exponent on x in the tenth term is $n - 9$. From this we have

$$
C(n, n - 9)x^{n-9}y^9 = \frac{n!}{(n - 9)!9!}x^{n-9}y^9
$$

for the tenth term of the expansion. Stated generally, we have Theorem 14.8.

THEOREM 14.8   The rth term of the binomial expansion of $(x + y)^n$, where $n \geq r - 1$, is

$$
C(n, n - (r - 1))x^{n-(r-1)}\, y^{r-1}.
$$

Example 20   Find the sixth term of $(a + 2b)^{10}$.

In the sixth term we see that $2b$ has an exponent of 5, while a has an exponent of $10 - 5$ or 5. The sixth term thus becomes

$$C(10, 5)a^5(2b)^5 = \frac{10!}{5!5!}a^5(2b)^5$$

$$= 252a^5(32b^5)$$

$$= 8064a^5b^5.$$

Example 21   Use the binomial theorem to find $(1.01)^8$ correct to three decimal places.

We can write 1.01 as $1 + .01$. Then, using the binomial theorem we have

$$
\begin{aligned}
(1.01)^8 &= (1 + .01)^8 \\
&= 1^8 + C(8, 7)(1)^7(.01) + C(8, 6)(1)^6(.01)^2 + \cdots + (.01)^8 \\
&= 1 + 8(1)(.01) + 28(1)(.0001) + \cdots + (.01)^8 \\
&= 1 + .08 + .0028 + \cdots + (.01)^8 \\
&\approx 1.083.
\end{aligned}
$$

Since we need only three decimal places there was no need to evaluate more terms of the expansion, although additional terms would give the answer to more decimal places of accuracy.

14.5 Exercises

Write out the binomial expansion for the following and simplify the terms.

1. $(x + y)^6$
2. $(m + n)^4$
3. $(p - q)^5$
4. $(r^2 - s)^7$
5. $(2x + t^3)^4$

6. $(3x - 2y)^6$
7. $\left(\dfrac{m}{2} - 3n\right)^5$
8. $\left(2p + \dfrac{q}{3}\right)^4$

For each of the following, write the indicated term of the binomial expansion.

9. 5th term of $(m - 2p)^{12}$
10. 7th term of $(3x + y)^{15}$
11. 17th term of $\left(p^2 + \dfrac{q}{2}\right)^{20}$
12. 8th term of $(x^3 + 2y)^{14}$

Use the binomial expansion to evaluate each of the following to four decimal places.

13. $(1.10)^{10}$
14. $(0.99)^{15}$

15. $(1.99)^8$
16. $(3.02)^6$

14.6 The Binomial Theorem and Probability

Many probability problems are concerned with an experiment in which an event is repeated many times. For example, we may wish to find the probability of getting 7 heads in 8 tosses of a coin, or hitting a target 6 times out of 6, or finding 1 defective item in a sample of 15 items. These problems all have certain things in common.

(1)   The same experiment is repeated several times.
(2)   The probability of the desired outcome remains the same for each trial.
(3)   The repeated trials are independent.

Probability problems of this kind are called **repeated trials** problems. In each case, some outcome is designated a success and any other outcome is considered a failure. Thus, if the probability of a success in a single trial is p, the probability of failure will be $1 - p$.

Let us consider the solution of a problem of this type. Suppose we want to find the probability of getting 6 ones on 6 rolls of a die. The probability of getting a one on 1 roll is $1/6$, while the probability of any other result is $5/6$. Using the product theorem, and noting that the throws are independent events we have

$$P(6 \text{ ones on 6 rolls of a die}) = P(1)P(1)P(1)P(1)P(1)P(1)$$

$$= \left(\frac{1}{6}\right)^6$$

$$\approx .00002.$$

Now, let us find the probability of getting exactly 5 ones in 6 rolls of the die. The desired outcome for this experiment can occur in more than one way as shown below where S represents success (getting a one) and F represents failure (any other result).

S	S	S	S	S	F
S	S	S	S	F	S
S	S	S	F	S	S
S	S	F	S	S	S
S	F	S	S	S	S
F	S	S	S	S	S

The probability of each of these six outcomes is

$$\left(\frac{1}{6}\right)^5\left(\frac{5}{6}\right).$$

Since the six outcomes represent mutually exclusive alternative events, we add the six probabilities to get

$$P(\text{5 ones in 6 rolls of a die}) = 6\left(\frac{1}{6}\right)^5\left(\frac{5}{6}\right) = \frac{5}{6^5} \approx .0006.$$

In the same way, we can compute the probability of 4 ones in 6 rolls of a die. The probability of 4 successes and 2 failures will be

$$\left(\frac{1}{6}\right)^4\left(\frac{5}{6}\right)^2.$$

Again, the desired outcomes can occur in more than one way. To find the number of alternative outcomes, we can use combinations. We want to find the number of ways in which we can combine 4 successes and 2 failures–that is, we want $C(6, 4)$ (or $C(6, 2)$). Since $C(6, 4) = 6!/(4!2!) = 15$, we have

$$P(\text{4 ones in 6 rolls of a die}) = 15\left(\frac{1}{6}\right)^4\left(\frac{5}{6}\right)^2$$

$$= \frac{15 \cdot 25}{6^6}$$

$$\approx .008.$$

In general, for experiments of this kind where the same experiment is repeated with the probability of success in a single trial the same for all trials, we have the following theorem.

THEOREM 14.9   If p is the probability of success in a single trial, the probability of x successes and $n - x$ failures in n independent repeated trials of an experiment is

$$C(n, x)p^x(1 - p)^{n-x}.$$

In Theorem 14.9, if we let $1 - p = q$ and $x = r$, we have

$$C(n, r)p^rq^{n-r},$$

the expression for one of the terms of the binomial expansion of $(p + q)^n$. For example, in the problem discussed above, if we expand $\left(\frac{1}{6} + \frac{5}{6}\right)^6$, we have

$$\left(\frac{1}{6} + \frac{5}{6}\right)^6 = \left(\frac{1}{6}\right)^6 + C(6, 5)\left(\frac{1}{6}\right)^5\left(\frac{5}{6}\right) + C(6, 4)\left(\frac{1}{6}\right)^4\left(\frac{5}{6}\right)^2$$

$$+ C(6, 3)\left(\frac{1}{6}\right)^3\left(\frac{5}{6}\right)^3 + C(6, 2)\left(\frac{1}{6}\right)^2\left(\frac{5}{6}\right)^4$$

$$+ C(6, 1)\left(\frac{1}{6}\right)\left(\frac{5}{6}\right)^5 + \left(\frac{5}{6}\right)^6.$$

The terms of this binomial expansion give respectively the probabilities of 6 ones, 5 ones, 4 ones, 3 ones, 2 ones, 1 one, and 0 ones in 6 rolls of a die. This result can be generalized as a theorem.

THEOREM 14.10   The rth term of the binomial expansion of $(p + q)^n$ (where $q = 1 - p$) gives the probability of getting $n - (r - 1)$ successes and $r - 1$ failures in n repeated trials.

Example 22   Find the probability of getting 7 heads in 8 tosses of a coin.

The probability of success (getting a head) in a single toss is $1/2$. Thus the probability of failure (getting a tail) is $1/2$. We have

$$P(\text{7 heads in 8 tosses of a coin}) = C(8, 7)\left(\frac{1}{2}\right)^7\left(\frac{1}{2}\right)$$

$$= 8\left(\frac{1}{2}\right)^8$$

$$\approx .0312.$$

Example 23   Assuming that all selections of items for a sample are independent trials, find the probability of one defective item in a sample of 15 items from a production line, if the probability that an item will be defective is .01.

The probability of success (a defective item) is .01, while the probability of failure (a good item), is .99. Thus,

$$P(\text{1 defective item in 15 items}) = C(15, 1)(.01)(.99)^{14}$$

$$= 15(.01)(.99)^{14}$$

$$\approx .130.$$

If it seems wrong to call a defective item a success, one could call it a failure and find the probability of 14 successes in 15 trials. In that case we have $p = .99$ and $q = .01$.

$$P(\text{14 good items in 15 items}) = C(15, 14)(.99)^{14}(.01)$$

$$= 15(.99)^{14}(.01)$$

$$\approx .130.$$

14.6 Exercises

1. Find the probability that a family with seven children will have exactly five boys. (Assume the probability of having a boy is $1/2$.)
2. Find the probability of hitting a bull's eye eight out of ten times if the probability of getting a bull's eye in a single trial is .6.

3. A die is rolled twelve times. Find the probability of
   (a) a 1 twelve times.
   (b) six 1's out of twelve tries.
   (c) exactly one 1 in the twelve tries.
4. A coin is tossed five times. Find the probability of
   (a) all heads.
   (b) exactly three heads.
   (c) at least three heads.
5. A factory tests a sample of twenty bulbs for defectives. The probability that a particular bulb will be defective has been established by past experience to be .05.
   (a) What is the probability that there are no defectives in the sample?
   (b) What is the probability that the number of defectives in the sample is at most two?
6. In a ten question multiple choice test with five choices for each question, a student who was not prepared guessed on each item. Find the probability that he gets
   (a) six questions correct.
   (b) at least eight correct.
   (c) less than eight correct.

14.7 Mathematical Induction

Most of the reasoning used in this course has been **deductive reasoning.** By deductive reasoning, we mean the kind of reasoning which procedes from a hypothesis to a conclusion, forming a chain of reasoning using previously established facts as shown below.

To prove: Given that A is true, show that C is true.
Hypothesis: A is true.
Previously known fact: If A is true, then B is true.
Conclusion: B is true.
Previously known fact: If B is true, then C is true.
Conclusion: C is true.

There is another kind of reasoning called **inductive reasoning.** Inductive reasoning is used whenever a generalization is made based on a number of observations. This kind of reasoning is often used by scientists. On the basis of the outcome of experiments in the laboratory, a scientist may theorize that a certain fact is generally true. Further repetitions of the experiment may tend to strengthen the claim or may refute it. However, no matter how many times experimental results confirm the theory, it is never absolutely certain. There is always the chance that it may be wrong.

In mathematics, this type of reasoning is used to generalize from examples. Any conclusion arrived at in this manner must then be proved by deductive methods. One such method, called **mathematical induction,** depends on the **Axiom of Mathematical Induction:**

▶ If a given open sentence with the variable n is true for $n = 1$, and if the truth of the sentence for $n = k$, k any natural number, implies its truth for $n = k + 1$, then the sentence is true for every natural number.

The method of mathematical induction is particularly useful in proving theorems which claim that some property is true for all natural numbers. A typical example of the kind of result best proved by mathematical induction is given below.

Prove   The sum of the first n natural numbers is $\dfrac{n(n + 1)}{2}$.

To prove this, we must show that the statement is true when n is replaced by any of the natural numbers 1, 2, 3, ... That is, we must show

$$1 + 2 + 3 + \cdots + n = \frac{n(n + 1)}{2},$$

for all natural number values of n.

Using inductive reasoning we could easily establish the truth of the statement for the first few natural numbers.

$$1 = \frac{1(1 + 1)}{2}$$

$$1 + 2 = \frac{2(2 + 1)}{2}$$

$$1 + 2 + 3 = \frac{3(3 + 1)}{2}$$

$$1 + 2 + 3 + 4 = \frac{4(4 + 1)}{2},$$

and so on. No matter how many such examples we show to be true, we still have not *proved* the result to be true for *all* natural numbers. However, the result can be proved by mathematical induction.

In a proof by mathematical induction, two steps are required.

(1)   Show that the statement is true for $n = 1$.

(2)   Assume that the statement is true for some natural number k. Then show that this implies that the statement is true for the next natural number, $k + 1$.

We show in step 2 that *if* the statement is true for the natural number k, *then* it will also be true for the next natural number, $k + 1$. In step 1 we show that the statement is true for the natural number 1. Then, by step 2, it is true for 2, and hence for 3, and so on, for any value of n.

Now, let us use mathematical induction to prove the statement discussed above.

Prove $1 + 2 + 3 + \cdots + n = \dfrac{n(n + 1)}{2}$ for any natural number n.

Proof

STEP 1. We must show that $1 = \dfrac{1(1 + 1)}{2}$.

Since
$$\frac{1(1 + 1)}{2} = \frac{1(2)}{2} = 1,$$

the statement is true for $n = 1$.

STEP 2. Assume the statement is true for some natural number k, so that

$$1 + 2 + 3 + \cdots + k = \frac{k(k + 1)}{2}.$$

Now we must show that this implies

$$1 + 2 + 3 + \cdots + (k + 1) = \frac{(k + 1)[(k + 1) + 1]}{2}.$$

Our hypothesis is

$$1 + 2 + 3 + \cdots + k = \frac{k(k + 1)}{2}.$$

By adding the quantity $(k + 1)$ to both sides of the equation, we have

$$1 + 2 + 3 + \cdots + (k + 1) = \frac{k(k + 1)}{2} + (k + 1).$$

Then, factoring on the right, we have

$$= (k + 1)\left(\frac{k}{2} + 1\right)$$

$$= (k + 1)\left(\frac{k + 2}{2}\right)$$

$$1 + 2 + 3 + \cdots + (k + 1) = \frac{(k + 1)[(k + 1) + 1]}{2}.$$

The final result is the statement we wished to establish for $n = (k + 1)$. Thus, we have shown that if the statement is true for $n = k$, it is also true for $n = (k + 1)$ and hence for any natural number value of n.

Example 24 Use the method of mathematical induction to prove that for all natural numbers n, $n < 2^n$.

Proof

STEP 1. $$1 < 2^1$$

STEP 2. Assuming $k < 2^k$, we want to show that $(k + 1) < 2^{(k+1)}$. If we multiply both sides of the hypothesis, $k < 2^k$, by 2, we have

$$2k < 2 \cdot 2^k$$
$$2k < 2^{k+1}.$$

Also, $$2k = k + k,$$
and $$k + 1 \leq k + k \text{ (since } 1 \leq k)$$
Then, $$k + 1 \leq 2k < 2^{(k+1)},$$
or $$k + 1 < 2^{(k+1)}.$$

Thus, the statement is true for $(k + 1)$ whenever it is true for k, and the proof is completed.

The binomial theorem of the previous sections

$$(x + y)^n = x^n + \frac{n!}{(n - 1)!1!}x^{n-1}y + \frac{n!}{(n - 2)!2!}x^{n-2}y^2 + \cdots + y^n,$$

can be proved by the method of mathematical induction. To begin, we show that the theorem holds for $n = 1$.

$$(x + y)^1 = x^1 + y^1 = x + y$$

Now, assume that the theorem is true for $n = k$. We wish to show that the theorem will then be true for $n = k + 1$, or that

$$(x + y)^{k+1} = x^{k+1} + \frac{(k + 1)!}{k!1!}x^k y + \frac{(k + 1)!}{(k - 1)!2!}x^{k-1}y^2 + \cdots + y^{k+1}.$$

From the assumption that the theorem holds for $n = k$, we have

$$(x + y)^k = x^k + \frac{k!}{(k - 1)!1!}x^{k-1}y + \frac{k!}{(k - 2)!2!}x^{k-2}y^2 + \cdots + y^k.$$

We can multiply both sides of this equation by $(x + y)$.

$$(x + y)^k(x + y) = \left(x^k + \frac{k!}{(k - 1)!1!}x^{k-1}y + \cdots + y^k \right)(x + y)$$

$$(x + y)^{k+1} = x^{k+1} + \frac{k!}{(k - 1)!1!}x^k y + \frac{k!}{(k - 2)!2!}x^{k-1}y^2$$

$$+ \cdots + xy^k + x^k y + \frac{k!}{(k - 1)!1!}x^{k-1}y^2$$

$$+ y^{k+1}$$

Collect like terms on the right-hand side.

$$= x^{k+1} + \left[\frac{k!}{(k-1)!1!} + 1\right]x^k y$$

$$+ \left[\frac{1}{2} + \frac{1}{k-1}\right]\left[\frac{k!}{(k-2)!1!}x^{k-1}y^2\right] + \cdots + y^{k+1}$$

$$= x^{k+1} + \frac{(k+1)!}{k!1!}x^k y + \frac{(k+1)!}{(k-1)!2!}x^{k-1}y^2 + \cdots + y^{k+1}$$

This gives the desired expression for $n = k + 1$. (In the second term, the coefficient of $x^k y$ was simplified as follows.)

$$\frac{k!}{(k-1)!1!} + 1 = \frac{k! + (k-1)!}{(k-1)!1!}$$

$$= \frac{k(k-1)! + (k-1)!}{(k-1)!1!}$$

$$= \frac{(k+1)(k-1)!}{(k-1)!1!}$$

$$= \frac{(k+1)(k)(k-1)!}{k(k-1)!1!}$$

$$= \frac{(k+1)!}{k!1!}$$

Thus, by the axiom of mathematical induction, the theorem is true for any natural number n.

14.7 Exercises

Use the method of mathematical induction to prove the following statements. Assume n is a positive integer.

1. $2 + 4 + 6 + \cdots + 2n = n(n + 1)$.

2. $1 + 3 + 5 + \cdots + (2n - 1) = n^2$.

3. $3 + 6 + 9 + \cdots + 3n = \dfrac{3n(n + 1)}{2}$.

4. $5 + 10 + 15 + \cdots + 5n = \dfrac{5n(n + 1)}{2}$

5. $2 + 4 + 8 + \cdots + 2^n = 2^{n+1} - 2$.

6. $1^2 + 2^2 + 3^2 + \cdots + n^2 = \dfrac{n(n + 1)(2n + 1)}{6}$.

7. $1^3 + 2^3 + 3^3 + \cdots + n^3 = \dfrac{n^2(n + 1)^2}{4}$.

8. $(a^m)^n = a^{mn}$, (assume m is a constant).

9. $\dfrac{1}{2} + \dfrac{1}{2^2} + \dfrac{1}{2^3} + \cdots + \dfrac{1}{2^n} = 1 - \dfrac{1}{2^n}$.

10. $(a + d) + (a + 2d) + (a + 3d) + \cdots + (a + nd) = \dfrac{n}{2}[2a + (n + 1)d]$,
    (assume a and d are constant).

11. $ar + ar^2 + ar^3 + \cdots + ar^n = \dfrac{ar(1 - r^n)}{1 - r}$, (assume a and r are constant).

12. $(1 + a)^n \geq 1 + a^n$, $(a \geq 0)$, (assume a is a constant).

13. What is wrong with the following proof by mathematical induction?
    Prove: Any natural number equals the next natural number.
    To begin, we assume the statement true for some natural number k:

    $$k = k + 1.$$

    We must now show that the statement is true for $n = k + 1$. If we add 1 to both sides, we have

    $$k + 1 = k + 1 + 1$$
    $$k + 1 = k + 2.$$

    Hence, if the statement is true for $n = k$, it is also true for $n = k + 1$. Thus, the theorem is proved.

14. Prove De Moivre's theorem (page 263) for all positive integer values of n.

Appendix

Table 1 Squares and Square Roots

n	n^2	$\sqrt{n}$	$\sqrt{10n}$	n	n^2	$\sqrt{n}$	$\sqrt{10n}$
1	1	1.000	3.162	51	2601	7.141	22.583
2	4	1.414	4.472	52	2704	7.211	22.804
3	9	1.732	5.477	53	2809	7.280	23.022
4	16	2.000	6.325	54	2916	7.348	23.238
5	25	2.236	7.071	55	3025	7.416	23.452
6	36	2.449	7.746	56	3136	7.483	23.664
7	49	2.646	8.367	57	3249	7.550	23.875
8	64	2.828	8.944	58	3364	7.616	24.083
9	81	3.000	9.487	59	3481	7.681	24.290
10	100	3.162	10.000	60	3600	7.746	24.495
11	121	3.317	10.488	61	3721	7.810	24.698
12	144	3.464	10.954	62	3844	7.874	24.900
13	169	3.606	11.402	63	3969	7.937	25.100
14	196	3.742	11.832	64	4096	8.000	25.298
15	225	3.873	12.247	65	4225	8.062	25.495
16	256	4.000	12.649	66	4356	8.124	25.690
17	289	4.123	13.038	67	4489	8.185	25.884
18	324	4.243	13.416	68	4624	8.246	26.077
19	361	4.359	13.784	69	4761	8.307	26.268
20	400	4.472	14.142	70	4900	8.367	26.458
21	441	4.583	14.491	71	5041	8.426	26.646
22	484	4.690	14.832	72	5184	8.485	26.833
23	529	4.796	15.166	73	5329	8.544	27.019
24	576	4.899	15.492	74	5476	8.602	27.203
25	625	5.000	15.811	75	5625	8.660	27.386
26	676	5.099	16.125	76	5776	8.718	27.568
27	729	5.196	16.432	77	5929	8.775	27.749
28	784	5.292	16.733	78	6084	8.832	27.928
29	841	5.385	17.029	79	6241	8.888	28.107
30	900	5.477	17.321	80	6400	8.944	28.284
31	961	5.568	17.607	81	6561	9.000	28.460
32	1024	5.657	17.889	82	6724	9.055	28.636
33	1089	5.745	18.166	83	6889	9.110	28.810
34	1156	5.831	18.439	84	7056	9.165	28.983
35	1225	5.916	18.708	85	7225	9.220	29.155
36	1296	6.000	18.974	86	7396	9.274	29.326
37	1369	6.083	19.235	87	7569	9.327	29.496
38	1444	6.164	19.494	88	7744	9.381	29.665
39	1521	6.245	19.748	89	7921	9.434	29.833
40	1600	6.325	20.000	90	8100	9.487	30.000
41	1681	6.403	20.248	91	8281	9.539	30.166
42	1764	6.481	20.494	92	8464	9.592	30.332
43	1849	6.557	20.736	93	8649	9.644	30.496
44	1936	6.633	20.976	94	8836	9.695	30.659
45	2025	6.708	21.213	95	9025	9.747	30.822
46	2116	6.782	21.448	96	9216	9.798	30.984
47	2209	6.856	21.679	97	9409	9.849	31.145
48	2304	6.928	21.909	98	9604	9.899	31.305
49	2401	7.000	22.136	99	9801	9.950	31.464
50	2500	7.071	22.361	100	10000	10.000	31.623

Table 2 Common Logarithms

n	0	1	2	3	4	5	6	7	8	9
1.0	.0000	.0043	.0086	.0128	.0170	.0212	.0253	.0294	.0334	.0374
1.1	.0414	.0453	.0492	.0531	.0569	.0607	.0645	.0682	.0719	.0755
1.2	.0792	.0828	.0864	.0899	.0934	.0969	.1004	.1038	.1072	.1106
1.3	.1139	.1173	.1206	.1239	.1271	.1303	.1335	.1367	.1399	.1430
1.4	.1461	.1492	.1523	.1553	.1584	.1614	.1644	.1673	.1703	.1732
1.5	.1761	.1790	.1818	.1847	.1875	.1903	.1931	.1959	.1987	.2014
1.6	.2041	.2068	.2095	.2122	.2148	.2175	.2201	.2227	.2253	.2279
1.7	.2304	.2330	.2355	.2380	.2405	.2430	.2455	.2480	.2504	.2529
1.8	.2553	.2577	.2601	.2625	.2648	.2672	.2695	.2718	.2742	.2765
1.9	.2788	.2810	.2833	.2856	.2878	.2900	.2923	.2945	.2967	.2989
2.0	.3010	.3032	.3054	.3075	.3096	.3118	.3139	.3160	.3181	.3201
2.1	.3222	.3243	.3263	.3284	.3304	.3324	.3345	.3365	.3385	.3404
2.2	.3424	.3444	.3464	.3483	.3502	.3522	.3541	.3560	.3579	.3598
2.3	.3617	.3636	.3655	.3674	.3692	.3711	.3729	.3747	.3766	.3784
2.4	.3802	.3820	.3838	.3856	.3874	.3892	.3909	.3927	.3945	.3962
2.5	.3979	.3997	.4014	.4031	.4048	.4065	.4082	.4099	.4116	.4133
2.6	.4150	.4166	.4183	.4200	.4216	.4232	.4249	.4265	.4281	.4298
2.7	.4314	.4330	.4346	.4362	.4378	.4393	.4409	.4425	.4440	.4456
2.8	.4472	.4487	.4502	.4518	.4533	.4548	.4564	.4579	.4594	.4609
2.9	.4624	.4639	.4654	.4669	.4683	.4698	.4713	.4728	.4742	.4757
3.0	.4771	.4786	.4800	.4814	.4829	.4843	.4857	.4871	.4886	.4900
3.1	.4914	.4928	.4942	.4955	.4969	.4983	.4997	.5011	.5024	.5038
3.2	.5051	.5065	.5079	.5092	.5105	.5119	.5132	.5145	.5159	.5172
3.3	.5185	.5198	.5211	.5224	.5237	.5250	.5263	.5276	.5289	.5302
3.4	.5315	.5328	.5340	.5353	.5366	.5378	.5391	.5403	.5416	.5428
3.5	.5441	.5453	.5465	.5478	.5490	.5502	.5514	.5527	.5539	.5551
3.6	.5563	.5575	.5587	.5599	.5611	.5623	.5635	.5647	.5658	.5670
3.7	.5682	.5694	.5705	.5717	.5729	.5740	.5752	.5763	.5775	.5786
3.8	.5798	.5809	.5821	.5832	.5843	.5855	.5866	.5877	.5888	.5899
3.9	.5911	.5922	.5933	.5944	.5955	.5966	.5977	.5988	.5999	.6010
4.0	.6021	.6031	.6042	.6053	.6064	.6075	.6085	.6096	.6107	.6117
4.1	.6128	.6138	.6149	.6160	.6170	.6180	.6191	.6201	.6212	.6222
4.2	.6232	.6243	.6253	.6263	.6274	.6284	.6294	.6304	.6314	.6325
4.3	.6335	.6345	.6355	.6365	.6375	.6385	.6395	.6405	.6415	.6425
4.4	.6435	.6444	.6454	.6464	.6474	.6484	.6493	.6503	.6513	.6522
4.5	.6532	.6542	.6551	.6561	.6571	.6580	.6590	.6599	.6609	.6618
4.6	.6628	.6637	.6646	.6656	.6665	.6675	.6684	.6693	.6702	.6712
4.7	.6721	.6730	.6739	.6749	.6758	.6767	.6776	.6785	.6794	.6803
4.8	.6812	.6821	.6830	.6839	.6848	.6857	.6866	.6875	.6884	.6893
4.9	.6902	.6911	.6920	.6928	.6937	.6946	.6955	.6964	.6972	.6981
5.0	.6990	.6998	.7007	.7016	.7024	.7033	.7042	.7050	.7059	.7067
5.1	.7076	.7084	.7093	.7101	.7110	.7118	.7126	.7135	.7143	.7152
5.2	.7160	.7168	.7177	.7185	.7193	.7202	.7210	.7218	.7226	.7235
5.3	.7243	.7251	.7259	.7267	.7275	.7284	.7292	.7300	.7308	.7316
5.4	.7324	.7332	.7340	.7348	.7356	.7364	.7372	.7380	.7388	.7396
n	0	1	2	3	4	5	6	7	8	9

Table 2 Common Logarithms (continued)

n	0	1	2	3	4	5	6	7	8	9
5.5	.7404	.7412	.7419	.7427	.7435	.7443	.7451	.7459	.7466	.7474
5.6	.7482	.7490	.7497	.7505	.7513	.7520	.7528	.7536	.7543	.7551
5.7	.7559	.7566	.7574	.7582	.7589	.7597	.7604	.7612	.7619	.7627
5.8	.7634	.7642	.7649	.7657	.7664	.7672	.7679	.7686	.7694	.7701
5.9	.7709	.7716	.7723	.7731	.7738	.7745	.7752	.7760	.7767	.7774
6.0	.7782	.7789	.7796	.7803	.7810	.7818	.7825	.7832	.7839	.7846
6.1	.7853	.7860	.7868	.7875	.7882	.7889	.7896	.7903	.7910	.7917
6.2	.7924	.7931	.7938	.7945	.7952	.7959	.7966	.7973	.7980	.7987
6.3	.7993	.8000	.8007	.8014	.8021	.8028	.8035	.8041	.8048	.8055
6.4	.8062	.8069	.8075	.8082	.8089	.8096	.8102	.8109	.8116	.8122
6.5	.8129	.8136	.8142	.8149	.8156	.8162	.8169	.8176	.8182	.8189
6.6	.8195	.8202	.8209	.8215	.8222	.8228	.8235	.8241	.8248	.8254
6.7	.8261	.8267	.8274	.8280	.8287	.8293	.8299	.8306	.8312	.8319
6.8	.8325	.8331	.8338	.8344	.8351	.8357	.8363	.8370	.8376	.8382
6.9	.8388	.8395	.8401	.8407	.8414	.8420	.8426	.8432	.8439	.8445
7.0	.8451	.8457	.8463	.8470	.8476	.8482	.8488	.8494	.8500	.8506
7.1	.8513	.8519	.8525	.8531	.8537	.8543	.8549	.8555	.8561	.8567
7.2	.8573	.8579	.8585	.8591	.8597	.8603	.8609	.8615	.8621	.8627
7.3	.8633	.8639	.8645	.8651	.8657	.8663	.8669	.8675	.8681	.8686
7.4	.8692	.8698	.8704	.8710	.8716	.8722	.8727	.8733	.8739	.8745
7.5	.8751	.8756	.8762	.8768	.8774	.8779	.8785	.8791	.8797	.8802
7.6	.8808	.8814	.8820	.8825	.8831	.8837	.8842	.8848	.8854	.8859
7.7	.8865	.8871	.8876	.8882	.8887	.8893	.8899	.8904	.8910	.8915
7.8	.8921	.8927	.8932	.8938	.8943	.8949	.8954	.8960	.8965	.8971
7.9	.8976	.8982	.8987	.8993	.8998	.9004	.9009	.9015	.9020	.9025
8.0	.9031	.9036	.9042	.9047	.9053	.9058	.9063	.9069	.9074	.9079
8.1	.9085	.9090	.9096	.9101	.9106	.9112	.9117	.9122	.9128	.9133
8.2	.9138	.9143	.9149	.9154	.9159	.9165	.9170	.9175	.9180	.9186
8.3	.9191	.9196	.9201	.9206	.9212	.9217	.9222	.9227	.9232	.9238
8.4	.9243	.9248	.9253	.9258	.9263	.9269	.9274	.9279	.9284	.9289
8.5	.9294	.9299	.9304	.9309	.9315	.9320	.9325	.9330	.9335	.9340
8.6	.9345	.9350	.9355	.9360	.9365	.9370	.9375	.9380	.9385	.9390
8.7	.9395	.9400	.9405	.9410	.9415	.9420	.9425	.9430	.9435	.9440
8.8	.9445	.9450	.9455	.9460	.9465	.9469	.9474	.9479	.9484	.9489
8.9	.9494	.9499	.9504	.9509	.9513	.9518	.9523	.9528	.9533	.9538
9.0	.9542	.9547	.9552	.9557	.9562	.9566	.9571	.9576	.9581	.9586
9.1	.9590	.9595	.9600	.9605	.9609	.9614	.9619	.9624	.9628	.9633
9.2	.9638	.9643	.9647	.9652	.9657	.9661	.9666	.9671	.9675	.9680
9.3	.9685	.9689	.9694	.9699	.9703	.9708	.9713	.9717	.9722	.9727
9.4	.9731	.9736	.9741	.9745	.9750	.9754	.9759	.9763	.9768	.9773
9.5	.9777	.9782	.9786	.9791	.9795	.9800	.9805	.9809	.9814	.9818
9.6	.9823	.9827	.9832	.9836	.9841	.9845	.9850	.9854	.9859	.9863
9.7	.9868	.9872	.9877	.9881	.9886	.9890	.9894	.9899	.9903	.9908
9.8	.9912	.9917	.9921	.9926	.9930	.9934	.9939	.9943	.9948	.9952
9.9	.9956	.9961	.9965	.9969	.9974	.9978	.9983	.9987	.9991	.9996
n	0	1	2	3	4	5	6	7	8	9

Table 3 Natural Trigonometric Functions

From MODERN TRIGONOMETRY: A Functional Approach by David G. Crowdis and Brandon W. Wheeler. Copyright © 1971 by McGraw-Hill Book Company. Reprinted by permission.

Angle θ

Degrees	Radians	cos θ	sin θ	tan θ	sec θ	csc θ	cot θ	Radians	Degrees
0°00'	0.0000	1.0000	0.0000	0.0000	1.000	No value	No value	1.5708	90°00'
10	029	000	029	029	000	343.8	343.8	679	50
20	058	000	058	058	000	171.9	171.9	650	40
30	087	000	087	087	000	114.6	114.6	621	30
40	116	0.9999	116	116	000	85.95	85.94	592	20
50	145	999	145	145	000	68.76	68.75	563	10
1°00'	0.0175	0.9998	0.0175	0.0175	1.000	57.30	57.29	1.5533	89°00'
10	204	998	204	204	000	49.11	49.10	504	50
20	233	997	233	233	000	42.98	42.96	475	40
30	262	997	262	262	000	38.20	38.19	446	30
40	291	996	291	291	000	34.38	34.37	417	20
50	320	995	320	320	001	31.26	31.24	388	10
2°00'	0.0349	0.9994	0.0349	0.0349	1.001	28.65	28.64	1.5359	88°00'
10	378	993	378	378	001	26.45	26.43	330	50
20	407	992	407	407	001	24.56	24.54	301	40
30	436	990	436	437	001	22.93	22.90	272	30
40	465	989	465	466	001	21.49	21.47	243	20
50	495	988	494	495	001	20.23	20.21	213	10
3°00'	0.0524	0.9986	0.0523	0.0524	1.001	19.11	19.08	1.5184	87°00'
10	553	985	552	553	002	18.10	18.07	155	50
20	582	983	581	582	002	17.20	17.17	126	40
30	611	981	610	612	002	16.38	16.35	097	30
40	640	980	640	641	002	15.64	15.60	068	20
50	669	978	669	670	002	14.96	14.92	039	10
		sin θ	cos θ	cot θ	csc θ	sec θ	tan θ	Radians	Degrees

Angle θ

Angle θ

Table 3

Degrees	Radians	cos θ	sin θ	tan θ	sec θ	csc θ	cot θ		
4°00'	0.0698	0.9976	0.0698	0.0699	1.002	14.34	14.30	1.5010	86°00'
10	727	974	727	729	003	13.76	13.73	981	50
20	765	971	756	758	003	13.23	13.20	952	40
30	785	969	785	787	003	12.75	12.71	923	30
40	814	967	814	816	003	12.29	12.25	893	20
50	844	964	843	846	004	11.87	11.83	864	10
5°00'	0.0873	0.9962	0.0872	0.0875	1.004	11.47	11.43	1.4835	85°00'
10	902	959	901	904	004	11.10	11.06	806	50
20	931	957	929	934	004	10.76	10.71	777	40
30	960	954	958	963	005	10.43	10.39	748	30
40	0.0989	951	0.0987	0.0992	005	10.13	10.08	719	20
50	0.1018	948	0.1016	0.1022	005	9.839	9.788	690	10
6°00'	0.1047	0.9945	0.1045	0.1051	1.006	9.567	9.514	1.4661	84°00'
10	076	942	074	080	006	9.309	9.255	632	50
20	105	939	103	110	006	9.065	9.010	603	40
30	134	936	132	139	006	8.834	8.777	573	30
40	164	932	161	169	007	8.614	8.556	544	20
50	193	929	190	198	007	8.405	8.345	515	10
7°00'	0.1222	0.9925	0.1219	0.1228	1.008	8.206	8.144	1.4486	83°00'
10	251	922	248	257	008	8.016	7.953	457	50
20	280	918	276	287	008	7.834	7.770	428	40
30	309	914	305	317	009	7.661	7.596	399	30
40	338	911	334	346	009	7.496	7.429	370	20
50	367	907	363	376	009	7.337	7.269	341	10

Natural Trigonometric Functions (continued)

Angle θ (Degrees)	Radians	sin θ	cos θ	cot θ	csc θ	sec θ	tan θ	Radians	Degrees
8°00'	0.1396	0.9903	0.1392	0.1405	1.010	7.185	7.115	1.4312	82°00'
10	425	899	421	435	010	7.040	6.968	283	50
20	454	894	449	465	011	6.900	827	254	40
30	484	890	478	495	011	765	691	224	30
40	513	886	507	524	012	636	561	195	20
50	542	881	536	554	012	512	435	166	10
9°00'	0.1571	0.9877	0.1564	0.1584	1.012	6.392	6.314	1.4137	81°00'
10	600	872	593	614	013	277	197	108	50
20	629	868	622	644	013	166	6.084	079	40
30	658	863	650	673	014	6.059	5.976	050	30
40	687	858	679	703	014	5.955	871	1.4021	20
50	716	853	708	733	015	855	769	1.3992	10
10°00'	0.1745	0.9848	0.1736	0.1763	1.015	5.759	5.671	1.3963	80°00'
10	774	843	765	793	016	665	576	934	50
20	804	838	794	823	016	575	485	904	40
30	833	833	822	853	017	487	396	875	30
40	862	827	851	883	018	403	309	846	20
50	891	822	880	914	018	320	226	817	10
11°00'	0.1920	0.9816	0.1908	0.1944	1.019	5.241	5.145	1.3788	79°00'
10	949	811	937	0.1974	019	164	5.066	759	50
20	0.1978	805	965	0.2004	020	089	4.989	730	40
30	0.2007	799	0.1994	035	020	5.016	915	701	30
40	036	793	0.2022	065	021	4.945	843	672	20
50	065	787	051	095	022	876	773	643	10
		sin θ	cos θ	cot θ	csc θ	sec θ	tan θ	Radians	Degrees
								|— Angle θ —|	

Table 3

Angle

Degrees	Radians	cos θ	sin θ	tan θ	sec θ	csc θ	cot θ	Radians	Degrees
12°00'	0.2094	0.9781	0.2079	0.2126	1.022	4.810	4.705	1.3614	78°00'
10	123	775	108	156	023	745	638	584	50
20	153	769	136	186	024	682	574	555	40
30	182	763	164	217	024	620	511	526	30
40	211	757	193	247	025	560	449	497	20
50	240	750	221	278	026	502	390	468	10
13°00'	0.2269	0.9744	0.2250	0.2309	1.026	4.445	4.331	1.3439	77°00'
10	298	737	278	339	027	390	275	410	50
20	327	730	306	370	028	336	219	381	40
30	356	724	334	401	028	284	165	352	30
40	385	717	363	432	029	232	113	323	20
50	414	710	391	462	030	182	061	294	10
14°00'	0.2443	0.9703	0.2419	0.2493	1.031	4.134	4.011	1.3265	76°00'
10	473	696	447	524	031	086	3.962	235	50
20	502	689	476	555	032	4.039	914	206	40
30	531	681	504	586	033	3.994	867	177	30
40	560	674	532	617	034	950	821	148	20
50	589	667	560	648	034	906	776	119	10
15°00'	0.2618	0.9659	0.2588	0.2679	1.035	3.864	3.732	1.3090	75°00'
10	647	652	616	711	036	822	689	061	50
20	676	644	644	742	037	782	647	032	40
30	705	636	672	773	038	742	606	1.3003	30
40	734	628	700	805	039	703	566	1.2974	20
50	763	621	728	836	039	665	526	945	10

Natural Trigonometric Functions (continued)

Angle θ Degrees	Radians	sin θ	cos θ	tan θ	cot θ	sec θ	csc θ	Radians	Degrees
16°00'	0.2793	0.2756	0.9613	0.2867	3.487	1.040	3.628	1.2915	74°00'
10	822	784	605	899	450	041	592	886	50
20	851	812	596	931	412	042	556	857	40
30	880	840	588	962	376	043	521	828	30
40	909	868	580	0.2944	340	044	487	799	20
50	938	896	572	0.3026	305	045	453	770	10
17°00'	0.2967	0.2924	0.9563	0.3057	3.271	1.046	3.420	1.2741	73°00'
10	0.2996	952	555	089	237	047	388	712	50
20	0.3025	0.2979	546	121	204	048	357	683	40
30	054	0.3007	537	153	172	048	326	654	30
40	083	035	528	185	140	049	295	625	20
50	113	062	520	217	108	050	265	595	10
18°00'	0.3142	0.3090	0.9511	0.3249	3.078	1.051	3.236	1.2566	72°00'
10	171	118	502	281	047	052	207	537	50
20	200	145	492	314	3.018	053	179	508	40
30	229	173	483	346	2.989	054	152	479	30
40	258	201	474	378	960	056	124	450	20
50	287	228	465	411	932	057	098	421	10
19°00'	0.3316	0.3256	0.9455	0.3443	2.904	1.058	3.072	1.2392	71°00'
10	345	283	446	476	877	059	046	363	50
20	374	311	436	508	850	060	3.021	334	40
30	403	338	426	541	824	061	2.996	.305	30
40	432	365	417	574	798	062	971	275	20
50	462	393	407	607	773	063	947	246	10

Bottom (complementary) column headers: sin θ | cos θ | cot θ | csc θ | sec θ | tan θ | Radians | Degrees — Angle θ

Table 3

Angle θ

Degrees	Radians	cos θ	sin θ	tan θ	sec θ	csc θ	cot θ		
20°00′	0.3491	0.9397	0.3420	0.3640	1.064	2.924	2.747	1.2217	70°00′
10	520	387	448	673	065	901	723	188	50
20	549	377	475	706	066	878	699	159	40
30	578	367	502	739	068	855	675	130	30
40	607	356	529	772	069	833	651	101	20
50	636	346	557	805	070	812	628	072	10
21°00′	0.3665	0.9336	0.3584	0.3839	1.071	2.790	2.605	1.2043	69°00′
10	694	325	611	872	072	769	583	1.2014	50
20	723	315	638	906	074	749	560	1.1985	40
30	752	304	665	939	075	729	539	956	30
40	782	293	692	0.3973	076	709	517	926	20
50	811	283	719	0.4006	077	689	496	897	10
22°00′	0.3840	0.9272	0.3746	0.4040	1.079	2.669	2.475	1.1868	68°00′
10	869	261	773	074	080	650	455	839	50
20	898	250	800	108	081	632	434	810	40
30	927	239	827	142	082	613	414	781	30
40	956	228	854	176	084	595	394	752	20
50	985	216	881	210	085	577	375	723	10
23°00′	0.4014	0.9205	0.3907	0.4245	1.086	2.559	2.365	1.1694	67°00′
10	043	194	934	279	088	542	337	665	50
20	072	182	961	314	089	525	318	636	40
30	102	171	0.3987	348	090	508	300	606	30
40	131	159	0.4014	383	092	491	282	577	20
50	160	147	041	417	093	475	264	548	10

Natural Trigonometric Functions (continued)

Angle θ (Degrees)	Radians							Radians	Angle θ (Degrees)
24°00'	0.4189	0.9135	0.4067	0.4452	1.095	2.459	2.246	1.1519	66°00'
10	218	124	094	487	096	443	229	490	50
20	247	112	120	522	097	427	211	461	40
30	276	100	147	557	099	411	194	432	30
40	305	088	173	592	100	396	177	403	20
50	334	075	200	628	102	381	161	374	10
25°00'	0.4363	0.9063	0.4226	0.4663	1.103	2.366	2.145	1.1345	65°00'
10	392	051	253	699	105	352	128	316	50
20	422	038	279	734	106	337	112	286	40
30	451	026	305	770	108	323	097	257	30
40	480	013	331	806	109	309	081	228	20
50	509	0.9001	358	841	111	295	066	199	10
26°00'	0.4538	0.8988	0.4384	0.4877	1.113	2.281	2.050	1.1170	64°00'
10	567	975	410	913	114	268	035	141	50
20	596	962	436	950	116	254	020	112	40
30	625	949	462	0.4986	117	241	2.006	083	30
40	654	936	488	0.5022	119	228	1.991	054	20
50	683	924	514	059	121	215	977	1.1025	10
27°00'	0.4712	0.8910	0.4540	0.5095	1.122	2.203	1.963	1.0996	63°00'
10	741	897	566	132	124	190	949	966	50
20	771	884	592	169	126	178	935	937	40
30	800	870	617	206	127	166	921	908	30
40	829	857	643	243	129	154	907	879	20
50	858	843	669	280	131	142	894	850	10
		sin θ	cos θ	cot θ	csc θ	sec θ	tan θ	Radians	Degrees

Angle θ

Table 3

Angle θ

Degrees	Radians	cos θ	sin θ	tan θ	sec θ	csc θ	cot θ	Radians	Degrees
28°00'	0.4887	0.8829	0.4695	0.5317	1.133	2.130	1.881	1.0821	62°00'
10	916	816	720	354	134	118	868	792	50
20	945	802	746	392	136	107	855	763	40
30	0.4974	788	772	430	138	096	842	734	30
40	0.5003	774	797	467	140	085	829	705	20
50	032	760	823	505	142	074	816	676	10
29°00'	0.5061	0.8746	0.4848	0.5543	1.143	2.063	1.804	1.0647	61°00'
10	091	732	874	581	145	052	792	617	50
20	120	718	899	619	147	041	780	588	40
30	149	704	924	658	149	031	767	559	30
40	178	689	950	696	151	020	756	530	20
50	207	675	0.4975	735	153	010	744	501	10
30°00'	0.5236	0.8660	0.5000	0.5774	1.155	2.000	1.732	1.0472	60°00'
10	265	646	025	812	157	1.990	720	443	50
20	294	631	050	851	159	980	709	414	40
30	323	616	075	890	161	970	698	385	30
40	352	601	100	930	163	961	686	356	20
50	381	587	125	0.5969	165	951	675	327	10
31°00'	0.5411	0.8572	0.5150	0.6009	1.167	1.942	1.664	1.0297	59°00'
10	440	557	175	048	169	932	653	268	50
20	469	542	200	088	171	923	643	239	40
30	498	526	225	128	173	914	632	210	30
40	527	511	250	168	175	905	621	181	20
50	556	496	275	208	177	896	611	152	10

Natural Trigonometric Functions (continued)

32°00′	0.5585	0.8480	0.5299	0.6249	1.179	1.887	1.600	1.0123	58°00′
10	614	465	324	289	181	878	590	094	50
20	643	450	348	330	184	870	580	065	40
30	672	434	373	371	186	861	570	036	30
40	701	418	398	412	188	853	560	1.0007	20
50	730	403	422	453	190	844	550	0.9977	10
33°00′	0.5760	0.8387	0.5446	0.6494	1.192	1.836	1.540	0.9948	57°00′
10	789	371	471	536	195	828	530	919	50
20	818	355	495	577	197	820	520	890	40
30	847	339	519	619	199	812	511	861	30
40	876	323	544	661	202	804	501	832	20
50	905	307	568	703	204	796	492	803	10
34°00′	0.5934	0.8290	0.5592	0.6745	1.206	1.788	1.483	0.9774	56°00′
10	0.5963	274	616	787	209	781	473	745	50
20	0.5992	258	640	830	211	773	464	716	40
30	0.6021	241	664	873	213	766	455	687	30
40	050	225	688	916	216	758	446	657	20
50	080	208	712	0.6959	218	751	437	628	10
35°00′	0.6109	0.8192	0.5736	0.7002	1.221	1.743	1.428	0.9599	55°00′
10	138	175	760	046	223	736	419	570	50
20	167	158	783	089	226	729	411	541	40
30	196	141	807	133	228	722	402	512	30
40	225	124	831	177	231	715	393	483	20
50	254	107	854	221	233	708	385	454	10
		sin θ	cos θ	cot θ	csc θ	sec θ	tan θ	Radians	Degrees
								Angle θ	

Table 3

Angle θ

Degrees	Radians	cos θ	sin θ	tan θ	sec θ	csc θ	cot θ		
36°00′	0.6283	0.8090	0.5878	0.7265	1.236	1.701	1.376	0.9425	54°00′
10	312	073	901	310	239	695	368	396	50
20	341	056	925	355	241	688	360	367	40
30	370	039	948	400	244	681	351	338	30
40	400	021	972	445	247	675	343	308	20
50	429	0.8004	0.5995	490	249	668	335	279	10
37°00′	0.6458	0.7986	0.6018	0.7536	1.252	1.662	1.327	0.9250	53°00′
10	487	966	041	581	255	655	319	221	50
20	516	951	065	627	258	649	311	192	40
30	545	934	088	673	260	643	303	163	30
40	574	916	111	720	263	636	295	134	20
50	603	898	134	766	266	630	288	105	10
38°00′	0.6632	0.7880	0.6157	0.7813	1.269	1.624	1.280	0.9076	52°00′
10	661	862	180	860	272	618	272	047	50
20	690	844	202	907	275	612	265	0.9018	40
30	720	826	225	0.7954	278	606	257	0.8988	30
40	749	808	248	0.8002	281	601	250	959	20
50	778	790	271	050	284	595	242	930	10
39°00′	0.6807	0.7771	0.6293	0.8098	1.287	1.589	1.235	0.8901	51°00′
10	836	753	316	146	290	583	228	872	50
20	865	735	338	195	293	578	220	843	40
30	894	716	361	243	296	572	213	814	30
40	923	698	383	292	299	567	206	785	20
50	952	679	406	342	302	561	199	756	10

Natural Trigonometric Functions (continued)

Degrees	Radians	sin θ	cos θ	cot θ	csc θ	sec θ	tan θ	Radians	Degrees
40°00'	0.6981	0.7660	0.6428	0.8391	1.305	1.556	1.192	0.8727	50°00'
10	0.7010	642	450	441	309	550	185	698	50
20	039	623	472	491	312	545	178	668	40
30	069	604	494	541	315	540	171	639	30
40	098	585	517	591	318	535	164	610	20
50	127	566	539	642	322	529	157	581	10
41°00'	0.7156	0.7547	0.6561	0.8693	1.325	1.524	1.150	0.8552	49°00'
10	185	528	583	744	328	519	144	523	50
20	214	509	604	796	332	514	137	494	40
30	243	490	626	847	335	509	130	465	30
40	272	470	648	899	339	504	124	436	20
50	301	451	670	0.8952	342	499	117	407	10
42°00'	0.7330	0.7431	0.6691	0.9004	1.346	1.494	1.111	0.8378	48°00'
10	359	412	713	057	349	490	104	348	50
20	389	392	734	110	353	485	098	319	40
30	418	373	756	163	356	480	091	290	30
40	447	353	777	217	360	476	085	261	20
50	476	333	799	271	364	471	079	232	10
43°00'	0.7505	0.7314	0.6820	0.9325	1.367	1.466	1.072	0.8203	47°00'
10	534	294	841	380	371	462	066	174	50
20	563	274	862	435	375	457	060	145	40
30	592	254	884	490	379	453	054	116	30
40	621	234	905	545	382	448	048	087	20
50	650	214	926	601	386	444	042	058	10
		sin θ	cos θ	cot θ	csc θ	sec θ	tan θ	Radians	Degrees

Angle θ

Table 3 Natural Trigonometric Functions (continued)

| Angle θ | | | | | | | | | |
Degrees	Radians	cos θ	sin θ	tan θ	sec θ	csc θ	cot θ	Radians	Degrees
44°00'	0.7679	0.7193	0.6947	0.9657	1.390	1.440	1.036	0.8029	46°00'
10	709	173	967	713	394	435	030	0.7999	50
20	738	153	0.6988	770	398	431	024	970	40
30	767	133	0.7009	827	402	427	018	941	30
40	796	112	030	884	406	423	012	912	20
50	825	092	050	0.9942	410	418	006	883	10
		sin θ	cos θ	cot θ	csc θ	sec θ	tan θ	Radians	Degrees
									Angle θ

Table 4 Logarithms of Trigonometric Functions*

angles	log sin	log cos	log tan	log cot	
0° 00′		10.0000			90° 00′
10	7.4637	.0000	7.4637	12.5363	50
20	.7648	.0000	.7648	.2352	40
30	7.9408	.0000	7.9409	12.0591	30
40	8.0658	.0000	8.0658	11.9342	20
50	.1627	10.0000	.1627	.8373	10
1° 00′	8.2419	9.9999	8.2419	11.7581	89° 00′
10	.3088	.9999	.3089	.6911	50
20	.3668	.9999	.3669	.6331	40
30	.4179	.9999	.4181	.5819	30
40	.4637	.9998	.4638	.5362	20
50	.5050	.9998	.5053	.4947	10
2° 00′	8.5428	9.9997	8.5431	11.4569	88° 00′
10	.5776	.9997	.5779	.4221	50
20	.6097	.9996	.6101	.3899	40
30	.6397	.9996	.6401	.3599	30
40	.6677	.9995	.6682	.3318	20
50	.6940	.9995	.6945	.3055	10
3° 00′	8.7188	9.9994	8.7194	11.2806	87° 00′
10	.7423	.9993	.7429	.2571	50
20	.7645	.9993	.7652	.2348	40
30	.7857	.9992	.7865	.2135	30
40	.8059	.9991	.8067	.1933	20
50	.8251	.9990	.8261	.1739	10
4° 00′	8.8436	9.9989	8.8446	11.1554	86° 00′
10	.8613	.9989	.8624	.1376	50
20	.8783	.9988	.8795	.1205	40
30	.8946	.9987	.8960	.1040	30
40	.9104	.9986	.9118	.0882	20
50	.9256	.9985	.9272	.0728	10
5° 00′	8.9403	9.9983	8.9420	11.0580	85° 00′
10	.9545	.9982	.9563	.0437	50
20	.9682	.9981	.9701	.0299	40
30	.9816	.9980	.9836	.0164	30
40	8.9945	.9979	8.9966	11.0034	20
50	9.0070	.9977	9.0093	10.9907	10
6° 00′	9.0192	9.9976	9.0216	10.9784	84° 00′
10	.0311	.9975	.0336	.9664	50
20	.0426	.9973	.0453	.9547	40
30	.0539	.9972	.0567	.9433	30
40	.0648	.9971	.0678	.9322	20
50	.0755	.9969	.0786	.9214	10
7° 00′	9.0859	9.9968	9.0891	10.9109	83° 00′
	log cos	log sin	log cot	log tan	angles

* Add −10 to all logarithms.

From ALGEBRA AND TRIGONOMETRY by Earl Swokowski. Copyright ©
1967 by Prindle, Weber & Schmidt, Inc. Reprinted by permission.

Table 4 Logarithms of Trigonometric Functions
(continued)

angles	log sin	log cos	log tan	log cot	
7° 00′	9.0859	9.9968	9.0891	10.9109	83° 00′
10	.0961	.9966	.0995	.9005	50
20	.1060	.9964	.1096	.8904	40
30	.1157	.9963	.1194	.8806	30
40	.1252	.9961	.1291	.8709	20
50	.1345	.9959	.1385	.8615	10
8° 00′	9.1436	9.9958	9.1478	10.8522	82° 00′
10	.1525	.9956	.1569	.8431	50
20	.1612	.9954	.1658	.8342	40
30	.1697	.9952	.1745	.8255	30
40	.1781	.9950	.1831	.8169	20
50	.1863	.9948	.1915	.8085	10
9° 00′	9.1943	9.9946	9.1997	10.8003	81° 00′
10	.2022	.9944	.2078	.7922	50
20	.2100	.9942	.2158	.7842	40
30	.2176	.9940	.2236	.7764	30
40	.2251	.9938	.2313	.7687	20
50	.2324	.9936	.2389	.7611	10
10° 00′	9.2397	9.9934	9.2463	10.7537	80° 00′
10	.2468	.9931	.2536	.7464	50
20	.2538	.9929	.2609	.7391	40
30	.2606	.9927	.2680	.7320	30
40	.2674	.9924	.2750	.7250	20
50	.2740	.9922	.2819	.7181	10
11° 00′	9.2806	9.9919	9.2887	10.7113	79° 00′
10	.2870	.9917	.2953	.7047	50
20	.2934	.9914	.3020	.6980	40
30	.2997	.9912	.3085	.6915	30
40	.3058	.9909	.3149	.6851	20
50	.3119	.9907	.3212	.6788	10
12° 00′	9.3179	9.9904	9.3275	10.6725	78° 00′
10	.3238	.9901	.3336	.6664	50
20	.3296	.9899	.3397	.6603	40
30	.3353	.9896	.3458	.6542	30
40	.3410	9893	.3517	.6483	20
50	.3466	.9890	.3576	.6424	10
13° 00′	9.3521	9.9887	9.3634	10.6366	77° 00′
10	.3575	.9884	.3691	.6309	50
20	.3629	.9881	.3748	.6252	40
30	.3682	.9878	.3804	.6196	30
40	.3734	.9875	.3859	.6141	20
50	.3786	.9872	.3914	.6086	10
14° 00′	9.3837	9.9869	9.3968	10.6032	76° 00′
	log cos	log sin	log cot	log tan	angles

Table 4 Logarithms of Trigonometric Functions (continued)

angles	log sin	log cos	log tan	log cot	
14° 00′	9.3837	9.9869	9.3968	10.6032	76° 00′
10	.3887	.9866	.4021	.5979	50
20	.3937	.9863	.4074	.5926	40
30	.3986	.9859	.4127	.5873	30
40	.4035	.9856	.4178	.5822	20
50	.4083	.9853	.4230	.5770	10
15° 00′	9.4130	9.9849	9.4281	10.5719	75° 00′
10	.4177	.9846	.4331	.5669	50
20	.4223	.9843	.4381	.5619	40
30	.4269	.9839	.4430	.5570	30
40	.4314	.9836	.4479	.5521	20
50	.4359	.9832	.4527	.5473	10
16° 00′	9.4403	9.9828	9.4575	10.5425	74° 00′
10	.4447	.9825	.4622	.5378	50
20	.4491	.9821	.4669	.5331	40
30	.4533	.9817	.4716	.5284	30
40	.4576	.9814	.4762	.5238	20
50	.4618	.9810	.4808	.5192	10
17° 00′	9.4659	9.9806	9.4853	10.5147	73° 00′
10	.4700	.9802	.4898	.5102	50
20	.4741	.9798	.4943	.5057	40
30	.4781	.9794	.4987	.5013	30
40	.4821	.9790	.5031	.4969	20
50	.4861	.9786	.5075	.4925	10
18° 00′	9.4900	9.9782	9.5118	10.4882	72° 00′
10	.4939	.9778	.5161	.4839	50
20	.4977	.9774	.5203	.4797	40
30	.5015	.9770	.5245	.4755	30
40	.5052	.9765	.5287	.4713	20
50	.5090	.9761	.5329	.4671	10
19° 00′	9.5126	9.9757	9.5370	10.4630	71° 00′
10	.5163	.9752	.5411	.4589	50
20	.5199	.9748	.5451	.4549	40
30	.5235	.9743	.5491	.4509	30
40	.5270	.9739	.5531	.4469	20
50	.5306	.9734	.5571	.4429	10
20° 00′	9.5341	9.9730	9.5611	10.4389	70° 00′
10	.5375	.9725	.5650	.4350	50
20	.5409	.9721	.5689	.4311	40
30	.5443	.9716	.5727	.4273	30
40	.5477	.9711	.5766	.4234	20
50	.5510	.9706	.5804	.4196	10
21° 00′	9.5543	9.9702	9.5842	10.4158	69° 00′
	log cos	log sin	log cot	log tan	angles

* Add −10 to all logarithms.

Table 4 Logarithms of Trigonometric Functions (continued)

angles	log sin	log cos	log tan	log cot	
21° 00′	9.5543	9.9702	9.5842	10.4158	69° 00′
10	.5576	.9697	.5879	.4121	50
20	.5609	.9692	.5917	.4083	40
30	.5641	.9687	.5954	.4046	30
40	.5673	.9682	.5991	.4009	20
50	.5704	.9677	.6028	.3972	10
22° 00′	9.5736	9.9672	9.6064	10.3936	68° 00′
10	.5767	.9667	.6100	.3900	50
20	.5798	.9661	.6136	.3864	40
30	.5828	.9656	.6172	.3828	30
40	.5859	.9651	.6208	.3792	20
50	.5889	.9646	.6243	.3757	10
23° 00′	9.5919	9.9640	9.6279	10.3721	67° 00′
10	.5948	.9635	.6314	.3686	50
20	.5978	.9629	.6348	.3652	40
30	.6007	.9624	.6383	.3617	30
40	.6036	.9618	.6417	.3583	20
50	.6065	.9613	.6452	.3548	10
24° 00′	9.6093	9.9607	9.6486	10.3514	66° 00′
10	.6121	.9602	.6520	.3480	50
20	.6149	.9596	.6553	.3447	40
30	.6177	.9590	.6587	.3413	30
40	.6205	.9584	.6620	.3380	20
50	.6232	.9579	.6654	.3346	10
25° 00′	9.6259	9.9573	9.6687	10.3313	65° 00′
10	.6286	.9567	.6720	.3280	50
20	.6313	.9561	.6752	.3248	40
30	.6340	.9555	.6785	.3215	30
40	.6366	.9549	.6817	.3183	20
50	.6392	.9543	.6850	.3150	10
26° 00′	9.6418	9.9537	9.6882	10.3118	64° 00′
10	.6444	.9530	.6914	.3086	50
20	.6470	.9524	.6946	.3054	40
30	.6495	.9518	.6977	.3023	30
40	.6521	.9512	.7009	.2991	20
50	.6546	.9505	.7040	.2960	10
27° 00′	9.6570	9.9499	9.7072	10.2928	63° 00′
10	.6595	.9492	.7103	.2897	50
20	.6620	.9486	.7134	.2866	40
30	.6644	.9479	.7165	.2835	30
40	.6668	.9473	.7196	.2804	20
50	.6692	.9466	.7226	.2774	10
28° 00′	9.6716	9.9459	9.7257	10.2743	62° 00′
	log cos	log sin	log cot	log tan	angles

Table 4   Logarithms of Trigonometric Functions
(continued)

angles	log sin	log cos	log tan	log cot	
28° 00′	9.6716	9.9459	9.7257	10.2743	62° 00′
10	.6740	.9453	.7287	.2713	50
20	.6763	.9446	.7317	.2683	40
30	.6787	.9439	.7348	.2652	30
40	.6810	.9432	.7378	.2622	20
50	.6833	.9425	.7408	.2592	10
29° 00′	9.6856	9.9418	9.7438	10.2562	61° 00′
10	.6878	.9411	.7467	.2533	50
20	.6901	.9404	.7497	.2503	40
30	.6923	.9397	.7526	.2474	30
40	.6946	.9390	.7556	.2444	20
50	.6968	.9383	.7585	.2415	10
30° 00′	9.6990	9.9375	9.7614	10.2386	60° 00′
10	.7012	.9368	.7644	.2356	50
20	.7033	.9361	.7673	.2327	40
30	.7055	.9353	.7701	.2299	30
40	.7076	.9346	.7730	.2270	20
50	.7097	.9338	.7759	.2241	10
31° 00′	9.7118	9.9331	9.7788	10.2212	59° 00′
10	.7139	.9323	.7816	.2184	50
20	.7160	.9315	.7845	.2155	40
30	.7181	.9308	.7873	.2127	30
40	.7201	.9300	.7902	.2098	20
50	.7222	.9292	.7930	.2070	10
32° 00′	9.7242	9.9284	9.7958	10.2042	58° 00′
10	.7262	.9276	.7986	.2014	50
20	.7282	.9268	.8014	.1986	40
30	.7302	.9260	.8042	.1958	30
40	.7322	.9252	.8070	.1930	20
50	.7342	.9244	.8097	.1903	10
33° 00′	9.7361	9.9236	9.8125	10.1875	57° 00′
10	.7380	.9228	.8153	.1847	50
20	.7400	.9219	.8180	.1820	40
30	.7419	.9211	.8208	.1792	30
40	.7438	.9203	.8235	.1765	20
50	.7457	.9194	.8263	.1737	10
34° 00′	9.7476	9.9186	9.8290	10.1710	56° 00′
10	.7494	.9177	.8317	.1683	50
20	.7513	.9169	.8344	.1656	40
30	.7531	.9160	.8371	.1629	30
40	.7550	.9151	.8398	.1602	20
50	.7568	.9142	.8425	.1575	10
35° 00′	9.7586	9.9134	9.8452	10.1548	55° 00′
	log cos	log sin	log cot	log tan	angles

* Add −10 to all logarithms.

Table 4 Logarithms of Trigonometric Functions
(continued)

angles	log sin	log cos	log tan	log cot	
35° 00′	9.7586	9.9134	9.8452	10.1548	55° 00′
10	.7604	.9125	.8479	.1521	50
20	.7622	.9116	.8506	.1494	40
30	.7640	.9107	.8533	.1467	30
40	.7657	.9098	.8559	.1441	20
50	.7675	.9089	.8586	.1414	10
36° 00′	9.7692	9.9080	9.8613	10.1387	54° 00′
10	.7710	.9070	.8639	.1361	50
20	.7727	.9061	.8666	.1334	40
30	.7744	.9052	.8692	.1308	30
40	.7761	.9042	.8718	.1282	20
50	.7778	.9033	.8745	.1255	10
37° 00′	9.7795	9.9023	9.8771	10.1229	53° 00′
10	.7811	.9014	.8797	.1203	50
20	.7828	.9004	.8824	.1176	40
30	.7844	.8995	.8850	.1150	30
40	.7861	.8985	.8876	.1124	20
50	.7877	.8975	.8902	.1098	10
38° 00′	9.7893	9.8965	9.8928	10.1072	52° 00′
10	.7910	.8955	.8954	.1046	50
20	.7926	.8945	.8980	.1020	40
30	.7941	.8935	.9006	.0994	30
40	.7957	.8925	.9032	.0968	20
50	.7973	.8915	.9058	.0942	10
39° 00′	9.7989	9.8905	9.9084	10.0916	51° 00′
10	.8004	.8895	.9110	.0890	50
20	.8020	.8884	.9135	.0865	40
30	.8035	.8874	.9161	.0839	30
40	.8050	.8864	.9187	.0813	20
50	.8066	.8853	.9212	.0788	10
40° 00′	9.8081	9.8843	9.9238	10.0762	50° 00′
10	.8096	.8832	.9264	.0736	50
20	.8111	.8821	.9289	.0711	40
30	.8125	.8810	.9315	.0685	30
40	.8140	.8800	.9341	.0659	20
50	.8155	.8789	.9366	.0634	10
41° 00′	9.8169	9.8778	9.9392	10.0608	49° 00′
10	.8184	.8767	.9417	.0583	50
20	.8198	.8756	.9443	.0557	40
30	.8213	.8745	.9468	.0532	30
40	.8227	.8733	.9494	.0506	20
50	.8241	.8722	.9519	.0481	10
42° 00′	9.8255	9.8711	9.9544	10.0456	48° 00′
	log cos	log sin	log cot	log tan	angles

Table 4 Logarithms of Trigonometric Functions
(continued)

angles	log sin	log cos	log tan	log cot	
42° 00′	9.8255	9.8711	9.9544	10.0456	48° 00′
10	.8269	.8699	.9570	.0430	50
20	.8283	.8688	.9595	.0405	40
30	.8297	.8676	.9621	.0379	30
40	.8311	.8665	.9646	.0354	20
50	.8324	.8653	.9671	.0329	10
43° 00′	9.8338	9.8641	9.9697	10.0303	47° 00′
10	.8351	.8629	.9722	.0278	50
20	.8365	.8618	.9747	.0253	40
30	.8378	.8606	.9772	.0228	30
40	.8391	.8594	.9798	.0202	20
50	.8405	.8582	.9823	.0177	10
44° 00′	9.8418	9.8569	9.9848	10.0152	46° 00′
10	.8431	.8557	.9874	.0126	50
20	.8444	.8545	.9899	.0101	40
30	.8457	.8532	.9924	.0076	30
40	.8469	.8520	.9949	.0051	20
50	.8482	.8507	.9975	.0025	10
45° 00′	9.8495	9.8495	10.0000	10.0000	45° 00′
	log cos	log sin	log cot	log tan	angles

* Add −10 to all logarithms.

Answers to Selected Exercises

1.1 Exercises (page 3)

1. $\{1, 2, 3, 4\}$ **3.** $\{-2, -1, 0, 1, 2\}$ **5.** $\{7, 9, 11, 13, \ldots\}$ **7.** $\{-1, 0, 1, 2, 3, 4, 6, 8\}$ **9.** $\{x \mid x$ is an integer$\} = U$ **11.** $\{-1\}$ **13.** P **15.** $\{-3, -2, -1\}$
17. $\{-1, \ 0, \ 2, \ 4, \ 6, \ 8\}$ **19.** $\{-1\}$ **21.** $\subset$ **23.** $\cup$ **25.** $\cap$ **27.** $\cup$
29. $=$ **31.** X and Y are disjoint   **33.** $X \subset Y$ **35.** $X = \emptyset$ **37.** $X \subset Y$
39. no special requirements

1.2 Exercises (page 7)

1. true   **3.** false $(0 - 1 = -1)$ **5.** true   **7.** false $(0 \notin N)$ **9.** false $(1/0$ does not exist)   **11.** commutative axiom   **13.** distributive axiom   **15.** associative axiom   **17.** inverse axiom   **19.** substitution axiom   **21.** subset of, equals
23. Set axioms include commutative axioms, associative axioms, identity axioms, closure axioms and distributive axioms.   **25.** definition of subtraction; identity axiom of addition; substitution   **27.** commutative axiom; associative axiom; inverse axiom; identity axiom of multiplication; substitution   **29.** associative axiom; associative axiom; commutative axiom; associative axiom; associative axiom; substitution

1.3 Exercises (page 12)

1. commutative and associative axioms; additive inverse axiom; identity axiom; definition; definition; additive inverse is unique   **3.** addition property of equality; associative axiom; additive inverse axiom; identity axiom; definition of subtraction   **5.** definition of division; Exercise 4 above; definition of division   **7.** reflexive property; commutative and associative axioms; Exercise 6 above, part II

1.4 Exercises (page 17)

1. false   **3.** false   **5.** true   **7.** transitive axiom   **9.** Theorem 1.8(b)
11. Theorem 1.8(a)

	Greatest Lower bound	*Least Upper bound*
13.	none	6
15.	1	4
17.	none	none

19. definition of $<$; Theorem 1.2; distributive axiom; given; closure axiom; definition of $<$
21. given; Theorem 1.8(c); Exercise 9 of Section 1.3; substitution

PART II

$(-1)(-a) > (-1)(-b)$	Exercise 20
$1 \cdot a > 1 \cdot b$	Theorem 1.7(d)
$a > b$	identity

23. definition of $<$; definition of $<$; addition property of equality and substitution; associative and commutative axioms; closure axiom; definition of $<$

1.5 Exercises (page 22)

1. $=$ **3.** $\leq$ **5.** $=$ **7.** $=$ **9.** $\leq, \leq$ **11.** $\geq, \leq$ **13.** $=$ **15.** $=$
17. $\leq$ **19.** $=$ **21.** $=$

23.

$$5$$

25.

$$3 \quad\quad 6$$

27.

$$1 \quad\quad 3$$

29.

$$2 \quad\quad 5$$

31. $\varnothing$ **33.** $\{1, -1\}$ **35.** $\{-4, -1\}$ **37.** $\{3, -9\}$

2.1 Exercises (page 27)

1. 30   **3.** 3   **5.** $-3/2$ **7.** $2m + 1$ **9.** $2m^2 + 4m$ **11.** $70 + 24h + 2h^2$
13. $8x^2 - 16x + 6$ **15.** $2x^2 + 4xr + 2r^2 - 4x - 4r$ **17.** 1   **19.** 448
21. $x^2 - x + 3$ **23.** $9y^2 - 4y + 4$ **25.** $-14q^2 + 11q - 14$
27. $28r^2 + r - 2$ **29.** $18p^2 - 27p - 35$ **31.** $25r^2 - 4$ **33.** $4k^2 - 12k + 9$
35. $12k^4 + 21k^3 - 5k^2 + 3k + 2$ **37.** $8z^3 - 12z^2 + 6z - 1$ **39.** $8m^3 + 1$
41. $216m^9$ **43.** $\dfrac{-96}{m^7}$ **45.** y^{1+2m} **47.** $\dfrac{-2}{m^{n-2}}$
49. If $n = 0$, we have $a^m \cdot a^0 = a^m \cdot 1 = a^m$. Also, $a^{m+n} = a^{m+0} = a^m$.
Hence, $a^m a^n = a^{m+n}$.

2.2 Exercises (page 31)

1. $2m^2 - 4m + 8$ **3.** $3y + 2$ **5.** $5p - 1$ **7.** $3x + 2 + \dfrac{-2}{4x - 5}$

9. $4z - 3 + \dfrac{-5}{2z + 5}$ **11.** $x^2 - 3x + 2$ **13.** $5x^2 - 3x + 4 + \dfrac{4}{3x + 4}$

15. $2z^3 - z^2 - \dfrac{1}{2}z - \dfrac{5}{4} + \dfrac{(-\frac{13}{4})z + \frac{17}{4}}{2z^2 + z + 1}$ **17.** $m^3 - m^2 - 6m$

19. $3x^2 + 4x + \dfrac{3}{x - 5}$ **21.** $4m^2 - 4m + 1 + \dfrac{-3}{m + 1}$

23. $x^4 + x^3 + 2x - 1 + \dfrac{3}{x + 2}$ **25.** $\dfrac{1}{3}x^2 - \dfrac{1}{9}x + \dfrac{1}{x - \frac{1}{3}}$

27. $x^3 + x^2 + x + 1$ **29.** $x^3 + x^2 + x + 1 + \dfrac{2}{x - 1}$

2.3 Exercises (page 34)

1. $4m(3n - 2)$ **3.** $2px(3x - 4x^2 - 6)$ **5.** $(4p - 1)(p + 1)$
7. $(2z - 5)(z + 6)$ **9.** $(3q + 4)(2q - 3)$ **11.** $3(2r - 1)(2r + 5)$
13. $(6r - 5s)(3r + 2s)$ **15.** $(8pq + 11mn)(8pq - 11mn)$
17. $(9m^2 + 4n^4)(3m + 2n^2)(3m - 2n^2)$ **19.** $x^2(3 - x)^2$
21. $(m + 1)(m^2 - m + 1)(m - 1)(m^2 + m + 1)$
23. $(2m^2 - 3n)(4m^4 + 6m^2n + 9n^2)$ **25.** $(2x - y)y$
27. $(3x - y)(19y^2 - 24xy + 9x^2)$ **29.** $(x + y + 5z)(x + y - 3z)$
31. $[3(a + b) + 8c][2(a + b) - 5c]$ **33.** $4pq$ **35.** $(m^n + 4)(m^n - 4)$
37. $(x^n - y^{2n})(x^{2n} + x^ny^{2n} + y^{4n})$ **39.** $(2x^n + 3y^n)(x^n - 13y^n)$

2.4 Exercises (page 37)

1. $\dfrac{4}{x - 1}$ **3.** $a + 2$ **5.** 1 **7.** $\dfrac{m - 4}{m + 4}$ **9.** $\dfrac{x^2 - 1}{x^2}$ **11.** $\dfrac{x + y}{x - y}$

13. $\dfrac{x^2 - xy + y^2}{x^2 + xy + y^2}$ **15.** 1 **17.** $\dfrac{2y + 1}{y(y + 1)}$ **19.** $\dfrac{3}{2(a + b)}$ **21.** $\dfrac{-2}{(a + 1)(a - 1)}$

23. $\dfrac{2m^2 + 2}{(m - 1)(m + 1)}$ **25.** $\dfrac{4}{a - 2}$ **27.** $\dfrac{5}{(a - 2)(a - 3)(a + 2)}$

29. $\dfrac{x - 11}{(x + 4)(x - 4)(x - 3)}$ **31.** $\dfrac{p + 5}{p(p + 1)}$ **33.** $\dfrac{-1}{x(x + h)}$ **35.** $\dfrac{x + 1}{x - 1}$

37. $\dfrac{(2 - b)(1 + b)}{b(1 - b)}$ **39.** $\dfrac{2x + 1}{x + 1}$ **41.** $\dfrac{3}{8}$

2.5 Exercises (page 41)

1. $\dfrac{1}{2^3} = \dfrac{1}{8}$ **3.** $\dfrac{1}{(-4)^3} = \dfrac{-1}{64}$ **5.** 8 **7.** $-\dfrac{1}{8}$ **9.** $2^6 = 64$ **11.** $\dfrac{x^2}{2y}$

13. p^4r^5 **15.** $\dfrac{x^6}{4y^4}$ **17.** $\dfrac{9m^5}{n^3}$ **19.** $a + b$ **21.** -1 **23.** $\dfrac{1}{ab}$

25. $\dfrac{y(xy - 9)}{x^2y^2 - 9}$

2.6 Exercises (page 45)

1. 2 **3.** 4 **5.** $3^{-2} = \frac{1}{9}$ **7.** $(\frac{2}{3})^{-3} = \frac{27}{8}$ **9.** $5^{-2} = \frac{1}{25}$ **11.** $4p^2$

13. $9x^4$ **15.** $\dfrac{4z^2}{x^5y}$ **17.** $\dfrac{r^2 - 2r + 1}{r}$ **19.** 1 **21.** $x^{11/12}$ **23.** $\dfrac{16}{y^{11/12}}$

25. $m^{3/2}$ **27.** r^{6+p} **29.** $\dfrac{x+1}{x^{1/2}}$ **31.** $12^{1/2}$ **33.** $x^{2/3}$ **35.** $y^{4/3}$

37. $y^{9/2}$ **39.** $m^{5/6}$ **41.** (a) Since $1^2 < 2$, $1 \in A$. (b) If $x = 2$, we have $x^2 = 4$. If $x^2 < 2$, then $x^2 < 4$, and 2 is an upper bound. (c) A has a least upper bound (d) $\sqrt{2}$

2.7 Exercises (page 47)

1. $3\sqrt[3]{3}$ **3.** $\dfrac{2\sqrt[4]{2}}{3}$ **5.** $72\sqrt{2}$ **7.** $3\sqrt{5}$ **9.** $9\sqrt{3}$ **11.** $\sqrt{2}$ **13.** $2\sqrt[3]{2}$

15. $2\sqrt{3}$ **17.** $7x^2z^4\sqrt{2x}$ **19.** $3py^2\sqrt[3]{3p^2x^2}$ **21.** $y\sqrt[3]{x^{2/3}}$ **23.** $\dfrac{\sqrt{6x}}{3x}$

25. $\dfrac{3\sqrt{2y}}{2y}$ **27.** $\dfrac{2\sqrt[3]{x}}{x}$ **29.** $\dfrac{x^2y\sqrt{xy}}{z}$ **31.** $\dfrac{a^2b\sqrt[4]{bc}}{c}$ **33.** $\dfrac{xy^2\sqrt[3]{9x^2zw}}{z^2w}$

35. $\dfrac{r\sqrt[4]{rs^3t^2}}{t}$ **37.** $\dfrac{\sqrt[4]{x^2y^3}}{y^2}$ **39.** $\dfrac{m\sqrt[3]{n^2}}{n}$ **41.** $ab\sqrt{ab}(b - 2a^2 + b^3)$

43. $\dfrac{x\sqrt{x+1}}{x+1}$ **45.** $\dfrac{2\sqrt{1+r}}{1+r}$ **47.** $-2x - 2\sqrt{x(x+1)} - 1$ **49.** $\dfrac{3}{2\sqrt{3}}$

51. $\dfrac{-1}{2(1 - \sqrt{2})}$ **53.** $\dfrac{x}{\sqrt{x} + x}$ **55.** $\dfrac{-1}{2x - 2\sqrt{x(x+1)} + 1}$

3.1 Exercises (page 53)

1. Domain: all reals; Range: all reals; Function **3.** Domain: all reals; Range: all reals; not a Function **5.** Domain: all reals; Range: nonnegative reals; Function **7.** Domain: $\{x \mid -4 \le x \le 4\}$; Range: $\{y \mid 0 \le y \le 4\}$; Function **9.** Domain: all reals; Range: $\{y \mid y = -2\}$: Function **11.** Domain: $\{x \mid x \ne 0\}$; Range: $\{y \mid y \ne 0\}$; Function **13.** -1 **15.** -10 **17.** 4 **19.** $3a - 1$ **21.** 11 **23.** 5 **25.** 2 **27.** 16 **29.** 1 **31.** 3 **33.** even **35.** even **37.** neither

3.2 Exercises (page 56)

1.

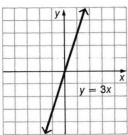

3.

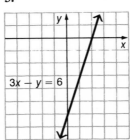

5.

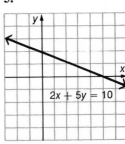

7.

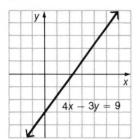

$4x - 3y = 9$

9.

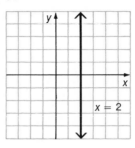

$x = 2$

11.

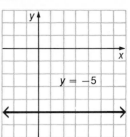

$y = -5$

13. (4, 2)   **15.** (5, −2)   **17.** Lines do not intersect.

3.3 Exercises (page 60)

1. $m = 2$ **3.** $m = 4$ **5.** $m = -\frac{3}{4}$ **7.** $m = \frac{1}{5}$ **9.** $m = 7$ **11.** $m = 0$
13. $m = 0$ **15.** $y - 2x = 5$ **17.** $2y + 3x = -7$ **19.** $4y + 3x = 6$
21. $y = 2$ **23.** $x = -8$ **25.** $x + 3y = 10$ **27.** $4y - x = 13$
29. $4y + 3x = 12$ **31.** $2x - 3y = 6$ **33.** $x = -5$

35.

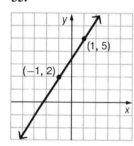

(1, 5)
(−1, 2)

37.

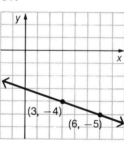

(3, −4)
(6, −5)

39.

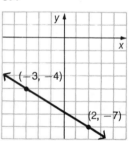

(−3, −4)
(2, −7)

41. $x + 3y = 11$ **43.** $x - y = 7$ **45.** no

3.4 Exercises (page 64)

1. $x = 4$ **3.** all real numbers   **5.** $x = 3$ **7.** $x = 12$ **9.** $m = -1\frac{2}{5}$
11. $\varnothing$ **13.** $x = -\frac{5}{13}$ **15.** $x \leq 4$ **17.** $p \geq -1$ **19.** $a = 1, a = 3$
21. $m = 1, m = -\frac{1}{3}$ **23.** $x = \frac{8}{3}, x = \frac{2}{3}$ **25.** $-2 < a < 10$
27. $m < \frac{1}{3}$ or $m > 1$ **29.** $b \leq -\frac{3}{2}$ or $b \geq \frac{1}{4}$ **31.** $V = k/P$
33. $g = \dfrac{V - V_0}{t}$ **35.** $R = \dfrac{r_1 r_2}{r_1 + r_2}$ **37.** $\frac{18}{5}$ hours   **39.** $\frac{37}{8}$ hours

3.5 Exercises (page 67)

1.

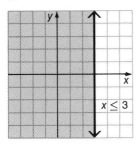

$x \leq 3$

3.

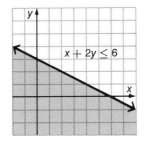

$x + 2y \leq 6$

5.

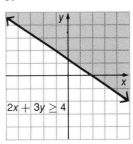

$2x + 3y \geq 4$

7.

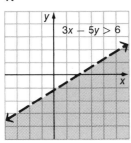

$3x - 5y > 6$

9.

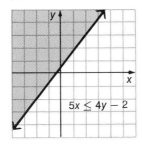

$5x \leq 4y - 2$

11.

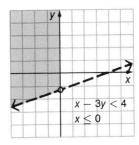

$x - 3y < 4$
$x \leq 0$

13.

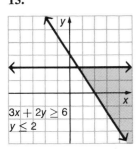

$3x + 2y \geq 6$
$y \leq 2$

15.

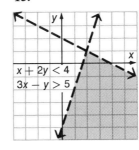

$x + 2y < 4$
$3x - y > 5$

17.

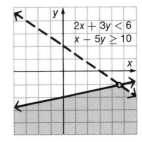

$2x + 3y < 6$
$x - 5y \geq 10$

3.6 Exercises (page 71)

1.

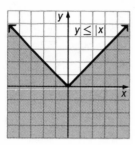

3.

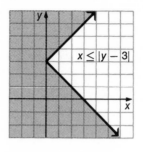

5.

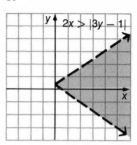

7.

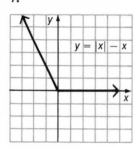

9.

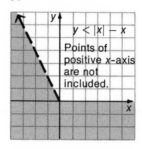

11.

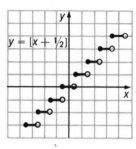

13.

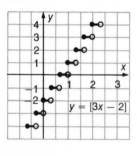

15.

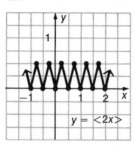

17.

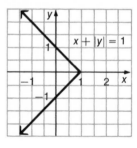

19.

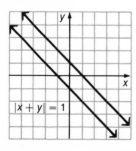

21.

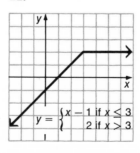

23.

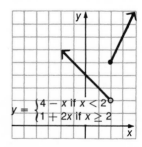

3.7 Exercises (page 74)

1. 19   **3.** 10   **5.** 7   **7.** -6 **9.** -9 **11.** -3 **13.** $x + 21$
15. $4m$ **17.** 215   **19.** 30   **21.** 125   **23.** 230   **25.** $\frac{205}{2}$ **27.** $\frac{155}{2}$
29. 6   **31.** 75   **33.** 231   **35.** 1281   **37.** 4680

4.1 Exercises (page 79)

1.

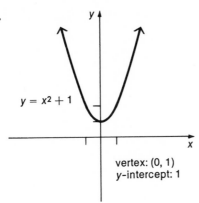

$y = x^2 + 1$

vertex: (0, 1)
y-intercept: 1

3.

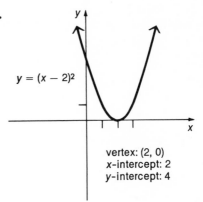

$y = (x - 2)^2$

vertex: (2, 0)
x-intercept: 2
y-intercept: 4

5.

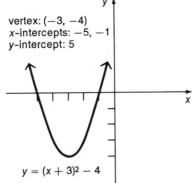

vertex: $(-3, -4)$
x-intercepts: $-5, -1$
y-intercept: 5

$y = (x + 3)^2 - 4$

7.

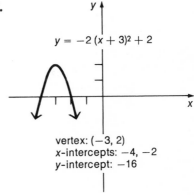

$y = -2(x + 3)^2 + 2$

vertex: $(-3, 2)$
x-intercepts: $-4, -2$
y-intercept: -16

9.

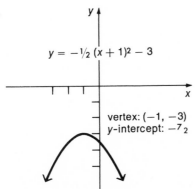

$y = -\frac{1}{2}(x + 1)^2 - 3$

vertex: $(-1, -3)$
y-intercept: $-7\frac{1}{2}$

11.

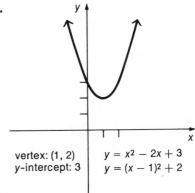

vertex: (1, 2)
y-intercept: 3

$y = x^2 - 2x + 3$
$y = (x - 1)^2 + 2$

13.

vertex: $(-4, -3)$
no rational x-intercepts
y-intercept: 13

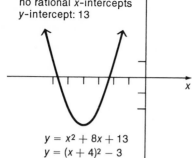

$y = x^2 + 8x + 13$
$y = (x + 4)^2 - 3$

15.

vertex: $(-2, 6)$
no rational x-intercepts
y-intercept: 2

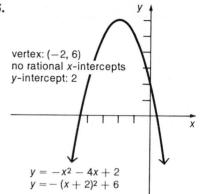

$y = -x^2 - 4x + 2$
$y = -(x + 2)^2 + 6$

17.

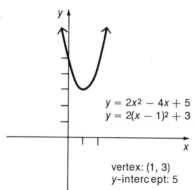

$y = 2x^2 - 4x + 5$
$y = 2(x - 1)^2 + 3$

vertex: $(1, 3)$
y-intercept: 5

19.

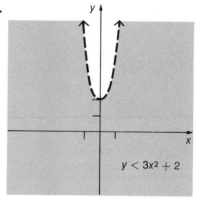

$y < 3x^2 + 2$

21.

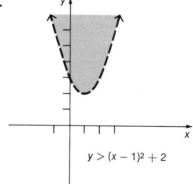

$y > (x - 1)^2 + 2$

23. 80×160 **25.** 10 and 10

4.2 Exercises (page 82)

1. 0, 2, or -3 **3.** 2 or 3   **5.** $\frac{10}{3}$ or $-\frac{5}{2}$ **7.** $-\frac{3}{2}$ or $-\frac{1}{4}$

9. $\dfrac{1 + \sqrt{5}}{2}$ or $\dfrac{1 - \sqrt{5}}{2}$ **11.** $\dfrac{-1 + \sqrt{7}}{2}$ or $\dfrac{-1 - \sqrt{7}}{2}$

13. $\dfrac{-7 + \sqrt{73}}{6}$ or $\dfrac{-7 - \sqrt{73}}{6}$ **15.** $3 + \sqrt{2}$ or $3 - \sqrt{2}$ **17.** $-\frac{3}{4}$ or $\frac{1}{3}$

19. $\dfrac{5 + 5\sqrt{2}}{3}$ or $\dfrac{5 - 5\sqrt{2}}{3}$ **21.** $-\frac{1}{4}$ or 3 **23.** $\frac{3}{2}$ or 1

25. $\dfrac{\sqrt{2} + \sqrt{6}}{2}$ or $\dfrac{\sqrt{2} - \sqrt{6}}{2}$ **27.** $\sqrt{2}$ or $\sqrt{2}/2$

29. $\dfrac{-n + \sqrt{n^2 - 8m}}{2m}$ or $\dfrac{-n - \sqrt{n^2 - 8m}}{2m}$ **31.** 0; one real zero

33. 1; 2 real zeros   **35.** 84; 2 real zeros   **37.** -23; no real zeros

4.3 Exercises (page 85)

We give only the real solutions.

1. 1 **3.** $\sqrt{3}$ or $-\sqrt{3}$ **5.** $\dfrac{\sqrt{2}}{2}$ or $\dfrac{-\sqrt{2}}{2}$ **7.** $\dfrac{\sqrt{10}}{2}$, $\dfrac{-\sqrt{10}}{2}$, 1, or -1

9. 4 or 6 **11.** $\dfrac{-5 + \sqrt{21}}{2}$ or $\dfrac{-5 - \sqrt{21}}{2}$

13. $\dfrac{-6 + 2\sqrt{3}}{3}$ or $\dfrac{-6 - 2\sqrt{3}}{3}$ or $\dfrac{-4 + \sqrt{2}}{2}$ or $\dfrac{-4 - \sqrt{2}}{2}$

15. 1 or $\dfrac{-5 + \sqrt{21}}{2}$ or $\dfrac{-5 - \sqrt{21}}{2}$ **17.** -63 or 28 **19.** 2 or 3

21. $-\frac{5}{2}$ **23.** 9 **25.** 2

4.4 Exercises (page 89)

1. $-3 \le x \le 3$ **3.** $0 < y < 10$ **5.** $r \le -2 - \sqrt{3}$ or $r \ge -2 + \sqrt{3}$
7. $-2 \le x \le 3$ **9.** $k < \frac{1}{2}$ or $k > 4$ **11.** $x \ne 0$ **13.** $x < 6$ or $x \ge \frac{15}{2}$
15. $k < 2$ or $k > 5$ **17.** $x \le -2$ or $0 \le x \le 2$

4.5 Exercises (page 91)

1. $\sqrt{26}$ **3.** $\sqrt{53}$ **5.** the distance between $(1,1)$ and $(2, -2)$ is $\sqrt{10}$. The distance between $(2, -2)$ and $(3, -5)$ is $\sqrt{10}$. The distance between $(1, 1)$ and $(3, -5)$ is $2\sqrt{10}$. Therefore the three points lie on the same straight line.

7.

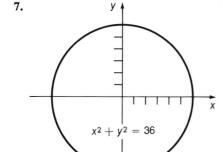

$x^2 + y^2 = 36$

9.

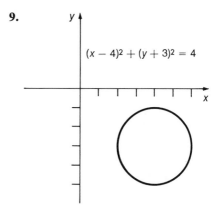

$(x - 4)^2 + (y + 3)^2 = 4$

11.

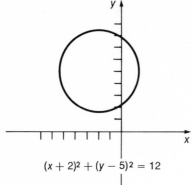

$(x + 2)^2 + (y - 5)^2 = 12$

13.

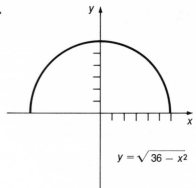

$y = \sqrt{36 - x^2}$

15.

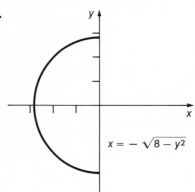

$x = -\sqrt{8 - y^2}$

17.

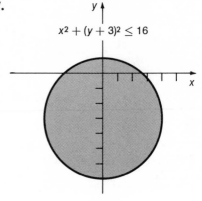

$x^2 + (y + 3)^2 \leq 16$

19.

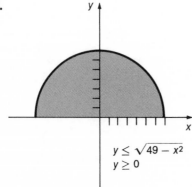

$y \leq \sqrt{49 - x^2}$
$y \geq 0$

	center	radius
21.	$(-3, -4)$	4
23.	$(6, -5)$	6
25.	$(-4, 7)$	0

4.6 Exercises (page 98)

1.

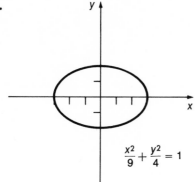

$$\frac{x^2}{9} + \frac{y^2}{4} = 1$$

3.

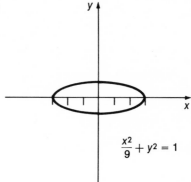

$$\frac{x^2}{9} + y^2 = 1$$

5.

$$\frac{x^2}{6} + \frac{y^2}{9} = 1$$

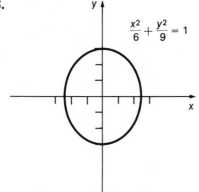

7.

$$x^2 + 4y^2 = 16$$
$$\frac{x^2}{16} + \frac{y^2}{4} = 1$$

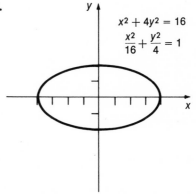

9.

$$x^2 = 9 + y^2$$
$$\frac{x^2}{9} - \frac{y^2}{9} = 1$$

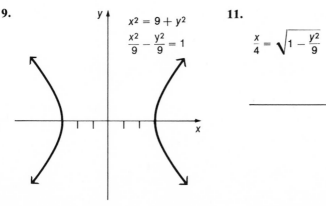

11.

$$\frac{x}{4} = \sqrt{1 - \frac{y^2}{9}}$$

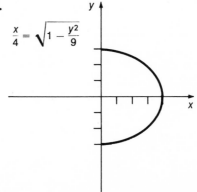

13.

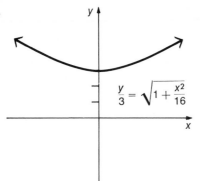

$$\frac{y}{3} = \sqrt{1 + \frac{x^2}{16}}$$

15.

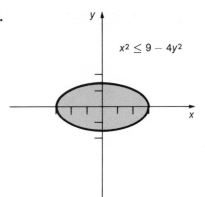

$$x^2 \leq 9 - 4y^2$$

17.

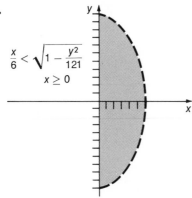

$$\frac{x}{6} < \sqrt{1 - \frac{y^2}{121}}$$
$$x \geq 0$$

19. $y = \frac{5}{3}x$ or $y = -\frac{5}{3}x$ **21.** $y = \frac{1}{2}x$ or $y = -\frac{1}{2}x$

4.7 Exercises (page 101)

1. $\frac{70}{3}$ **3.** 2304 feet   **5.** .0444 ohms   **7.** 250 pounds

5.1 Exercises (page 106)

1.

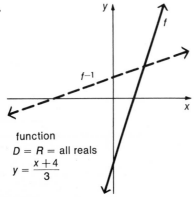

function
$D = R =$ all reals
$y = \frac{x + 4}{3}$

3.

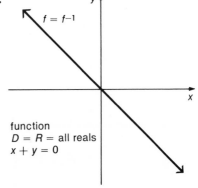

function
$D = R =$ all reals
$x + y = 0$

5.

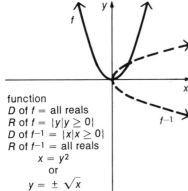

function
D of f = all reals
R of f = $\{y|y \geq 0\}$
D of f^{-1} = $\{x|x \geq 0\}$
R of f^{-1} = all reals
$x = y^2$
or
$y = \pm \sqrt{x}$

7.

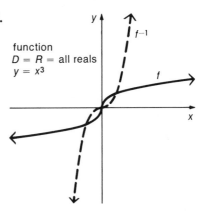

function
$D = R$ = all reals
$y = x^3$

9.

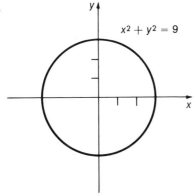

$x^2 + y^2 = 9$

11.

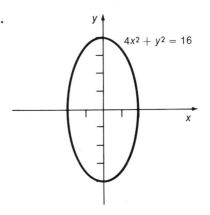

$4x^2 + y^2 = 16$

13.

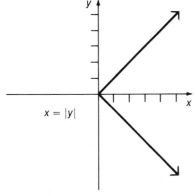

$x = |y|$

15.

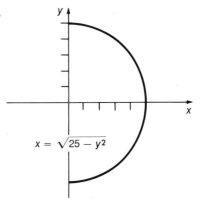

$x = \sqrt{25 - y^2}$

17.

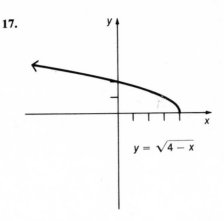

$y = \sqrt{4 - x}$

5.2 Exercises (page 109)

1.

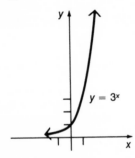

$y = 3^x$

3.

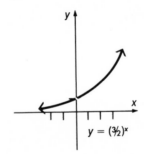

$y = (\tfrac{3}{2})^x$

5.

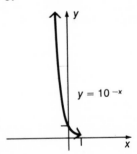

$y = 10^{-x}$

7.

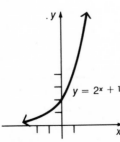

$y = 2^x + 1$

9.

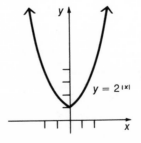

$y = 2^{|x|}$

11. $y = 2^x + 2^{-x}$

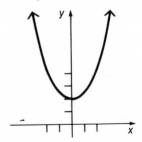

13.   (a) 1.4   (b) 2.1

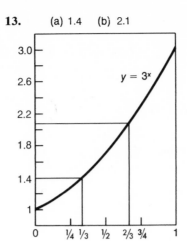

15. A series of points, one for each rational number on the number line.

5.3 Exercises (page 112)

1. $\log_3 81 = 4$ **3.** $\log_{10} 10000 = 4$ **5.** $\log_{1/2} 16 = -4$ **7.** $\log_{10} .0001 = -4$
9. $6^2 = 36$ **11.** $(\sqrt{3})^8 = 81$ **13.** $m^n = k$ **15.** $\log_a \left(\dfrac{xy}{m} \right)$
17. $\log_m \left(\dfrac{a^2}{b^6} \right)$ **19.** $\log_x a^{3/2} b^{-4}$

5.4 Exercises (page 116)

1. 2   **3.** 6   **5.** −4   **7.** −5   **9.** 2.9420   **11.** 4.1072   **13.** 0.8825
15. 6.9509 − 10   **17.** 0.5884   **19.** 4.8341   **21.** 6.5825 − 10   **23.** 34.4
25. .0747   **27.** 3124   **29.** 636.7   **31.** 81.6   **33.** 713   **35.** 2.01
37. .735   **39.** 36.3   **41.** 2.77   **43.** 31.2 pounds per cubic feet
45. (a) 350 years   (b) 3990 years   (c) 2300 years

5.5 Exercises (page 120)

1. 1.63   **3.** 1.16   **5.** ∅   **7.** 2.26   **9.** 1.08   **11.** −0.20   **13.** 11
15. 5   **17.** 4   **19.** 8   **21.** 10^{-2} or $1/100$ **23.** ∅   **25.** ∅   **27.** 17.5

5.6 Exercises (page 124)

1. 4.83   **3.** −.11   **5.** 4.59   **7.** 1.43   **9.** .59   **11.** .90   **13.** −1.59

15. .96   **17.** 495   **19.** 960,800   **21.** 17.3 years   **23.** We get $\dfrac{R}{r} = 1$.

25. $t = \dfrac{(5600)(\ln R/r)}{\ln 2}$ **27.** 378 feet   **29.** 18.9

5.7 Exercises (page 128)

1. $a_5 = 48; S_5 = 93$ **3.** $a_5 = \frac{3}{4}; S_5 = \frac{33}{4}$ **5.** $a_5 = \frac{16}{9}; S_5 = \frac{55}{9}$
7. $a_5 = 64; S_5 = 124$ **9.** $a_5 = 9; S_5 = \frac{121}{9}$ **11.** $a_8 = \frac{64}{243}$ **13.** 8 or -8
15. 4, 8, 16 or $-4, 8, -16$ **17.** 30   **19.** $\frac{255}{256}$ **21.** 372   **23.** $\frac{650}{81}$

5.8 Exercises (page 133)

1. $\frac{3}{2}$ **3.** 2   **5.** $\frac{1}{5}$ **7.** cannot be found by the formula of this section
9. $\frac{1}{3}$ **11.** the sum does not exist   **13.** $\frac{1}{20}$ **15.** $\frac{1}{4}$ **17.** $\frac{5}{9}$ **19.** $\frac{31}{99}$
21. $\frac{311}{900}$ **23.** 70 feet

6.1 Exercises (page 138)

s	$\sin s$	$\cos s$
1. -3π	0	-1
3. 17π	0	-1
5. $-\pi/2$	-1	0
7. $11\pi/2$	-1	0
9. $-7\pi/4$	$\sqrt{2}/2$	$\sqrt{2}/2$
11. -2.25π	$-\sqrt{2}/2$	$\sqrt{2}/2$

13. $1/2$ **15.** $\sqrt{2}/2$ **17.** $3/4$
19. $\sqrt{39/64} = \sqrt{39}/8$ **21.** I   **23.** II

6.2 Exercises (page 141)

1. $90°$ **3.** $45°$ **5.** $135°$ **7.** $100°$ **9.** $120°$ **11.** $300°$ **13.** $4\pi/3$
15. 3π **17.** $5\pi/4$ **19.** $13\pi/18$ **21.** $5\pi/2$ **23.** $7\pi/6$

	sine	cosine
25.	$1/2$	$\sqrt{3}/2$
27.	$\sqrt{2}/2$	$-\sqrt{2}/2$
29.	$-1/2$	$-\sqrt{3}/2$
31.	$-\sqrt{3}/2$	$1/2$
33.	$\sqrt{2}/2$	$\sqrt{2}/2$
35.	$-\sqrt{5}/3$	
37.	$-2\sqrt{6}/7$	

6.3 Exercises (page 145)

	$\sin s$	$\cos s$	$\tan s$	$\cot s$	$\csc s$	$\sec s$
1.	$\sqrt{3}/2$	$1/2$	$\sqrt{3}$	$\sqrt{3}/3$	$2\sqrt{3}/3$	2
3.	$-1/2$	$-\sqrt{3}/2$	$\sqrt{3}/3$	$\sqrt{3}$	-2	$-2\sqrt{3}/3$
5.	$-\sqrt{2}/2$	$\sqrt{2}/2$	-1	-1	$-\sqrt{2}$	$\sqrt{2}$
7.	$\sqrt{2}/2$	$\sqrt{2}/2$	1	1	$\sqrt{2}$	$\sqrt{2}$
9.	$3/4$	$\sqrt{7}/4$	$3\sqrt{7}/7$	$\sqrt{7}/3$	$4/3$	$4\sqrt{7}/7$
11.	$\sqrt{65}/9$	$-4/9$	$-\sqrt{65}/4$	$-4\sqrt{65}/65$	$9\sqrt{65}/65$	$-9/4$
13.	$3\sqrt{13}/13$	$2\sqrt{13}/13$	$3/2$	$2/3$	$\sqrt{13}/3$	$\sqrt{13}/2$
15.	$7\sqrt{74}/74$	$-5\sqrt{74}/74$	$-7/5$	$-5/7$	$\sqrt{74}/7$	$-\sqrt{74}/5$

6.4 Exercises (page 148)

1. 35°   **3.** 37°   **5.** 60°   **7.** π/5   **9.** .07   **11.** −.5736   **13.** .1793
15. .5727   **17.** −3.071   **19.** −.9110   **21.** 191°   **23.** 51°   **25.** 313° 30′
27. 3.4936   **29.** 2.4027

	by formula	by table	difference
31.	.8417	.8415	.0002
33.	.0100	.0100	.0000
35.	1.0000	1.0000	.0000

6.5 Exercises (page 154)

1.

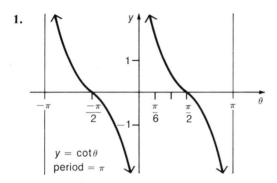

$y = \cot\theta$
period $= \pi$

3.

$y = 2\cos\theta$
period $= 2\pi$

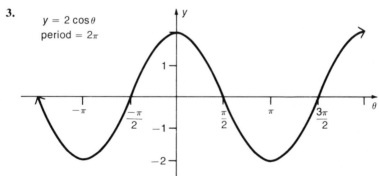

5.

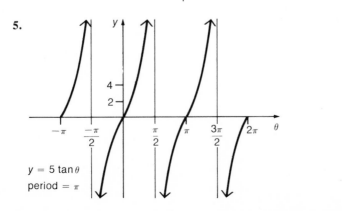

$y = 5\tan\theta$
period $= \pi$

7.

$y = \sin \dfrac{2}{3}\theta$

period = 3π

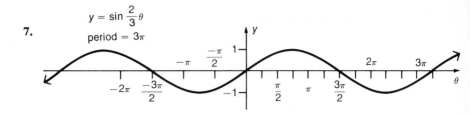

9.

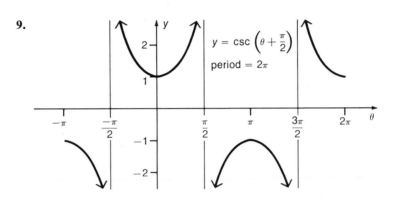

$y = \csc\left(\theta + \dfrac{\pi}{2}\right)$

period = 2π

11.

$y = \tan\left(\theta - \dfrac{\pi}{4}\right)$

period = π

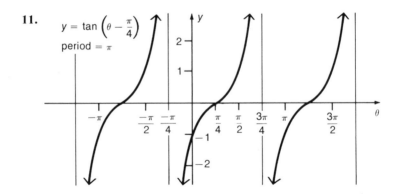

13.

$y = \sin \tfrac{1}{2}\theta - 1$

period = 4π

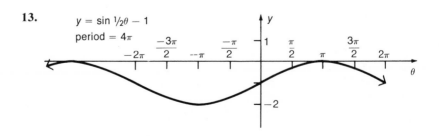

15.

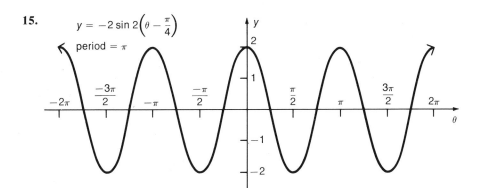

$y = -2 \sin 2\left(\theta - \frac{\pi}{4}\right)$

period $= \pi$

17.

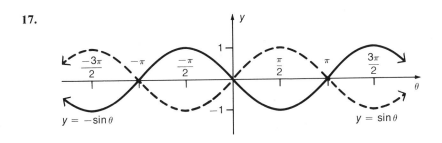

$y = -\sin \theta$

$y = \sin \theta$

6.6 Exercises (page 157)

	(in degrees)	s (in radians)
1.	-60	$-\pi/3$
3.	45	$\pi/4$
5.	-90	$-\pi/2$
7.	60	$\pi/3$
9.	135	$3\pi/4$
11.	150	$5\pi/6$
13.	$-7° \, 40'$	$-.1338$
15.	$113° \, 30'$	1.9810
17.	$24° \, 30'$	$.4276$
19.	48	$.8378$

21. $1/2$ **23.** $-\sqrt{3}/3$ **25.** $\sqrt{2}$ **27.** $.6$ **29.** 3 **31.** $-2\sqrt{5}/5$
33. $-\sqrt{10}/10$ **35.** -1

37.

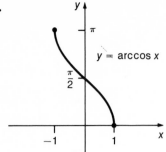

$y = \arccos x$

39.

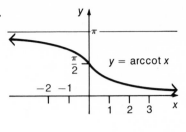

$y = \text{arccot } x$

7.1 Exercises (page 163)

1. c **3.** d **5.** g **7.** b

	sin	cos	tan	cot	sec	csc
9.	$\sqrt{5}/3$	$2/3$	$\sqrt{5}/2$	$2\sqrt{5}/5$	$3/2$	$3\sqrt{5}/5$
11.	$5/13$	$12/13$	$5/12$	$12/5$	$13/12$	$13/5$

7.2 Exercises (page 168)

1. $\dfrac{\sqrt{2}}{4} - \dfrac{\sqrt{6}}{4}$ **3.** $2 + \sqrt{3}$ **5.** $2 + \sqrt{3}$ **7.** -1 **9.** $-\sqrt{2}/2$

11. $-\cos\theta$ **13.** $\dfrac{\sqrt{3}}{2}\sin\theta - \dfrac{1}{2}\cos\theta$ **15.** $\dfrac{\tan\theta - 1}{1 + \tan\theta}$

7.3 Exercises (page 172)

1. $\sin 2\pi/5$ **3.** $\sin 20°$ **5.** $2\cos 2x$ **7.** $\tan 2\theta$

	sin	cos	tan	cot	sec	csc
9.	$-\dfrac{\sqrt{2-\sqrt{2}}}{2}$	$\dfrac{\sqrt{2+\sqrt{2}}}{2}$	$\sqrt{3-2\sqrt{2}}$	$-\sqrt{3+2\sqrt{2}}$	$\sqrt{4-2\sqrt{2}}$	$-\sqrt{4+2\sqrt{2}}$
11.	$-\dfrac{\sqrt{2-\sqrt{3}}}{2}$	$-\dfrac{\sqrt{2+\sqrt{3}}}{2}$	$\sqrt{7-4\sqrt{3}}$	$\sqrt{7+4\sqrt{3}}$	$-\sqrt{8-4\sqrt{3}}$	$-\sqrt{8+4\sqrt{3}}$
13.	$\sqrt{5}/5$	$2\sqrt{5}/5$	$1/2$	2	$\sqrt{5}/2$	$\sqrt{5}$
15.	$\sqrt{6}/6$	$-\sqrt{30}/6$	$-\sqrt{5}/5$	$-\sqrt{5}$	$-\sqrt{30}/5$	$\sqrt{6}$
17.	$20/101$	$-99/101$	$-20/99$	$-99/20$	$-101/99$	$101/20$

19. $4\cos^3 x - 3\cos x$ **21.** $4\sin x \cos x - 8\sin^3 x \cos x$ or $8\cos^3 x \sin x - 4\cos x \sin x$

23. $\dfrac{1 - \cos x}{1 + \cos x}$

7.4 Exercises (page 176)

1. $\frac{1}{2}(\sin 60° - \sin 10°)$ **3.** $\frac{3}{2}(\sin 8x - \sin 2x)$ **5.** $\frac{1}{2}(\cos 2\theta - \cos 4\theta)$

7. $2\cos 45° \sin 15°$ **9.** $2\cos 95° \sin 53°$ **11.** $2\cos\dfrac{15}{2}\theta \sin\dfrac{9}{2}\theta$

7.5 Exercises (page 178)

1. π **3.** $0, \pi, 2\pi, \pi/3, 5\pi/3$ **5.** $\pi/2$ **7.** $\pi/8, 7\pi/8, 9\pi/8, 15\pi/8$
9. no solution   **11.** $22\frac{1}{2}°, 67\frac{1}{2}°, 112\frac{1}{2}°, 157\frac{1}{2}°, 202\frac{1}{2}°, 247\frac{1}{2}°, 292\frac{1}{2}°, 337\frac{1}{2}°$
13. $30°, 150°, 270°$ **15.** $135°, 315°, 108°, 288°$ **17.** $0°, 120°, 240°, 360°$
19. $90°$

8.1 Exercises (page 185)   The answers in this chapter may vary slightly, due to rounding errors.

1. $B = 62°, b = 28.2, c = 31.9$ **3.** $A = 32°, c = 34.0, b = 28.8$
5. $A = 21° 50', B = 68° 10', c = 32.3$ **7.** $B = 47° 10', b = 20.1, c = 27.4$
9. $c = 502.8, a = 170.6, B = 70° 10'$ **11.** 20.3   **13.** 50.2 feet   **15.** $34° 20'$
17. 41120 feet   **19.** 114 feet   **21.** 156.2

8.2 Exercises (page 190)

1. $C = 99°, a = 8.13, b = 10.2$ **3.** $C = 101° 10', a = 116.7, c = 158.7$
5. $B = 43° 50', a = 17.1, b = 27.3$ **7.** $AB = 118$ feet   **9.** 1.92 miles
11. not solvable (see law of cosines in a later section)   **13.** 10.7   **15.** 308.3

8.3 Exercises (page 194)

1. $B = 20° 30', C = 117°, c = 20.6$ **3.** impossible ($\sin C = 1.7230$, which is
impossible)   **5.** $B = 34°, C = 33° 40', c = 3.5$
7. $B_1 = 49° 30', C_1 = 91° 50', c_1 = 15.5$
$B_2 = 130° 30', C_2 = 10° 50', c_2 = 2.9$
9. $A_1 = 51° 40', C_1 = 95° 40', c_1 = 95.5$
$A_2 = 128° 20', C_2 = 19°, c_2 = 31.2$
11. A triangle with these dimensions cannot exist.

8.4 Exercises (page 198)

1. $B = 84°, C = 67° 10', a = 2.91$ **3.** $B = 76° 20', A = 58°, c = 5.9$
5. $a = 117.5, B = 57° 10', C = 42° 10'$ **7.** $b = 12.5, A = 31° 30', C = 23°$
9. $a = 15.4, B = 21°, C = 46° 10'$ **11.** $A = 29°, B = 46° 30', C = 104° 30'$
13. $A = 36° 40', B = 46° 40', C = 96° 40'$ **15.** 257 yards   **17.** 5.2, 8.8
19. 741 miles   **21.** about 3:15 pm, $152°$

8.5 Exercises (page 202)

1. 34.8   **3.** 346.1   **5.** 719.1   **7.** 16.2   **9.** 850.9   **11.** 1074
13. 20 cans

9.1 Exercises (page 206)

1. False—not all corresponding elements are equal.   **3.** True   **5.** True

7. $\begin{pmatrix} 9 & 12 & 0 & 2 \\ 1 & -1 & 2 & -4 \end{pmatrix}$

9. Not possible—only matrices of the same dimension can be added.

11. $\begin{pmatrix} -2 & -3 & 3 \\ 4 & 3 & -1 \end{pmatrix}$

13. $X + T = \begin{pmatrix} x+r & y+s \\ z+t & w+u \end{pmatrix}$, and so is $T + X$.

15. $X + (-X) = \begin{pmatrix} x+(-x) & y+(-y) \\ z+(-z) & w+(-w) \end{pmatrix} = 0.$

17. All of the statements would still be true as long as 0, P, T, and X are of the same order.

9.2 Exercises (page 210)

1. True **3.** False—A could be 2×4 and B could be 4×2. Both BA and AB would then exist.

5. False—$\begin{pmatrix} 2 & 3 \\ 0 & 0 \end{pmatrix}\begin{pmatrix} -3 & 0 \\ 2 & 0 \end{pmatrix} = \begin{pmatrix} 0 & 0 \\ 0 & 0 \end{pmatrix}.$ **7.** $\begin{pmatrix} 14 & 18 \\ 7 & 14 \end{pmatrix}$

9. $\begin{pmatrix} -6 & 6 & 6 \\ -5 & 5 & 5 \\ -4 & 4 & 4 \end{pmatrix}$

11. (3) **13.** It is not possible to find this product.

19. (a) $\begin{pmatrix} 17\frac{3}{4} & 23\frac{3}{4} \\ 9 & 12\frac{3}{4} \\ 33 & 42 \\ 6 & 8 \end{pmatrix}$

(b) (220   890   105   125   70)   (c) (4165   5605)

9.3 Exercises (page 217)

1. $\begin{pmatrix} -1/2 & 3/2 \\ 1 & -2 \end{pmatrix}$ **3.** $\begin{pmatrix} 1 & 0 \\ 0 & 1/2 \end{pmatrix}$

5. Does not exist.

7. $\begin{pmatrix} -1/13 & 7/26 \\ 2/13 & -1/26 \end{pmatrix}$ **9.** $\begin{pmatrix} 3 \\ 5 \end{pmatrix}$ **11.** $\begin{pmatrix} -5 \\ -18 \end{pmatrix}$

9.4 Exercises (page 221)

1. -36 **3.** 7 **5.** 0 **7.** 17 **9.** 166 **11.** 0 **13.** 0 **15.** 1

9.5 Exercises (page 226)

12. (a) -32 (b) -6 **13.** $\frac{5}{2}$ **15.** 7 **17.** 8

10.1 Exercises (page 233)

1. $x = 1, y = 3$ **3.** $p = 4, q = -2$ **5.** $x = 2, y = -2$
7. $x = 12, y = 6$ **9.** $x = 2, y = 2$ **11.** $x = 1, y = 2, z = 3$

13. $m = -1, n = 2, p = 1$ **15.** $a = 4, b = 1, c = 2$ **17.** $a = -3, b = -1$
19. 3000 at 8%, 6000 at 9%, and 1000 at 5%

10.2 Exercises (page 238)

1. $x = 2, y = 0$ **3.** $x = \frac{9}{7}, y = \frac{2}{7}$ **5.** $x = \frac{1}{2}, y = \frac{3}{8}$ **7.** no solution
9. $x = \frac{197}{91}, y = -\frac{118}{91}, z = -\frac{23}{91}$ **11.** $x = \frac{31}{5}, y = \frac{19}{10}, z = -\frac{29}{10}$

10.3 Exercises (page 242)

1. $x = 2, y = 0$ **3.** $x = \frac{7}{2}, y = -1$ **5.** $x = 5, y = 0$
7. $x = -1, y = 23, z = 16$ **9.** $x = 3, y = 2, z = -4$
11. $x = 2, y = 1, z = -1$ **13.** $x = -1, y = 2, z = 5, w = 1$

10.4 Exercises (page 244)

1. $x = \frac{1}{2}, y = \frac{1}{2}$ **3.** $x = \frac{53}{29}, y = \frac{-7}{29}$ **5.** $x = \frac{55}{41}, y = \frac{-19}{82}$ **7.** $x = \frac{55}{7}, y = \frac{15}{7}$
9. $x = \frac{7}{2}, y = -1$

10.5 Exercises (page 249)

1. $(-6/5, 17/5), (2, -3)$ **3.** $(-3/5, 7/5), (-1, 1)$
5. $(3, 6), (-3, 6), (3, -6), (-3, -6)$ **7.** any x and y such that $x^2 + y^2 = 10$
9. no solution   **11.** $(3, 5), (-3, -5)$ **13.** $(3, 2), (-3, -2), (4, 3/2), (-4, -3/2)$
15. $(\sqrt{5}, 0), (-\sqrt{5}, 0), (\sqrt{5}, \sqrt{5}), (-\sqrt{5}, -\sqrt{5})$ **17.** $(1, 3), (31/14, -9/14)$
19. $\left(\dfrac{1 + \sqrt{13}}{2}, \dfrac{-1 + \sqrt{13}}{2} \right), \left(\dfrac{-1 - \sqrt{21}}{2}, \dfrac{3 + \sqrt{21}}{2} \right)$

10.6 Exercises (page 252)

1.

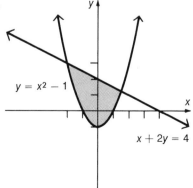

3.

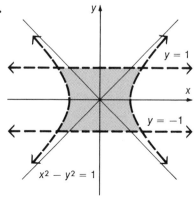

5.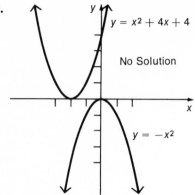

$y = x^2 + 4x + 4$

No Solution

$y = -x^2$

7.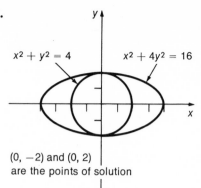

$x^2 + y^2 = 4$ $x^2 + 4y^2 = 16$

$(0, -2)$ and $(0, 2)$
are the points of solution

9.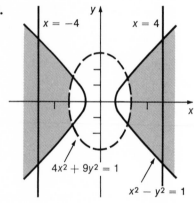

$x = -4$ $x = 4$

$4x^2 + 9y^2 = 1$

$x^2 - y^2 = 1$

11.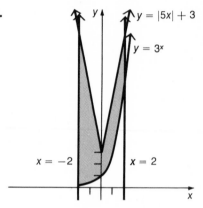

$y = |5x| + 3$

$y = 3^x$

$x = -2$ $x = 2$

11.1 Exercises (page 256)

1. $7 - i$ **3.** $3 - 6i$ **5.** $2 + i$ **7.** $8 - i$ **9.** $-14 + 2i$ **11.** $31 - 5i$

13. $3 + 4i$ **15.** 5 **17.** $25i$ **19.** $24 - 7i$ **21.** i **23.** $\dfrac{7}{25} - \dfrac{24}{25}i$

25. $\dfrac{13}{20} - \dfrac{1}{20}i$ **27.** $\dfrac{32}{37} - \dfrac{7}{37}i$ **29.** $-1 - 2i$ **31.** -1 **33.** 1 **35.** i

37. $3i$ **39.** $3i\sqrt{2}$ **41.** $5i\sqrt{6}$

11.2 Exercises (page 261)

1. – 9.

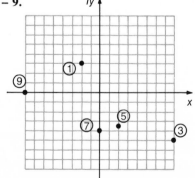

iy

x

11. $1 + i$ **13.** $-2 + 2i$ **15.** $-2 + 4i$ **17.** $(2\sqrt{2}, -45°)$ **19.** $(3, 90°)$
21. $(2, 60°)$ **23.** $\sqrt{2} + \sqrt{2}i$ **25.** $\dfrac{-\sqrt{2}}{2} + \dfrac{\sqrt{2}}{2}i$ **27.** $0 + 3i$

23. – 27.

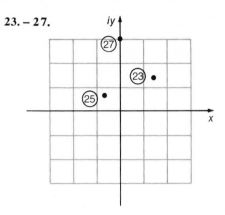

29. $\sqrt{2} + \sqrt{2}i$ **31.** $0 + 10i$ **33.** $-2 - 2\sqrt{3}i$
35. $3\sqrt{2}(\cos 315° + i \sin 315°)$ **37.** $6(\cos 240° + i \sin 240°)$

11.3 Exercises (page 264)

1. $0 + 6i$ **3.** $4 + 0i$ **5.** $-48 + 0i$ **7.** $-8 + 0i$ **9.** $1 + 0i$
11. $\dfrac{125}{2} - \dfrac{125\sqrt{3}}{2}i$ **13.** $-16\sqrt{3} + 16i$ **15.** $-128 + 128\sqrt{3}i$
17. $\dfrac{3\sqrt{3}}{2} + \dfrac{3}{2}i$ **19.** $-1 - \sqrt{3}i$ **21.** $2 + 0i$

11.4 Exercises (page 267)

1.

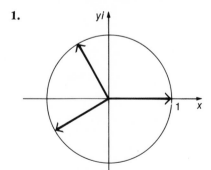

$\cos 0° + i\sin 0° = 1$

$\cos 120° + i\sin 120°$

$\cos 240° + i\sin 240°$

3.

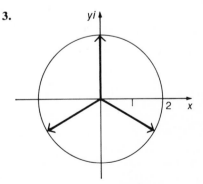

$2 (\cos 90° + i\sin 90°)= 2i$

$2 (\cos 210° + i\sin 210°)$

$2 (\cos 330° + i\sin 330°)$

5.

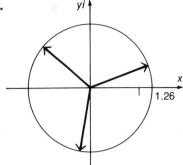

$\sqrt[3]{2}\,(\cos 20° + i\sin 20°)$

$\sqrt[3]{2}\,(\cos 140° + i\sin 140°)$

$\sqrt[3]{2}\,(\cos 260° + i\sin 260°)$

7.

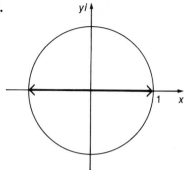

$\cos 0° + i\sin 0° = 1$

$\cos 180° + i\sin 180° = -1$

9.

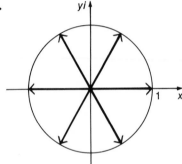

$\cos 0° + i\sin 0° = 1$

$\cos 60° + i\sin 60°$

$\cos 120° + i\sin 120°$

$\cos 180° + i\sin 180° = -1$

$\cos 240° + i\sin 240°$

$\cos 300° + i\sin 300°$

11.

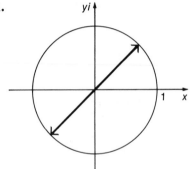

$\cos 45° + i\sin 45°$

$\cos 225° + i\sin 225°$

13. $1, -\dfrac{1}{2} + \dfrac{\sqrt{3}}{2}i, -\dfrac{1}{2} - \dfrac{\sqrt{3}}{2}i$

15. $\dfrac{\sqrt{2}}{2} + \dfrac{\sqrt{2}}{2}i, \dfrac{-\sqrt{2}}{2} + \dfrac{\sqrt{2}}{2}i, \dfrac{-\sqrt{2}}{2} - \dfrac{\sqrt{2}}{2}i, \dfrac{\sqrt{2}}{2} - \dfrac{\sqrt{2}}{2}i$

17. $.924 + .383i, -.383 + .924i, -.924 - .383i, .383 - .924i$

19. $1.88 + .684i, -1.53 + 1.29i, -.35 - 1.97i$

12.1 Exercises (page 271)

1. $3i, -3i$ **3.** $\dfrac{5}{2}i, \dfrac{-5}{2}i$ **5.** $2 + 2i, 2 - 2i$ **7.** $\dfrac{1}{2} + i, \dfrac{1}{2} - i$

9. $-1 + \sqrt{2}i, -1 - \sqrt{2}i$ **11.** $\dfrac{1 + \sqrt{7}i}{4}, \dfrac{1 - \sqrt{7}i}{4}$

13. $\dfrac{1 + \sqrt{23}i}{6}, \dfrac{1 - \sqrt{23}i}{6}$ **15.** $\dfrac{-i + \sqrt{5}i}{2}, \dfrac{-i - \sqrt{5}i}{2}$

17. $\dfrac{1 + \sqrt{2}}{i}, \dfrac{1 - \sqrt{2}}{i}$ **19.** $\dfrac{1 + \sqrt{7}i}{2 + 2i}, \dfrac{1 - \sqrt{7}i}{2 + 2i}$ **21.** $x^2 + 1 = 0$

23. $x^2 - 4x + 13 = 0$ **25.** $x^2 + 3 = 0$ **27.** $x^2 - 2x + 3 = 0$

29. $x^2 + (2i - 3)x + (5 - i) = 0$ **31.** $x^2 + (-5 - 2i)x + (14 + 8i) = 0$

12.2 Exercises (page 273)

1. yes   **3.** yes   **5.** no   **7.** yes   **9.** yes   **11.** yes   **13.** no   **15.** 2

17. $-6 - i$ **19.** $-2 - 3i$ **21.** -20 **23.** 4

25. $(0, 1)$

12.3 Exercises (page 277)

1. $x^2 - 10x + 26$ **3.** $x^3 - 4x^2 + 6x - 4$ **5.** $x^3 - 3x^2 + x + 1$

7. $x^4 - 6x^3 + 10x^2 + 2x - 15$ **9.** $x^3 - 8x^2 + 22x - 20$

11. $x^4 - 4x^3 + 5x^2 - 2x - 2$ **13.** $-1 + i, -1 - i$ **15.** $3, -2 - 2i$

17. $-\dfrac{1}{2} + \dfrac{\sqrt{5}}{2}i, -\dfrac{1}{2} - \dfrac{\sqrt{5}}{2}i$ **19.** $i, 2i, -2i$ **21.** $3, -2, 1 - 3i$

12.4 Exercises (page 279)

1. $-1, -2, 5$ **3.** $2, -3, -5$ **5.** no rational solutions   **7.** $1, -2, -3, -5$

9. $\frac{3}{2}, -\frac{1}{3}, -4$ **11.** $\frac{1}{2}, -\frac{2}{3}, -\frac{3}{2}$ **13.** $\frac{1}{2}$ **15.** no rational solutions

12.5 Exercises (page 283)

9. $-3, -1.4, 1.4$ **11.** $-2, 1.6, 4.4$ **13.** 1.6 **15.** $-.7, 1.6, 2.7, 4.4$

12.6 Exercises (page 287)

1.

$P(x) = (x + 1)^3$

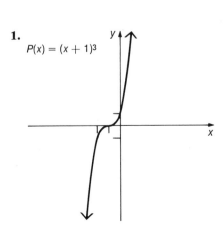

3. $P(x) = x^3 - 7x - 6$

	1	0	−7	−6
−3	1	−3	2	−12
−2	1	−2	−3	0
−3/2	1	−3/2	−19/4	9/8
−1	1	−1	−6	0
1	1	1	−6	−12
2	1	2	−3	−12
3	1	3	2	0

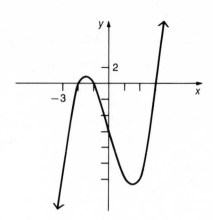

5.

$P(x) = x^4 - 5x^2 + 6$

	1	0	−5	0	6
−3	1	−3	4	−12	42
−2	1	−2	−1	2	2
−1	1	−1	−4	4	2
1	1	1	−4	−4	2
2	1	2	−1	−2	2
3/2	1	3/2	−11/4	−33/8	−3/16

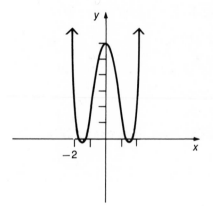

7.

$P(x) = 6x^3 + 11x^2 - x - 6$

	6	11	−1	−6
−3	6	−7	20	−66
−2	6	−1	1	−8
−1	6	5	−6	0
1	6	17	16	10

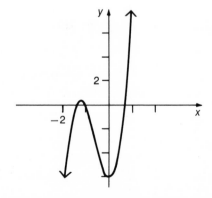

9.

$P(x) = x^4 + x^3 - 2$

	1	1	0	0	-2
-3	1	-2	6	-18	52
-2	1	-1	2	-4	6
-1	1	0	0	0	-2
1	1	2	2	2	0
2	1	3	6	12	22
-1/2	1	1/2	-1/4	1/8	-33/16
1/2	1	3/2	3/4	3/8	-29/16

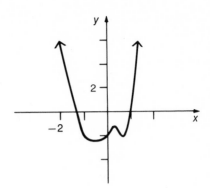

11.

$P(x) = 8x^4 - 2x^3 - 47x^2 - 52x - 15$

	8	-2	-47	-52	-15
-3	8	-26	31	-145	420
-2	8	-18	-11	-30	45
-1	8	-10	-37	-15	0
1	8	6	-41	-93	-108
2	8	14	-19	-90	-195
3	8	22	19	5	0
-1/2	8	-6	-44	-30	0
-5/4	8	-12	-32	-12	0

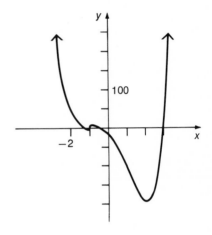

12.7 Exercises (page 289)

1.

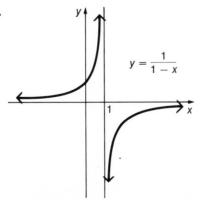

$$y = \frac{1}{1 - x}$$

3.

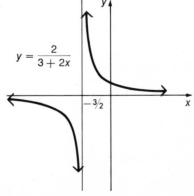

$$y = \frac{2}{3 + 2x}$$

5.

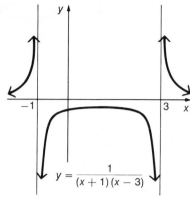

$$y = \frac{1}{(x+1)(x-3)}$$

7.

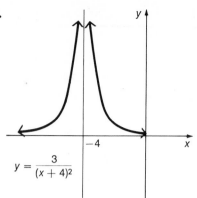

$$y = \frac{3}{(x+4)^2}$$

9.

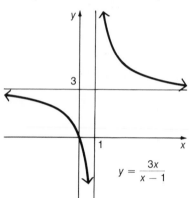

$$y = \frac{3x}{x-1}$$

11.

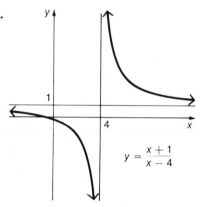

$$y = \frac{x+1}{x-4}$$

13.

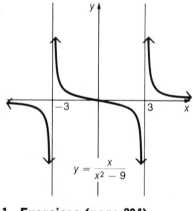

$$y = \frac{x}{x^2 - 9}$$

15.

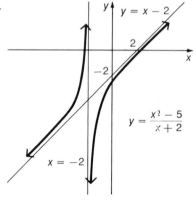

$$y = x - 2$$

$$y = \frac{x^2 - 5}{x + 2}$$

$$x = -2$$

13.1 Exercises (page 294)

1. $45°$ **3.** $60°$ **5.** $108° \, 30'$ **7.** $115°$ **9.** 2 **11.** $-2/3$ **13.** 0
15. $-3/2$ **17.** $-1/3$ **19.** $1/2$ **21.** no slope   **23.** 1 **25.** $x + 4y = 0$
27. $-2x + 3y = 11$ **29.** (a) $k = 6$ (b) $k = 11$

13.2 Exercises (page 298)

1. $(12, 1)$ **3.** $(15/2, 7)$ **5.** $(1/2, 1/2)$ **7.** $(13, 10)$ **9.** $(-10, 11)$

13.3 Exercises (page 302)

1. 5   **3.** 2/5   **5.** 72°   **7.** 48°   **9.** 90°, 45°, 45°   **11.** 90°, 60°, 30°
13. $m = -3.732$, $\alpha = 105°$, or $m = .268$, $\alpha = 15°$ **15.** 4.50
17. $d = \sqrt{(x-p)^2 + y^2}$ **19.** $y^2 = 4px$ **21.** $(1/64, 0)$ **23.** $(1/4, 0)$
25. $(0, 1)$

13.4 Exercises (page 306)

1.

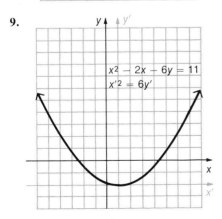

$(x - 5)^2 = 4y$
$x' = 4y'$

3.

$(y - 5)^2 = 3(x + 2)$
$y'^2 = 3x'$

5.

$\dfrac{(x + 3)^2}{9} + \dfrac{(y - 5)^2}{4} = 1$

$\dfrac{x'^2}{9} + \dfrac{y'^2}{4} = 1$

7.

$\dfrac{(x - 6)^2}{9} - \dfrac{(y + 2)^2}{25} = 1$

$\dfrac{x'^2}{9} - \dfrac{y'^2}{25} = 1$

$x' = 3/5y'$

$x' = -3/5y'$

9.

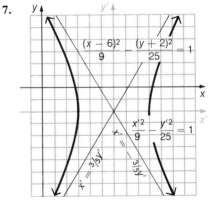

$x^2 - 2x - 6y = 11$
$x'^2 = 6y'$

11.

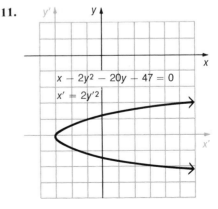

$x - 2y^2 - 20y - 47 = 0$

$x' = 2y'^2$

13.

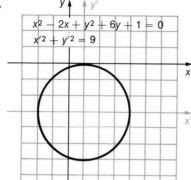

$$x^2 - 2x + y^2 + 6y + 1 = 0$$
$$x'^2 + y'^2 = 9$$

15. no such graph

17.

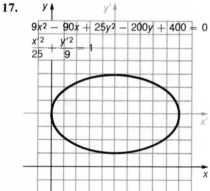

$$9x^2 - 90x + 25y^2 - 200y + 400 = 0$$
$$\frac{x'^2}{25} + \frac{y'^2}{9} = 1$$

19.

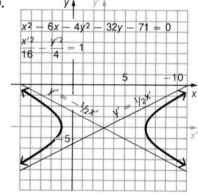

$$x^2 - 6x - 4y^2 - 32y - 71 = 0$$
$$\frac{x'^2}{16} - \frac{y'^2}{4} = 1$$

13.5 Exercises (page 313)

1. $\dfrac{x'^2}{12} + \dfrac{y'^2}{4} = 1$

3–11 See Art

3.

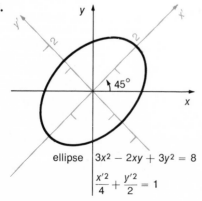

ellipse $3x^2 - 2xy + 3y^2 = 8$

$$\frac{x'^2}{4} + \frac{y'^2}{2} = 1$$

5.

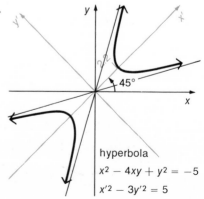

hyperbola

$x^2 - 4xy + y^2 = -5$

$x'^2 - 3y'^2 = 5$

7.

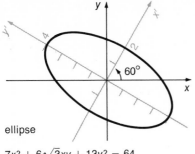

ellipse

$7x^2 + 6\sqrt{3}xy + 13y^2 = 64$

$\dfrac{x'^2}{4} + \dfrac{y'^2}{16} = 1$

9.

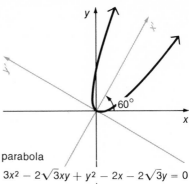

parabola

$3x^2 - 2\sqrt{3}xy + y^2 - 2x - 2\sqrt{3}y = 0$

$x' = y'^2$

13.6 Exercises (page 317)

1.–7.

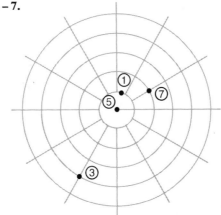

9.

circle

$x^2 + (y - 1)^2 = 1$

center at $(0,1)$

radius $= 1$

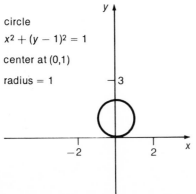

11.

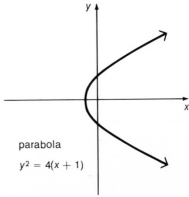

parabola

$y^2 = 4(x + 1)$

13.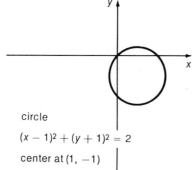

circle

$(x - 1)^2 + (y + 1)^2 = 2$

center at $(1, -1)$

radius $= \sqrt{2} = 1.41$

15.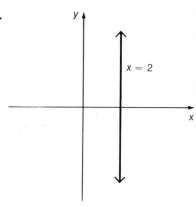

$x = 2$

17.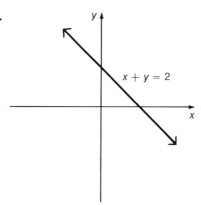

$x + y = 2$

19.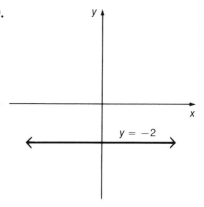

$y = -2$

21. $r = 2 + 2 \cos \theta$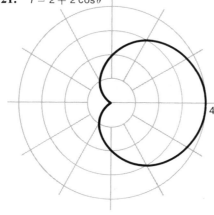

23. $r = 3 + \cos \theta$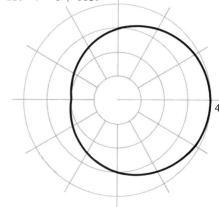

25. $r = \sin 4\theta$

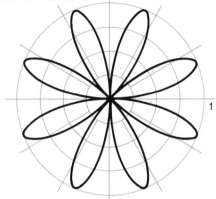

27. $r^2 = 4\cos 2\theta$

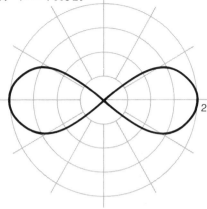

29. $r = 4(1 - \cos\theta)$

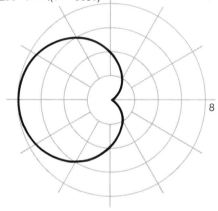

31. $r = 5\theta$

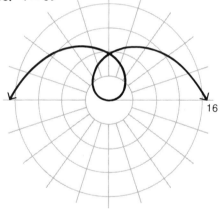

33. $r\theta = \pi$

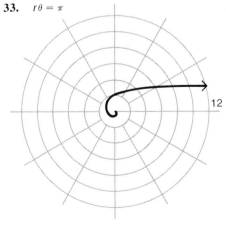

35. $\ln r = \theta$

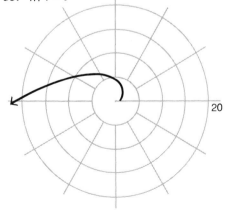

37.

$r = \dfrac{\cos 2\theta}{\cos \theta}$

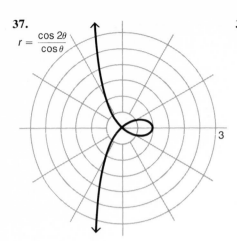

3

39. $r = \sin \theta \cos^2 \theta$

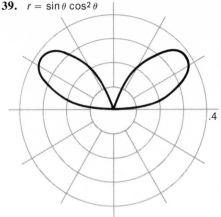

.4

14.1 Exercises (page 322)

1. (a) 5040 (b) 60 (c) 720 (d) 12   **3.** 720   **5.** 120   30,240   **7.** 30
9. 120   **11.** 4   **13.** 552   **15.** $26 \cdot 25 \cdot 24 \cdot 10 \cdot 9 \cdot 8$, $2(26 \cdot 25 \cdot 24 \cdot 10 \cdot 9 \cdot 8)$

14.2 Exercises (page 326)

1. (a) 6 (b) 6 (c) 56 (d) 45   **3.** 27,405   **5.** 220, 220   **7.** 1326   **9.** 10
11. 28   **13.** (a) 84 (b) 10 (c) 40 (d) 28

14.3 Exercises (page 331)

1. $\{H\}$ **3.** $\{HHH, HHT, HTH, HTT, THH, THT, TTH, TTT\}$
5. $\{FFF, FFT, FTF, FTT, TFF, TFT, TTF, TTT\}$
7. (a) $E_1 = \{HH, TT\}$, $P(E_1) = \frac{1}{2}$ (b) $E_2 = \{HH, HT, TH\}$, $P(E_2) = \frac{3}{4}$
9. (a) $E_1 = \{2 \text{ and } 4\}$; $P(E_1) = \dfrac{1}{10}$

     (b) $E_2 = \{1 \text{ and } 3, 1 \text{ and } 5, 3 \text{ and } 5\}$; $P(E_2) = \dfrac{3}{10}$

     (c) $E_3 = \varnothing$; $P(E_3) = 0$

     (d) $E_4 = \{1 \text{ and } 2, 1 \text{ and } 4, 2 \text{ and } 3, 2 \text{ and } 5, 3 \text{ and } 4, 4 \text{ and } 5\}$; $P(E_4) = \dfrac{3}{5}$

11. (a) $\frac{1}{5}$ (b) $\frac{8}{15}$ (c) 0 (d) 1 to 4 (e) 7 to 8   **13.** 1 to 4   **15.** $\frac{2}{5}$
17. (a) $\frac{1}{9}$ (b) $\frac{1}{36}$ (c) $\frac{35}{36}$ (d) $\frac{1}{2}$ (e) $\frac{1}{6}$

14.4 Exercises (page 336)

1. (a) $\frac{1}{2}$ (b) $\frac{7}{10}$ (c) $\frac{2}{5}$ **3.** (a) $\frac{1}{6}$ (b) $\frac{1}{3}$ (c) $\frac{1}{6}$ **5.** (a) $\frac{1}{8}$ (b) $\frac{3}{8}$ (c) $\frac{1}{4}$
7. (a) $\frac{1}{17}$ (b) $\frac{25}{102}$ (c) $\frac{1}{221}$ (d) $\frac{13}{102}$ (e) $\frac{25}{204}$ **9.** (a) $\frac{2}{3}$ (b) $(\frac{2}{3})^3$
11. (a) $\frac{9}{29}$ (b) $\frac{197}{203}$ **13.** $\frac{55}{221}$

14.5 Exercises (page 341)

1. $x^6 + 6x^5y + 15x^4y^2 + 20x^3y^3 + 15x^2y^4 + 6xy^5 + y^6$
3. $p^5 - 5p^4q + 10p^3q^2 - 10p^2q^3 + 5pq^4 - q^5$
5. $16x^4 + 32x^3t^3 + 24x^2t^6 + 8xt^9 + t^{12}$
7. $\dfrac{m^5}{32} - \dfrac{15m^4n}{16} + \dfrac{45m^3n^2}{4} - \dfrac{135m^2n^3}{2} + \dfrac{405mn^4}{2} - 243n^5$
9. $7920m^8p^4$ 11. $\dfrac{4845}{65536}p^8q^{16}$ 13. 2.5937 15. 245.9374

14.6 Exercises (page 344)

1. $C(7, 5)\left(\dfrac{1}{2}\right)^7 = \dfrac{21}{2^7}$ 3. (a) $\left(\dfrac{1}{6}\right)^{12}$ (b) $C(12, 6)\left(\dfrac{1}{6}\right)^6\left(\dfrac{5}{6}\right)^6$ (c) $C(12, 1)\left(\dfrac{1}{6}\right)\left(\dfrac{5}{6}\right)^{11}$
5. (a) $(.95)^{20}$ (b) $(.95)^{20} + C(20, 1)(.05)(.95)^{19} + C(20, 2)(.05)^2(.95)^{18}$

Index

Cancellation property
  of addition, 10
  of multiplication, 10
Carbon 14 dating, 124
Cardioid, 314
Cartesian coordinate system, 51
Center of a circle, 89
Characteristic, 114
Circle, 89
  center of, 89
  radius of, 89
Cissoid, 318
Closure axioms, 6
Closure axiom of order, 16
Coefficient, 24
Coefficient matrix, 239
Cofunctions, 167
Column matrix, 205
Combinations, 323
Common difference, 72
Common logarithms, 113
Common ratio, 126
Commutative axioms, 5
Complement, 328
Completeness, axiom of, 17
Completing the square, 76
Complex fraction, 37
Complex numbers, 254
  absolute value of, 261
  additive identity of, 255
  additive inverse of, 255
  conjugate of, 256
  equality of, 254
  field of, 256
  multiplicative identity of, 255
  multiplicative inverse of, 255
  negative of, 255
  nth root of, 265
  polar form of, 261
  product of, 254
  quotient of, 256
  standard form of, 254
  subtraction of, 255

  sum of, 254
  trigonometric form of, 261
Composite function, 51
Compound sentence, 20
Conditional equations, 159, 176ff
Conditional probability, 335
Conics
  circle, 89
  ellipse, 92
  hyperbola, 94
  parabola, 76
Conic sections, 98
  summary, 98
Constant of variation, 100
Conjugate, 256
Coordinate system
  Cartesian, 51
  polar, 260, 313
Cosecant function, 142
Cosine function, 136
Cotangent function, 142
Cramer's Rule, 235

Deductive reasoning, 345
Degree, 25
De Moivre's theorem, 263
Dependent events, 336
Dependent system, 230
Descartes, René, 52
Descartes' rule of signs, 282
Determinants, 218
  expansion of, 220
  properties of, 222
Difference of two cubes, 33
Difference of two squares, 33
Dimensions of a matrix, 204
Direct variation, 100
Discriminant, 81
Distance formula, 89
Distance from point to line, 300
Distributive axiom, 6
  extended, 6

Geometric progression, 126
    infinite, 131
    sum of n terms, 127
Geometric sequence, 126
Greatest integer function, 69
Greatest lower bound, 16

Half-plane, 65
Half-value identities, 170, 171
Heron's area formula, 201
Horizontal asymptote, 287
Hyperbola, 94
    asymptotes, 95
Hyperbolic spiral, 318

Identity, 159
    additive, 5
    axioms, 5
    matrix, 214
    multiplicative, 6
Imaginary axis, 258
Imaginary number, 254
Inconsistent equations, 230
Independent events, 336
Index of a radical, 43
Inductive reasoning, 345
Inequalities, 63
    absolute value, 64
    linear, 65
    quadratic, 86
Infinite geometric sequence, 129
Infinite geometric progression, 131
Initial side of an angle, 138
Integer exponents, 39
Integers, 3
Interpolation, 115
Intersection, 2
Inverse
    additive, 6
    axioms, 6
    multiplicative, 6

relations, 103
trigonometric functions, 154ff
variation, 100
Irrational numbers, 3

Joint variation, 101

Law of cosines, 196
Law of sines, 187
    ambiguous case, 192
Leading coefficient, 269
Least upper bound, 16
Lemniscate, 315
Less than, definition, 15
Limaçon, 318
Limits, 130
Linear
    equations, 229
    functions, 54
    interpolation, 115
    relations, 54
Logarithm, 110
    characteristic of, 114
    common, 113
    mantissa of, 114
    natural, 120
Logarithmic
    equations, 118
    functions, 110
    spiral, 318
Lower bound, 16

Mantissa, 114
Mathematical induction, 346
Mathematical system, 4
Matrices, 204
    additive inverses of, 206
    augmented, 239
    coefficients of, 239
    column, 205

	DATE DUE		
Carrey			
4-14-76			
8-14-78			
1/18/79 fried			
AG 2 '91			